RISK, SOCIETY, AND I
Edited by Ragnar

Risk, Uncertainty, and Rational Action

Carlo C. Jaeger, Ortwin Renn,
Eugene A. Rosa and Thomas Webler

London • Washington, DC

First published in 2001 by Earthscan
Moved to digital printing 2010

Earthscan Ltd, Dunstan House, 14a St Cross Street, London EC1N 8XA, UK
Earthscan LLC,1616 P Street, NW, Washington, DC 20036, USA
Earthscan publishes in association with the International Institute for Environment and Development

For more information on Earthscan publications, see www.earthscan.co.uk or write to earthinfo@earthscan.co.uk

ISBN: 978-1-85383-770-8

Typeset by JS Typesetting Ltd, Porthcawl, Mid Glamorgan
Cover design by Yvonne Booth

A catalogue record for this book is available from the British Library

At Earthscan we strive to minimize our environmental impacts and carbon footprint through reducing waste, recycling and offsetting our CO_2 emissions, including those created through publication of this book. For more details of our environmental policy, see www.earthscan.co.uk.

FSC
Mixed Sources
Product group from well-managed forests and other controlled sources
Cert no. SGS-COC-2953
www.fsc.org
© 1996 Forest Stewardship Council

Printed and bound in the UK by CPI Antony Rowe.
The paper used is FSC certified.

CONTENTS

FOREWORD

Thinking About Uncertainty and Risk

It has been suggested that *risk* is the quintessential characteristic of the modern world, socially, ecologically, and institutionally. It is so because of the ways in which we conceptualize and think about risk, quite as much as the nature of the risks with which we are confronted.

The ways in which we conceptualize and think about risk are, of course, related to the changing nature of risks, most notably because scientific discoveries have informed the nature of risks and made us aware of uncertainties yet to be understood, and because so many uncertainties and risks result from human activities.

This book is about how we think about risks and uncertainties, and how we might better do so in the interest of the common good. Jaeger, Renn, Rosa, and Webler take us on a journey that is both fascinating and frightening – fascinating as it traces the philosophical and scientific roots of the dominant paradigm of modern social thought, *rationality*; frightening because of the paradigm's limitations and vulnerability in an increasingly risky and uncertain world.

Relating risk to ontological security (confidence "in the continuity of ... self-identity and in the constancy of the surrounding social and material environments of action"; Giddens 1991, p. 92), brings into sharp relief the problem of achieving a balance *among rationalities* and between trust and acceptable risk. Paradoxically both types of balance appear today to be more fragile than ever before in human history; paradoxically, because despite advances in knowledge of and control over the demons of the past (pestilence, famine, disease, etc.), we have yet to come to terms with the limits or the sustainability of human ecology (Freese 1997) – surely the ultimate uncertainty and the ultimate risk.

By carefully identifying and explaining the rational actor paradigm (RAP), its strengths and its weaknesses, variations and alternative strategies, the stage is set for a radical shift of risk analysis thinking and practice. The National Research Council Committee on Risk Characterization (Stern and Fineberg 1996) calls for redefining traditional risk assessment, combining scientific approaches to hazard identification and characterization with deliberative strategies involving interested and affected parties; i.e., stakeholders. The risk decision process does not stop with risk characterization. It must include risk management, in a continuous deliberative and iterative process.

Jaeger et al. move beyond the NRC Committee's assessment and recommendations, to consider how the interests of stakeholders, their values and concerns, might best be taken into consideration; how, in particular, the risk decision process might involve the organizations and institutions that so dominate the macrosocial environments in which we live. Several metaphors inform this process: the "social fabric," intended to convey the complex bonds and relationships (micro and macro) that hold society together, or that, in weakened states, characterize its disintegration; the "social amplification of risk," a comprehensive approach to risk analysis that seeks to explain how risks become recognized, amplified and/or attenuated; and "social arenas" of behavior, a metaphor used to describe – and to advocate – a policy decision making process among competitors operating under conditions of structured rules of interaction. These metaphors build on recent theory and research from all the social and behavioral sciences, at once an indication of the richness of advances in these sciences and of theoretical weakness.

Moving beyond RAP, the authors suggest, is facilitated by recognition of the ambivalence of preferences and expectations (Smelser 1998), processes related to the evolution of rules (Jaeger 1994a), recent developments in network theory (White 1992), and the reflexivity of much recent theorizing concerning organizations and human collective behavior generally (Beck et al 1994).

There is hope in developments such as these that some theoretical synthesis may yet successfully overcome the rigidity of RAP and its lack of congruence with the real world, while retaining its advantages – and in the analyses and the discourse experiences reported in this book. Hope arises, as well, from Kempton et al's (1995) discovery of common values regarding environmental degradation, among quite varied groups; and Heimer's brilliant demonstration of mechanisms that are able to "bring self-interest and responsibility into alignment" (1998, p 84), in settings as varied as insurance companies' practices regarding "moral hazard" and among families with children who require the services of neonatal intensive care units.

The risk decision making processes advocated by Jaeger, Renn, Rosa, and Webler aspire to recognition of their common interests among those who search for solutions to problems of uncertainty and risk. The increasingly global character of many uncertainties and risks unites us all – whether or not we actively engage in the search for solutions – in communities of fate; challenge enough, surely, for all to ponder.

James F. Short, Jr.

PREFACE[1]

This book goes to press as the world has entered both a new century and a new millennium – a transition which has ignited the public imagination by making us look both at our past and our future. The timing of a book on risk could hardly be more propitious.

The 20th century began with an enchantment with the new technology of assembly line mass production and with an unprecedented techno-optimism. That optimism launched such monumental technological changes as the harnessing of the atom and of the genetic code. But, we are now beginning to understand the magnitude of the intended and unintended consequences of these and other massive technologies of the last century. So, that century has ended with, if not disenchantment with technology, at least a deep ambivalence about it. It has ended with the uneasy recognition that we live in a world replete with risks of our making.

While the idea of risk management can be traced to ancient times, to the 18th century BC code of Hammurabi for example (Dietz, Frey, and Rosa 1999), and while risk has been an essential underpinning to investment and insurance practices for centuries, the systematic application of risk to evaluate the technologies and products of high modernism is a child of the late 20th century. Risk, as we now know it, is a wholly new phenomenon. It evolved in response to the recognition that our technologies were not only getting larger and larger, but also more complex and, in some respects, more dangerous. It evolved, too, out of a paradox created by modern science. Contemporary technologies and products owe much of their existence to both basic and applied science. Yet, in service to public health and welfare, science has developed increased sophistication and precision – remarkably increased in many cases – in detecting the unwanted side effects, the risks, of the technologies it helped create.

Risk developed over the past several decades as the key analytical lens for attempting to anticipate the consequences of our purposive actions on the environment and ourselves. But, risk is much more than just an analytic orientation and it is much more than a suite of evaluation methodologies. Risk is also the mark of a new consciousness, a way of looking at a world of technological and environmental uncertainty. It is a consciousness born of such vast uncertainties as nuclear peace, the durability of the ozone protecting us, global warming, the extinction of species, and the possibility of creating species. It is a consciousness entirely foreign to the optimistic beginning of the 20th century. At the close of that century risk, not technological optimism, has become a dominant cultural theme.

[1] The book is the result of genuine teamwork. Therefore, its authors are listed alphabetically. For the preface, we considered a more personal style to be appropriate. Rosa wrote it, the others supported him with comments and amendments.

And at the beginning of the new millennium coping with risks of our own making has become one of the major challenges faced by humankind.

This book itself began under somewhat risky circumstances. Its inception took place over dinner at the Leoneck Hotel (now the Crazy Cow) in Zurich, Switzerland, in 1992, following a seminar given by Gene Rosa on global climate change organized by Carlo C. Jaeger, then of ETH: Eidgenössische Technische Hochschule Zürich (Swiss Federal Institute of Technology). At the time, Ortwin Renn and Tom Webler were both working at ETH on the Aargau project on risk management (see Section 5.1 of this book). While Ortwin Renn, first as dissertation supervisor at Clark University and then as co-researcher, had collaborated with Tom Webler, and while each of us knew each other by reputation, we had never worked together before. Indeed, the seminar was the very first occasion that brought us all together. Yet, bucking the odds, we audaciously agreed there and then to write a book on risk. But there was also a second element of risk in our book on risk. The writing would simultaneously take place in Stuttgart, Germany, Zurich, Switzerland, and at two ends of the United States: Massachusetts and Washington State. Despite both risks, and despite the sometimes frustrating experience of coordinating our efforts, we were able to combine our efforts and to complete the manuscript.

Our audacity is understandable, even in reconstructed retrospection. It was clear to us at the outset, from our separate and varied experiences in the risk field, that nowhere was there a book that systematically combined the analytic and management literature on risk with the rapidly growing literature on risk in the social sciences. It was especially the case that nowhere could we find a book that attempted to systematically integrate social scientific theory into risk. Similarly, we could find no structured critique of the fundamental theoretical foundation of the risk field, what in the book we call the rational actor paradigm (RAP), and which laid out both its strengths and weaknesses. Nor could we find where alternative general approaches, such as postmodernism, systems theory, and critical theory, had systematically made their way into the risk literature. We took it as our task to address all of these lacunae within an organized framework of analysis. This, we knew, was a daunting task. But we were not deterred, because each of us brought to the task skills and experiences that partly overlapped, and other skills that were quite different, but complementary and constitutive to the task.

All books begin with a mix of excitement and anticipation – the passion for an idea and the enthusiasm to craft it for others. This is the obvious element of a book's beginning. But, all books begin with a delusion – the delusion that the task is far less challenging, far less time-consuming, and far less frustrating than is ultimately the case. Perhaps it must be this way for books to be written at all. This book, like all the others, began with great excitement and, like all others, with a delusion. The excitement still exists for us. The delusion was that the task we set for ourselves was less challenging than it ultimately proved to be, that we could easily coordinate our efforts, and that the book would take a year or two at the most to complete – not the seven years that it actually took. We are no longer deluded.

Despite the unavoidable delusions, we believe that we have met our objectives. This book combines a critical evaluation of the extant risk literature, and its theoretical underpinnings, with a critical evaluation of social scientific theories, and some of their philosophical implications. Furthermore, it lays the groundwork for the integration of more broadly based theory into the field of risk. As such, it sets the stage for a whole host of future work – work that is not only rigorous, but theoretically informed. To our knowledge, the book is the first of its kind. Because of this, and because of its comprehensiveness, it should appeal to a broad range of professionals: government officials, public policymakers, risk professionals, risk managers, and social scientists studying risk. Furthermore, we believe that the book could find effective use in upper division undergraduate and graduate courses on risk, environmental assessment or management, technological assessment or management, and related courses.

One of the consequences of this book's long gestation period is the accumulation of a large number of intellectual debts, at various stages of its completion. We are especially indebted to Almut Beck, Robert Brulle, Tom Dietz, Hadi Dowlatabadi, Riley Dunlap, Ottmar Edenhofer, William F. Freudenburg, Silvio Funtovicz, Daniel Jackson, Valerie Jenness, Ragnar Löfstedt, Bernd Kasemir, Allan Mazur, Aaron McCright, Claudia Pahl, Kristin Shrader-Frechette, Paul Stern, James F. Short, Jr., Jerry Ravetz, Steve Rayner and Matthias Wächter for their help, encouragement and critical comments. Moreover, students participating in seminars at ETH, at Darmstadt University, Stuttgart University, the University of Surrey, Klagenfurt University, George Mason University, Washington State University and Antioch New England Graduate School provided the friendly curiosity and blunt criticism without which our task would have been impossible to accomplish.

The same holds for the audiences at scientific conferences where we presented portions of the book, in particular at the 1994 XIII World Congress of the International Sociological Association in Bielefeld, Germany, at the 1994 Sixth Annual Conference on Socio-Economics in Jouy-en-Josas, France, at the 1994 Annual Meetings of the Society for Human Ecology in East Lansing, Michigan, USA, at the 1995 XXXII Congress of the International Institute of Sociology in Trieste, Italy, at the 1995 Annual Meetings of the Society for Risk Analysis in Honolulu, Hawaii,

USA, at the 1997 Annual Meetings of the Risk Assessment and Policy Association in Alexandria, VA, USA, at the 1997 Conference, "Sociological Theory and the Environment," at Woudschoten, The Netherlands, and at the 1998 XIV World Congress of the International Sociological Association in Montréal, Canada. Portions of the work have also appeared as "Decision Analysis and Rational Action," in Steve Rayner and Elizabeth L. Malone, eds., 1998, *Human Choices & Climate Change*, Vol. 3, *Tools for Analysis*, Columbus, OH: Battelle Press and as "The Rational Actor Paradigm in Risk Theories: Analysis and Critique," in Maurie J. Cohen, ed., 1999, *Risk in the Modern Age: Social Theory, Science and the Environment*, New York: Macmillan.

We are also grateful to a number of people who helped retrieve a fugitive literature, who performed other work on the manuscript, who assisted in editing the text (particularly the parts written by non-native English speaking authors), and who helped with the formatting of the book. Included here are Sigrun Gmelin-Zudrell, Sabine Mücke, Meherangis Steinwandter and copy editor Jeanne Anderson.

To James F. Short, Jr. we owe a threefold thanks. He helped to pioneer the sociological study of risk with his 1983 presidential address to the American Sociological Association. His continued efforts in this endeavor have stimulated the thinking of many sociologists, but especially us. With his unrivaled editor's eye he read the manuscript in its entirey and did his best to save it from the pitfalls of multi-authored writing. Any remaining editorial flaws are our own. Finally, he wrote a Foreword that not only refocused our attention to the topic of this book, but refreshed our fascination with it.

Rosa's contribution was partially supported by the Edward R. Meyer Distinguished Professorship in Natural Resource and Environmental Policy in the Thomas S. Foley Institute for Public Policy and Public Service at Washington State University; Rosa and Webler's contributions were partially supported by the Center of Technology Assessment in Baden-Württemberg, and by EAWAG, The Swiss Federal Institute for Environmental Science and Technology.

Eugene A. Rosa · Carlo C. Jaeger · Ortwin Renn · Thomas Webler

1. GENERAL INTRODUCTION

1.1 Risk, Ontological Security, and Uncertainty

Automobile and plane crashes, toxic chemical spills and explosions, nuclear accidents, food contamination, genetic manipulation, the spread of AIDS, global climate change, ozone depletion, species extinction, and the persistence of nuclear weapon arsenals: the list goes on. Risks abound and people are increasingly aware that no one is entirely safe from the hazards of modern living. Risk reminds us of our dependency, interdependency, and vulnerability. Catastrophic risk is an even stronger reminder.

Untoward, sometimes catastrophic consequences of risk are newsworthy events. Catastrophic accidents attract the media. Technological disasters receive worldwide coverage. These events grab headlines and headlines grab public attention. Headlines across the globe testify to the pervasiveness of risk and to the growing embeddedness of risk in the public consciousness. So, at this, the mature stages of modernity, public awareness of risks common to people everywhere has, perhaps, never been as widespread. Risk, it seems, is embedded in a contemporary, worldwide culture (Rosa and Wong 1992).

Our contemporary world is clearly stretching the social fabric of modernity – making it more interdependent and, consequently, more vulnerable. Dominant patterns of the social fabric today include the worldwide spread of industrial production, called by some theorists "a new world manufacturing order" (Ross and Trachte 1990), and the internationalization of the division of labor, "the new international division of labor (NIDL)" (Fröbel, Heinrichs et al 1980). These labor and production patterns are accompanied by the availability everywhere of a wide assortment of consumer goods. The market baskets of people widely distanced around the globe are becoming increasingly similar, at least among the growing affluent classes.

These production and consumption patterns are also accompanied by another unshakable traveling companion: risk. The globalized interdependencies of production, consumption, and geopolitical arrangements mean that people everywhere are coming to share a common set of risks. No one could escape a nuclear holocaust, ozone depletion, the consequences of monoculture and species extinction. Toxic chemical exposure, industrial accidents, and global climate change pose increasing threats. Since risk is a central feature of the contemporary world, and since the imposition and management of many risks are the charge of others, there is a strong public interest in how societies manage or cope with risks.

Innovation can reduce risk. So can safety and vigilant management practices. Technical risk managers tell us that the world is measurably safer today than it was only one hundred years ago (Wilson 1979). They point to significant reductions in infant mortality and to the marked improvements in life expectancy as tangible indicators that life is getting less risky. But each wave of innovation and each new effort at safety also seem to bring with them the unintended consequences of new risks. The contemporary world continues on a trajectory of pervasive industrialization, underpinned by the application of technologies of increasing complexity and risk. It must, therefore, struggle to reconcile the fervent desire for progress and betterment for people everywhere with the consequences of technologically induced uncertainty and the possibility of disasters.

1.1.1 Risk and the Human Condition

Far from a transient annoyance, risk is constitutive of the human condition, as it has been from the beginning of human existence. While we worry about the risk of a nuclear disaster, our ancestors worried about the risk of being devoured by other species. We worry about the risk of global warming, while they worried about the risk of finding shelter to protect themselves from the Arctic cold. How have things changed?

There are three key differences between risks in the past and now: (1) in the past, most risks were proximate and local in impact, while today many risks are "eco-systemic" risks; (2) in the past, most risks were geographically circumscribed, while today many risks are global; and (3) in the past, many risks were likely thought of in terms of a group's unique circumstances, while today – as noted above – there is growing public awareness of common risks around the globe. The first type of risk refers to systemic environmental risks that can occur at a local level, but are cumulative (such as the destruction of moist forests in certain parts of the world), or that can occur at the level of entire ecosystems (such as air pollution over a wide area). The second type of risk implicates the globe as a whole (such as the risk of a nuclear disaster or risk of global warming). And the third change refers to a growing risk conscience in the contemporary world of high technology (Rosa 1994).

1.1.2 Imprimatur for the Age

The global changes outlined above represent the most recent of a number of fundamental "transitions" in the world (from feudalism to capitalism, for example). Our transitional age is witnessing social change that is huge, that is fundamental, and that is far-reaching – resulting in the evolution of worldwide institutions never seen before. These changes are profound because of their depth (they go to the foundation of how societies organize themselves) and because of their spread (societies nearly everywhere are experiencing these changes). The pervasiveness and profundity of this social change has not gone unnoticed by social theorists; quite the contrary. A central preoccupation of a number of social theorists has been to understand this transformation of the globe into a "new social world." Alexander (1994), for example, writes: "Recently, a new social world has come into being. We must try to make sense of it. For the task of intellectuals is not only to explain the world: they must interpret it as well." In a subtle way, these two tasks are related to a third, namely, to label this new world, to give it a suitable imprimatur.

In the view of two leading European social theorists, Anthony Giddens and Ulrich Beck, risk is that imprimatur. Both have developed – independently – theoretical frames that place risk at the core of the world transition. Risk is theorized as a fundamental of the emergent social order. For Giddens and Beck, *the spirit of our age is the universal concern with hazards* in the contemporary world, the vulnerability of the environment, and of the human species itself. Beck (1986/1992) calls the new social world – new age – the "Risk Society." The transition to this "other modernity", he argues, transforms our collective consciousness: we are now preoccupied with risk. The social world has become a world of risk, a world that makes transparent our vulnerabilities.

The adoption of "risk" as the imprimatur of our age marks a significant refocusing of social thought. The foundation of Western thought since the Enlightenment – from Comte, Spencer, Marx, Parsons, Habermas, and others – has been the expectation of progress, of continued improvement in the social world. The emergence of a "Risk Society," abruptly challenges that assumption. "Risk society means an epoch in which the dark sides of progress increasingly come to dominate social debate" (Beck 1991/1995, p 2). The focus of attention shifts from the "goods" of modernization, social and political, as well as economic, to its many – often unintended – "bads."

1.1.3 Ontological Security

Social fabric is a root metaphor for collective life. It is a webbing of interdependencies embedded in expectations, obligations, actions, and interactions – bonds and mutualities that are the essence of social life. They ensure a degree of regularity or orderliness to social action and are pivotal to a reflexive self-identity and to a weaving of the individual into the social fabric. They provide "ontological security," in the terminology of Anthony Giddens (1984; 1991). Ontological security refers to "the confidence that most human beings have in the continuity of their self-identity and in the constancy of the surrounding social and material environments of action" (1991, p 92). The reappearance of the sun each morning is a form of ontological security for humans everywhere.

We expect the astronomical, more generally: the material world to be more or less the same each day. For some prescientific societies an eclipse of the sun poses a serious threat to that security. Such threats are not typically taken lightly, for the eclipse often elicits worrisome responses, such as the rush to worry beads or to an appeasement of the gods. For modern societies, threats to ontological security increasingly take the form of uncertainties, and risks associated with them. Less and less do risks emanate from the natural world; more and more they are the result of human choice. As threats to ontological security, risks, then, threaten both environmental conditions and individual identity. Managing risks – and concomitantly threats to ontological security – has, therefore, become a major activity and a gauge of success for modern societies (Covello, Mumpower et al 1985; Royal Society 1992; Somers 1995).

Understanding how and why people and organizations and societies deal with risk by reducing uncertainties in order to maintain ontological security is clearly a task worthy of sociological investigation. In this book we approach such a study by essentially paraphrasing Kant's two central questions: "How did things get this way?" and "How can we understand what needs to be done about them?" Kant's questions invite sociological analysis because worries about risks are not just individual problems, but problems of a growing collective consciousness. Our task in this book is to exercise the sociological imagination in service to Kant's questions.

We begin by defining risk. Next, we sketch the dominant intellectual worldview – rational action – for understanding and managing risk. We then critically evaluate that position, pointing out both its strengths and weaknesses. Our critical evaluation leads to the recognition that there are other rationalities for explaining social life and for explaining risk. These other rationalities are the foundation of alternative approaches to understanding risk: reflexive modernization, critical theory, systems theory, and postmodernism. Key questions organize our inquiry: What is meant by risk? How do we perceive and respond to it? What means have evolved to analyze it and to manage it? What can be done about risks? What should be done? How do risks reproduce social structure? How is risk embedded in the social fabric?

1.2 What is Risk?

In common usage, risk has a wide range of connotations: fear of specific hazards, concern for the interdependency of humans and technological systems, uncertainty regarding financial gain or loss, fear of the malevolent forces of nature, or the thrill of adventure, or worry about the competence and trustworthiness of those who manage risks. Despite variation in usage, however, there are unifying features that ground the meaning of risk. All conceptions of risk presuppose a distinction between *predetermination* and *possibility* (Evers and Nowotny 1987; Markowitz 1990; Renn 1992a), for if the future were either predetermined or independent of present human activities, the notion of "risk" makes no sense.[2] Whatever the variation in connotation, risk implies the *possibility* of some outcome – possibility is, indeed, the first indispensable element of the idea of risk.

Risk implies both the possibility that an event or outcome can happen with the denial that either occurs with predetermined certainty. Risk thus necessarily implies *uncertainty*, the second indispensable element of risk. A further consideration is the fact that possible outcomes – consequences – are rarely neutral, but carry with them rewards or penalties, as the case may be. Humans evaluate the outcomes of their actions. As a result, humans try to make causal connections between present actions and future outcomes, and they exercise agency in attempting to shape the causes of future outcomes.

The foregoing considerations imply that certain states of the world which are possible but not predetermined can be identified objectively as risk. The lack of predetermination implies probability and, therefore, uncertainty. Not all uncertainty is risk, however; hence, a third essential feature is that risk is present only to the extent that uncertainty involves some feature of the world, stemming from natural events or human activities, that impacts human reality in some way. Risk, in human terms, exists only when humans have a *stake in outcomes*.

1.2.1 Defining Risk

With these considerations, we adopt, after Rosa (1998), the following definition (see also National Research Council 1983; Fischhoff, Watson et al 1984; Luhmann 1990):

Risk: A situation or event in which something of human value (including humans themselves) has been put at stake and where the outcome is uncertain.

In order to avoid certain confusions contained in other definitions of risk, it is important to point out several key features of this definition. First, it

[2] This view of risk contrasts sharply with more fatalistic views of nature and society, where fate is believed to predetermine outcomes. In this fated orientation, anticipating future outcomes in order to improve present choices is meaningless.

expresses an ontological state of the world. Risk captures the duality that humans are embedded in uncertain environments – natural and human-made. Second, it explicates states (uncertain but involving human stakes) that are properly conceptualized as risk. Third, it embeds the conventional definition of risk (as the probability of an occurrence or event multiplied by the value of the outcome of that event). Fourth, it is robust, in that it subsumes both undesirable risks (the dominant concern of the field) and desirable risks (Machlis and Rosa 1990), such as adventurous undertakings or investments in which risks are engaged in for thrills or similar satisfactions.

Although risk may be viewed as an ontological state of the world, humans neither ignore that world nor are they passive about it. By definition, stakes are involved. Individuals, collectivities, and institutions perceive some risks, but not others. Some risks engender concern, or alarm, while others are unconsciously or willfully ignored. Some attract professional attention, including management practices; some do not. Thus, while we define risk as an ontological entity, our understanding of risk is an epistemological matter. Humans everywhere seek to identify and understand risks. This involves perception, investigation, judgment, evaluation, and claims about our knowledge of risk.

Risk, as noted above, is closely tied to human agency in that it involves choices among various possibilities. Anticipating the consequences of alternative possibilities, evaluating their desirability, and choosing the most desirable option lie at the core of human agency. These volitions, too, lie at the core of the idea of *rational action*, the foundation of inquiries into risk and uncertainty. The first institution specifically devoted to risk, dating from the Italian Renaissance, was the insurance industry. The idea of rational action, too, has been the foundation of inquiries into risk and uncertainty since the beginning of probability theory in the seventeenth century by Pascal and others (Hacking 1975).

Because all risks carry with them either danger or opportunity – potential for loss or gain – the notion of risk adds incentive to make causal connections between present actions and future outcomes. The task of risk management is to anticipate outcomes of risk situations and to incorporate uncertainty into decision making. Risk management implies that undesirable outcomes can sometimes be avoided, and where unavoidable, can be mitigated if connections between cause and effect are made properly. Thus, risk typically is normative, as well as descriptive or analytic. That is, risk involves evaluative judgment about the desirability of outcomes.

Taken together, our ontological definition of risk, the epistemology of risk, and the normative and management aspects of risk, contain three key elements that distinguish the various conceptualizations of risk: type of outcome (typically, but not always, undesirable consequences); some gauging of the possibility of occurrence (typically, but not always, probability); and type of entity affected (individual, corporate, or institutional) that also judge outcomes, their possibility, and their desirability. Different risk perspectives conceptualize these three elements in quite different ways. They can be distinguished on the basis of the manner in which they address four questions that emerge from these elements:

- What concept of possibility is used?
- What types of outcomes and their consequences are considered?
- How are the concepts of possibility and outcome combined?
- Who is the actor that judges the three questions above?

In the following sections these four questions – the conceptualization of uncertainty, the scope of consequences, the combination rule, and the actor involved in making decisions – will provide the framework for explicating and evaluating a variety of perspectives on risk.

1.2.2 Analytical Perspectives

Because of its centrality to modern decision making, risk has developed an analytic infrastructure consisting of scientists, engineers, economists and other social scientists, decision theorists, and many agencies charged with managing risk. The roots and branches of this infrastructure are grounded in the idea of *rational action.* A variety of approaches have emerged to study risk from the perspective of rational action. Other approaches are offshoots of this perspective – either as an extension of rational action or as a critical reaction to it. In Section 1.3 we describe this pervasive version of rational action as the Rational Actor Paradigm (RAP).

A classification scheme for subsuming approaches to the study of risk is offered by Dietz, Frey, and Rosa (forthcoming). They distinguish between technical (subdivided into assessment, evaluation, and management), psychological (especially the psychometric approach), sociological (subdivided into contextual and organizational), anthropological, and geographical approaches. Based on this classification (see, also, Renn 1992a), we distinguish the following approaches to risk and uncertainty. All are either applications or elaborations of rational action, or critical responses to it:

- decision analysis
- quantitative risk assessment
- psychometrics

- insurance and portfolio investment
- natural hazards
- game theory
- risk communication
- social movements and resource mobilization
- bounded rationality.

As will be discussed in the following sections, these approaches vary in the selection of the underlying base unit, the choice of methodologies, the complexity of risk measures, the instrumental objective of risk, and the social function of risk. At the same time they share in common the underlying assumption of rationality. The idea of rationality is captured in their basic concepts and assumptions. This lends important strengths to them, not only in isolation, but even more so when taken in combination, since they all proceed from the same basic orientation. To what extent it also attracts important weaknesses remains to be seen, a focus of our critique in part II of the book.

1.3 The Rational Actor Paradigm (RAP)

The Italian Renaissance gave the world many gifts, among them, and no less awe-inspiring today, extraordinary accomplishments in science and the arts. Embedded within these tangible symbols was a new idea as fundamental, enduring, and influential as the art that embodied it. The idea is that humans are actors who strive to realize personal objectives by means of purposive action that is designed to achieve considered alternatives (Dawes 1988). Michelangelo's statue of David, for example,[3] not only embodies the artistic accomplishments of the period, but also

[3] Michelangelo's David, the largest marble statue that had been carved in Italy since the end of the Roman empire, was the apotheosis of a fascination with heroic-patriotic themes. David as slayer of Goliath, and as exemplar of those themes, began with a sculpture by Donatello a half-century earlier. Donatello's version, more life-sized and therefore less imposing than Michelangelo's, had its own monumental significance: it was the first sculpture of this size in Western Europe meant to be "seen in the round" (that is, from all sides) since antiquity (Cole and Gealt, 1989). In the Donatello original and all subsequent versions of the statue, David stands alone, divorced from any tangible or spiritual surroundings. He was no longer enmeshed in an artistic context of religious allegory, nor part of the grand procession of the faithful, nor embedded in the crust of fate, but was self-assured in this world. He was self-assured, self-reflective, and defiant.

All of these features symbolized the shifting of attention away from an other-worldly religion toward the presence of divine creativity in the secular world; humankind's gaze upward to the heavens, since the beginning of Christianity, now slowly but relentlessly pivoted down to its earthly surroundings. In a revival of the Protagorean dictum, man became the measure of all things. The apparently overmatched David defeated Goliath, not by brute strength and might,

this new idea. The enduring brilliance of this masterpiece is only partly due to its magnificence as a representation of the human body. It is equally brilliant as a tangible representation of the idea that man is the measure of all things via rational agency.

The elevation of human agency to a position of centrality in human thought has persisted throughout the ensuing centuries.[4] Though rooted in the Renaissance, the rational actor perspective was cultivated toward modern form by the scientific aspirations of the Enlightenment, a central concern of which was to develop a "moral science of the mind." So central was this objective that it attracted some of the leading lights not only of that age, but of all ages since. Chief among these were Hobbes, Hume, and Kant who, proceeding from deeply conflicting perspectives, sought to establish a scientific basis for deliberation, thinking, and acting on a rational basis.

The crucial step beyond the Renaissance idea of human agency was the attempt to define a notion of rationality independent of specific objectives. So effective were these efforts to shape the philosophical underpinnings of the rational actor perspective that all that remained for Bentham, the utilitarians, and those who followed, was to prune the growing sapling of the rational actor perspective. They did so by mapping the perspective onto a mathematical structure.

Hobbes argued that humans were governed by the twin forces of passion and reason. A universal feature of the human condition was that, in a state of nature, human life was governed by ruthless passion. It was a war of survival, and life was "solitary, nasty, brutish, and short." To bridle the excesses of passion, Hobbes sought "precepts" or "general rules" that could transcend and counter this natural state. Rules, he argued, could be found in the other governing force: reason. Reason would create "a common power to keep all in awe" (Hobbes 1651/1968, Ch. XI). Putting general rules of reason into action meant the application to human affairs of the same universal scientific spirit that was applicable to the natural world.

A century later, Hume would take up the banner of applying science to human affairs. Like Hobbes, he began with the presupposition that human action could be traced to the two motivating forces of passion and reason. Hume, however, assumed that all human action is the product of both

but by cunning and intelligence – by the use of reason and by purposive action. Look around you, said the Renaissance thinkers, and you will see everywhere the accomplishments and unlimited potential of the application of similarly rational thought and action: feats in architecture and engineering, in commerce, in art, in astronomical investigation, and geographical exploration – all visible measures of God's amazing gift to humankind, rationality.

[4] While seeds of the idea can be traced to classical Greece, in the works of Plato and Aristotle, they did not take root until the Italian Renaissance.

forces, not simply the reflection of one or the other. Hume viewed the state of nature with a much less jaundiced eye than Hobbes, allowing for passions that were beneficent as well as ruthless, and communal and altruistic as well as egocentric. As with Hobbes, Hume held that passions precede reason and are also universal to all people and constant over time. They are, therefore, the initiating force of action. It is the role of reason to be governed by passions: to establish rules and principles for satisfying desires that are common for all of humankind. Thus, both Hobbes and Hume – despite their fundamentally different views of the state of nature – are largely responsible for an "instrumental notion of rational action" (Hollis and Sugden 1993).

Kant also sought to develop a basis for rationality that followed universal rules. Unlike Hobbes and Hume, who began with supposed knowledge about human propensities or about empirical observations of human action, Kant's starting point was the moral duty within the individual actor. To Kant, moral action was rational action because it stemmed, not from conditions of the world, but entirely from *a priori* reason. Kant sought to uncover a "supreme principle of morality". If adopted by an actor, this principle would result in purely rational action, because the intention behind the action would be moral. Such action would also be expected of anyone in the actor's position. Moral duty compels the actor to act rationally. Kant argued for such a principle with his categorical imperative: "Act only according to that maxim by which you can at the same time will that it should become a universal law" (Kant 1785/1959, p 39). Thus, by shifting the basis of rational action from a social to a moral foundation, Kant shifted the criteria of rationality from the consequences of actions to the moral motives driving actions. In essence, Kant sought to establish rational action on a deontological foundation, while Hobbes and Hume – pursuing the same goal – chose a consequentialist foundation.

1.3.1 Rationality as Worldview, General Theory, and Special Theory

Seeded by classical Greece, rooted in Renaissance Italy, cultivated by philosophers of the Enlightenment, and pruned by the utilitarians and their intellectual offspring, the idea of rational action remains a central legacy of Western thought. It pervades our culture at all levels, from the most fundamental level of the worldview, through the level of general theories (we will denote them as paradigms), to specific theories.[5] At all

[5] Confusion is easily cultivated by a failure to make a distinction between these three levels. For example, as Arrow points out: "It is noteworthy that everyday usage of the term 'rationality' does not correspond to the economist's definition as transitivity and completeness, that is, maximizing something" (Arrow 1987, p 206). Furthermore, the distinction helps to identify the locus of

levels, however, rationality as a feature of individual actors – as distinguished from an understanding of rationality as a characteristic of social relations – is emphasized. At the center of the Western worldview is the idea that humans are rational organisms and that the world can be explained in terms of the interaction of atomistic entities: atomistic bodies for the natural world, atomistic minds for the world of human affairs. This worldview sees humans as rational beings motivated by self-interest and consciously evaluating alternative courses of action.

Thoroughly enmeshed in contemporary culture, the rational actor worldview is nearly ubiquitous. As a consequence, it provides the epistemological foundation for a wide variety of institutions in the modern world. It underpins the structure of markets, legislative government, education, international security, industrial management, health care administration, and countless others. Deeply submerged within these institutions, however, rationality operates largely below conscious awareness and is seldom scrutinized. Rationality, in this sense, is one of the most basic "givens" of Western thought; it shaped modernity and set one historical boundary that distinguished Western culture from others, and that increasingly characterizes a global culture shaped by a global economy.

At the next, narrower level is the idea of rational action as a general theory of human affairs. It has led to the most influential research tradition the social sciences have experienced so far. The emergence of the state and the functioning of markets, the definition of property rights and family structures, criminality, and human cognition have all been conceptualized in terms of rational action at this level. In this book, we denote the general theory of rational action as the rational actor paradigm, or simply by the acronym RAP.[6]

RAP is a two-sided coin. One side conceptualizes the social universe as an aggregate of atomistic actors. The other is normative – rational action as the basis of a social order that is appropriate and good. The theoretical power of RAP is due to the claim that interactions between actors can and often do lead to an overall equilibrium. This equilibrium is said to be highly desirable in the sense that it is satisfying to the atomistic actors. Thus, Hobbes argued that if people accept domination by some

intellectual debate. Seldom does debate take place at the broadest level of meaning: rational action as worldview. Rather, it takes place at the other two levels, but in different forms. At the paradigm level debates tend to be "disciplinary-level" debates with the focus on the fundamental question: To what extent is rationality a fruitful foundation for the discipline's principal domains of inquiry? At the level of special theories the focus tends to be on questions about the proper domains of applicability of rationality-based theories, and on questions about the cumulative growth of the theories.

[6] Using the word "paradigm" to refer to general theories is one of several uses introduced by Kuhn (1962/1970).

central government, they better their chances to realize individual desires. Hume developed a similar argument for stable property rights. Adam Smith (1776/1976) prepared the ground for the scientific analysis of market equilibria with the metaphor of the invisible hand.

As a paradigm, the idea of rational action "can be considered a heuristic device for interpreting behavior" (Hogarth and Reder 1987, p 3). Heuristics are intellectual tools – tools applicable to a wide variety of settings from a wide variety of perspectives or intellectual traditions. Thus, rational action as paradigm can just as readily be employed by the psychologist as by the political scientist, and even the novelist. Historically, however, rational action has been employed as the foundation of economics.

At its narrowest level of meaning, rational action is the basis for special theories in the social sciences. The concept is at the heart of formal models (axioms, propositions, derivations) for postulating interrelationships, and for drawing conclusions and making predictions.[7] It has a growing variety of manifestations, including expected utility theory and public choice theory in economics, decision theory in psychology, social choice theory in political science, rational choice theory in sociology, and others. At this level of specificity *a priori* behavioral predictions or *a posteriori* behavioral explanations are made on the basis of explicit expectations about purposive choices.

The three levels of rationality – worldview, general theory (RAP), and special theory – have a nested relationship with one another. They are nested with respect to their degree of ubiquity and embeddedness – their degree of spread and depth. Concomitantly, they are nested with respect to their malleability. Rationality as worldview is ubiquitous and deeply ingrained in our culture; the most important social institutions are epistemologically founded on this worldview. The idea of rationality at this level is, therefore, extremely difficult to change. General theory, RAP, while less ingrained, is nevertheless amazingly hard to change, too. Because paradigms such as RAP are broad orientations, they are resistant to the challenge of anomalous empirical evidence, as Kuhn (1962/1970) demonstrated so clearly for scientific theories. Special theories are the most malleable – more sensitive to the fit between theoretical prediction and empirical evidence. Thus, the idea of rational action as special theories

[7] Krimsky's (1992) distinction between "theoretic activity" and "theory" parallels our distinction between paradigm and special theory. He explains (p 6): "The former refers to certain methods of analysis, conceptualizations, or approaches to problems that are synthetic, abstract, and/or integrative and apply to large domains of phenomena. The latter consists of a set of principles (axioms, empirical generalizations, laws) that provide explanatory coherence to an empirical domain." Absent from Krimsky's formulation is rationality as worldview, the umbrella concept nesting the other two meanings.

can come and go, while general theory and worldview remain virtually unchanged.

1.3.2 Durability and Malleability: An Example

Early American criminology developed along the same rational lines as the existing American criminal justice system, when RAP was the basis for the scientific understanding of criminal acts. Eventually, owing to attacks from the emergent fields of psychology and sociology in the early part of this century, RAP gave theoretical ground to a "positivist" view of crime. That view emphasized the importance of psychological and situational factors in criminal behavior. By the mid-twentieth century, positivist criminology had become the dominant paradigm in the United States. Since the late 1970s, however, the positivist position has lost ground to a revival of classical thinking with its emphasis on rational action. Among RAP supporters debate centers on the most fruitful theoretical formulation of the paradigm – on the choice of special theory (for example, between retribution and rational choice theories).

So, while the rational foundation of traditional criminology waxed and waned as paradigm and theory, as worldview it waxed and waxed. Paradigm and theory changed. But the criminal justice system and legal processes were, by and large, unswervingly shaped by the principles of rational action (Siegel 1989).

1.3.3 Rationality in Economics

Nowhere has the idea of rational action enjoyed a more privileged position than in the field of economics. More than any other discipline, economics has elaborated the idea of rational action into a general theory for researchers studying a wide variety of human affairs. Thus, economics exhibits "a disciplinary unity that is lacking in the other social sciences" (Hogarth and Reder 1987, p 4). This accounts, in large part, for the privileged position of rationality in economic thinking. Because it is so thoroughly integrated at all levels, the rationality assumption is unshakable, even in the face of disconfirming empirical evidence. Falsifying evidence – the Popperian criterion of scientific knowledge claims – is embarrassing at the level of special theories only. It is rarely more than an anomaly to the broader domains – a vexing noise to the harmonious duet of worldview and paradigm. As general theory, RAP will only be superseded if a new paradigm can offer a superior ability to deal with empirical evidence.

The development of RAP from the time of Hobbes to the present documents the durability of rational action as a paradigm in economics, unparalleled in the history of any other scientific inquiry. This is not to deny ongoing debates about the rational action perspective. Over the past century there have been debates about the meaning and measurement of utility, for example. Yet, in spite of considerable heat, and occasional light, generated in the study of specific economic problems, these debates virtually never threaten the broader meanings of rationality: rationality as worldview and as paradigm. Indeed, where special theories fall under a growing weight of criticism the broader meanings – worldview and paradigm – reconstitute themselves around new theoretical formulations. Such reformulations have recently been finding their way into nearly every facet of human – and sometimes nonhuman – life. In the words of Herrnstein and Mazur (1987, p 41): "Today, utility maximization, also known as the theory of rational choice, is being used to analyze not only markets, but the law, the family, the state, and the behavior of animals in nature. Social scientists are employing it to explore how criminals come to commit crime, how judges hand down sentences, how couples decide between having BMWs and having babies, how legislatures enact law codes, even how squirrels forage for acorns." The rational perspective remains viable, even as its specific theoretical manifestations become threatened, discarded, replaced, or renewed.

Nevertheless, in the study and management of risk, RAP-based approaches have increasingly confronted difficult theoretical challenges. Questions have arisen concerning a growing number of empirical anomalies, as well as the internal logic, of RAP. Although several alternative approaches have been proposed to challenge its ascendancy, none has been able to supersede RAP. However, such challenges point the way toward alternatives that promise to provide a broader understanding of human action that may eventually lead to important advances not only in risk analysis, but also in the social sciences in general.

1.3.4 The Larger Sociological Critique of RAP

Neil Smelser, in his 1997 presidential address to the American Sociological Association, developed a view that in many ways resembles the core of our critical appraisal of RAP (Smelser 1998). Beginning with a premise similar to ours – that "The idea of 'the rational' holds a noble place in the history of philosophy...." and "the rational survives in the contemporary social sciences" (1998, p 1) – his critique parallels the substance of ours by focusing on the variety of social activity beyond the theoretical reach of RAP. It further parallels our critique in rationale: we seek to unpack both the strengths and weaknesses of RAP, with the goal of defining the conditions of RAP's applicability and that of alternative approaches (see Jaeger, Renn et al 1998 for an exemplary discussion referring to global environmental change). Smelser expresses a related goal when he writes: "I intend to neither applaud nor to bash the rational-choice approach yet again. Rather, I aim toward developing a supplement" (1998, p 3).

Smelser also reaches the same conclusion: that RAP is a powerful and useful theoretical tool – but only under a restricted set of conditions. RAP is applicable in settings that are socially-structured to prefigure the conditions and actions assumed within the logical structure of RAP: settings where actors are engaged strategically with preferences and actions whose gainful outcomes can be assessed. Well-established markets are the paradigm example of such settings. Political contests and other competitive situations often, too, meet the conditions presupposed by RAP. Smelser summarizes the governing principle: "...that rational-choice principles are *applicable* to situations in which choice is *institutionalized*" (1998, p 13, emphasis in the original).

The upshot of these critiques is to delimit the scope conditions of RAP. Scope conditions, too often neglected in social scientific theory, are important for explicating a theory's range of applicability. They permit a logical justification for interpreting evidence designed to test theory. In a loose sense, every *general* theoretical proposition is both true and false, since for every proposition tested we can always find supporting examples or embarrassing counter-examples. The specification of scope conditions is a logical route out of this dilemma, in that it "conditionalizes" theory by stating the conditions under which propositions are presumed to hold.

Despite our common goal and common conclusion, Smelser's critique and ours take different turns. In seeking to "supplement" RAP for all situations beyond its scope, Smelser looks to nonrational forces. In such situations, he claims, the variety of social behavior at all levels of aggregation can be better understood with nonrational models. A central underpinning of such models, in Smelser's framing, is a psychological postulate he calls "ambivalence." We often love and hate the same thing simultaneously. We often are compelled at once to approach or avoid the same thing. Ambivalence is typically "assumed away" in social science investigations due both to general cultural and to theoretical biases toward univalent explanations. RAP can be seen as a theory of motivation. Smelser supplements the limitations of RAP with an alternative theory of motivation. The net effect is an orientation that parallels RAP: a focus on the individual actor (as both rational and nonrational), a concomitant methodological individualism, and the parallel belief that larger social aggregations can be understood on the same terms.

To reach his microsocial orientation Smelser turns away from social theory, while we turn toward it. He dismisses the major theoretical alternatives to RAP – "neo-Marxism, critical theory, varieties of phenomenology, some parts of feminist and gender studies, cultural studies, and postmodernism" (1998, p 2) – as forms of *anti*-rationality. We, in contrast, argue that these theories can be conceptualized, not as anti-rational, but as *alternative* rationalities (compare with Chase et al 1998). Actors small and large rely on varieties of reasoning which go way beyond RAP. Accordingly, we critically examine not only RAP, but also critical theory, systems theory, some versions of phenomenology, and postmodernism as candidates to supplement it. We find that, like RAP, they have strengths and weaknesses, and delineate elements of an understanding of rationality which includes the one emphasized by RAP as a special case.

1.3.5 The Micro-Macro Link – and Markets

Few issues have so galvanized social theorists on both sides of the Atlantic as the theoretical challenge of linking micro and macro levels of explanation. For well over a decade, the principal preoccupation of sociological theory in Europe and America has been with the micro-macro linkage. How can we theoretically integrate the agency of individual social actors with the social and institutional structures that mediate or condition agency? Alexander and Giesen put the challenge this way: "The questions came to focus on whether action was rational or interpretive and whether social order was negotiated between individuals or imposed by collective, or emergent, forces" (1987, p 2). This focus, they note, was a preoccupation for classical sociology, in various stages of the work of Marx, Weber, and Durkheim, as it was for neoclassic sociology, especially in the work of Parsons.

Following World War II, social theorizing concerning the micro and macro became dichotomized – each orientation antagonistic to the other. The pendulum has now swung back and the key quest is the search for linkage.[8] Finding the linkage, however, is a hoary challenge that cuts across the boundaries of philosophy, the traditional sciences, as well as the social sciences. Attention to the challenge has varied across those boundaries, and has fluctuated within disciplinary boundaries. One discipline that has maintained a steady focus on the challenge is economics.

For over a century the foundation of economic theory and analysis has been individual rationality, manifest in strategic decisions that are coordinated by prices and markets. While this foundation has eroded somewhat under the weight of empirical scrutiny, the structure built on it

[8] The avant-garde of this effort was occupied by Anthony Giddens with his structuration theory, where social structure is conceptualized as a duality – at once both medium and outcome of social action. See, for example, Giddens (1984).

remains largely intact. In particular, markets and market forces are still constitutive of that foundation. The market remains a central feature in economic theorizing.

Markets are the focal point for the coordination of individual actors seeking to realize the benefits of their strategic decisions. In this role, the market enjoys a privileged theoretical position more generally because it is the mechanism for combining the concrete actions of individuals (e.g., buyers and sellers) to emergent institutional structures (e.g., communication networks). That is, the market is the mechanism that links the micro with the macro. If it is derived from a RAP-based foundation, the market achieves this link in a theoretically coherent fashion. Any acceptable theoretical alternative to RAP must, therefore, offer an equally elegant solution if it is to be a viable contender in the study of risk.

This theoretical proscription derives both from the formal demands of theory and from the hermeneutic accounts expected from theory. Markets are the context that provide the scope conditions under which RAP does its best theoretical work. Markets often provide settings where actor choices are institutionalized: settings where the principles governing RAP are most applicable because the context strategically orients actors to behave according to rational criteria. And markets provide some of the most effective and influential tools to deal with risk.

The market is fundamentally important on other grounds, too. The world is becoming market-based; this is a social fact that must be explained theoretically. This is the hermeneutic element expected of theory. The importance of the market after the fall of communism in the West, the reinterpretation of the market within traditional religious ideologies, such as Islam, and the reframing of its ideological meaning within remaining communist states, such as China, is strong testament to the power of the market. By its persistence and recently accelerated spread, the market remains a central institution in the contemporary world – perhaps the central institution – begging for theoretical interpretation.

For these reasons we provide a careful evaluation of the role of markets as part of our critique of RAP. We also include markets in our general argument. The essential point is that the theoretical structure of alternatives to RAP must, somehow, provide a logical place for markets while also providing an explanation of their origination and spread.

1.4 Plan of the Book

The appearance of social sciences other than economics in the field of risk is relatively recent. Before their entry, the field of risk studies was dominated by technical risk analysis performed by engineers and economists armed with tools taken from RAP. The pioneering article by

Chauncey Starr (1969), launching the modern era of risk analysis, is the application of a specific version of RAP-based theory: revealed preferences. Expected utility theory, one of the most explicated and mature of RAP's theories, continues to have considerable influence on risk research (Crouch and Wilson 1982; Kunreuther 1992; Rowe 1977). Indeed, the conventional definition of risk originating in engineering as probability times value of consequences is, in effect, a tool for constructing indifference relations between alternative options for decision. By telescoping risk down to a single metric, it produces the expectation that an individual should be indifferent toward low-consequence/high-probability hazards and high-consequence/low-probability hazards. It seems reasonable, therefore, to discuss the study of risk and uncertainties with explicit reference to RAP. The first part of this book is a critical evaluation of approaches based upon the idea of rational action, while the second part presents alternative approaches to RAP.

In Chapter 2 the importance of RAP for risk studies leads us to an in-depth look at the paradigm. Links between RAP and the rational actor idea as a worldview are explored in Section 2.1, where we discuss philosophical foundations of RAP, mainly with regard to the field of political philosophy. We do not move very far in these waters, to avoid distancing the discussion too much from the themes of risk and uncertainty. Rather, in the remaining sections of Chapter 2, we proceed to a discussion of RAP as a general theory. Here, we highlight its impact in economics, psychology and sociology.

RAP has inspired a wide range of special theories dealing with risk and uncertainty. These are covered in Chapters 3 and 4. Making up a major part of this volume, these chapters reflect on the paramount importance of RAP for studies of risk and uncertainty. We show that, arguably, RAP is a research tradition of such pervasive importance to the topic of risk and uncertainty that it cannot be ignored. Our critical review of these studies reveals that empirical findings obtained within this tradition have created stubborn anomalies embarrassing to RAP.

Revealing the shortcomings of RAP raises the question of alternatives. Part II is devoted to the search for alternative approaches to RAP. We take note in Chapter 5 of several special theories concerned with issues of risk and uncertainty that might successfully explain RAP's anomalies. Some of the most promising work in the sociology of risk belongs to this category. Empirical findings that are anomalous to the RAP framework can sometimes be treated in more satisfactory ways in these special theories. Some of these rely on broader conceptual schemes like cultural theory or social systems theory. Nevertheless, sociological approaches lack a general theory equivalent to RAP, thereby limiting their purchase of the topic. Given this limitation, we then ask: How can these theories be used together with the RAP-based theories which remain indispensable?

A more radical task is to ask: What approaches could successfully challenge the rationalist worldview that is shaping the global society of our times? Recently, several attempts, especially various versions of postmodernism that advocate the extirpation of the rationalist worldview, have generated considerable interest. In Chapter 6, therefore, we move to the level of worldviews. This leads to reconsideration of the dynamics of modernization, and especially the role of scientific rationality therein. Three current attempts to challenge the worldview which underpins RAP are discussed: postmodernism, fundamentalist revivals, and the development of critical theory.

In Chapters 6 and 7, we offer some hints at how a concept of social rationality – as opposed to the atomistic rationality of RAP – might be relevant for changes in what may be called the rationalist worldview. Such changes seem to be under way. These changes are emphasized by postmodernist authors. They claim that social reality can and should be investigated without relying on general theories. While there is much to be learned from postmodernism, we find that position self-defeating, and one that would lead to disastrous consequences in the field of risk management if it were adopted.

The consequences are no better for the contending position that argues for an approach to risk based upon a fundamentalist orientation which refuses the pluralistic outlook of RAP. One variant of this position comes from environmentalist quarters, where it is sometimes argued that environmental risks can and should simply be avoided by relying on some set of fundamental ecological values. We briefly discuss evidence suggesting that contemporary changes in worldviews may be strongly related to environmental problems. We conclude, however, that neither postmodernist relativism nor fundamentalist absolutism are acceptable orientations to deal with the risks with which humankind is confronted.

Instead, we argue for the stepwise development of a critical perspective which patiently tries to integrate and improve the insights offered by older traditions, listening to contradictory voices where conflicting interests clash. The danger of failing to deal with environmental and technical risks is real, but so is the failure to deal with these risks in a theoretically informed and responsible way. Dealing with such complexities means that the contributions of social scientists studying matters of risk are necessary, but not sufficient to properly comprehend the problem.

In the concluding Chapter 7 we will return to the level of general theory. We will not propose a successor to RAP, as this seems premature, but we will outline some requirements, elements, and implications of a plausible successor. Such a successor will need to include main insights of RAP while offering solutions to some of the paradoxes into which RAP has run both with regard to social phenomena in general and to issues of risk and uncertainty in particular.

A fresh understanding of markets is one indispensable element of any such integration. The theory will need to explain the operations of interdependent markets in which a multitude of economic agents interact under conditions of uncertainty, many of which lie beyond the control of market forces. With regard to the problems of risk management, we provide a list of likely issues that a successor of RAP might incorporate.

A core point can be illustrated with the example of financial markets. Often ignored is the fact that financial managers need to define thresholds of acceptable risk before they can start to maximize their profits. Without such thresholds, no well-defined decision problem exists. In fact, financial managers seem to define thresholds of acceptable risk in a complex process of social learning which, so far, has received surprisingly little attention. Future risk management will be required to facilitate similar processes involving a wide variety of stakeholders. We conclude by arguing that one promising way to foster such processes in the public realm is through the use of discourse methods as a means of public participation.

In sum, the plan of the book is to (1) outline the logic of RAP and its application to risk and uncertainty; (2) identify and explicate empirical anomalies embarrassing to RAP; (3) present alternative approaches that try to explain such anomalies; and (4) speculate as to an approach that may encompass RAP in a broader understanding of social rationality. The competing approaches are examined at the levels of worldview, of general, and of special theories. In Part I, we proceed in that order, while in Part II, we move from current exemplars of special theories and worldviews to the problems of general theory.

I. THE RATIONAL ACTOR WORLD

2. RAP: THE MONARCH AND HIS SHAKY KINGDOM

RAP provides the framework for many special theories of risk. Its strength is based on the fact that it combines *a model of individual action* and *a model of social interaction*. According to RAP, individual human agents are the atoms of the human universe. In this sense, RAP is based on an atomistic worldview. While atomistic ideas were first developed by the ancient Greek philosopher Democritus, they developed into a basic framework of thought, a worldview, only with the advent of modernity. In the realm of natural phenomena, Newtonian mechanics provided an atomistic explanation of an astonishing variety of phenomena. In the realm of human affairs, RAP played a similar role. Here we describe the worldview which provides the philosophical underpinnings of RAP as mechanistic thinking.[9]

The atomistic orientation of RAP assumes that human agents seek to maximize given utility functions under given constraints and to do so with the utmost privacy. This model obviously presents an incomplete picture of human existence. Birth and death, learning and talking, the quest for meaning, and the surprises of love and rejection are all reduced to minor details in the algorithmic machinery of utility maximization. At the same time, however, the RAP model has a truly amazing advantage over more subtle descriptions of human beings. When combined with a corresponding model for social interaction, it captures important aspects of how human beings interact in social settings, especially in a market economy. Much of the historical impact of RAP is due to the ability of these two combined models to describe some important aspects of what people actually do in a market economy.

Before looking at the use of RAP as a general theory, it is useful to consider the broader worldview of mechanistic thinking in which RAP is embedded. Toward this end, an examination of the origins of RAP in political philosophy offers a good starting point. In Section 2.1, we document how basic ideas of RAP first appeared in a philosophical context and how they are connected to the rise of mechanistic thinking.

RAP found its most fertile ground in the field of economics. Today, economics looms in stature over the other social sciences. The quantity of resources available for economic research outnumber those of any other social science discipline – probably all of the other social science disciplines combined. Economics is the only discipline in the social sciences for which Nobel prizes are awarded, a clear indication of the

---

[9] For philosophical studies of mechanistic thinking, see Whitehead (1929/1978), Dijksterhuis (1961), Einstein and Infeld (1966), Rust (1987).

relevance of RAP in contemporary culture. Thus, our discussion of RAP as a general theory looks first at economics. As we shall see in Section 2.2, economics has become the science *par excellence* of RAP, chiefly because of its success at formulating theorems and hypotheses in mathematical terms. The mathematics, and related quantitative empirical modeling of economics, has enabled the discipline to express the general theory of RAP with remarkable theoretical elegance.

It would be seriously misleading, however, to consider RAP as only an economic theory. By no means is the use of RAP restricted to the analysis of monetary phenomena. Mathematical formulation may be less prominent in other domains, but the basic ideas of RAP still provide powerful tools of analysis. For example, political institutions can be analyzed along similar lines as economic markets. In both cases, the interaction of self-interested actors may lead to desirable outcomes under suitable institutional conditions. With regard to political institutions, however, we will limit ourselves to discussion of the foundations of RAP in political philosophy (Section 2.1). After discussing economics (Section 2.2), we will look at the relevance of RAP for psychology and sociology (Section 2.3).

2.1 Philosophical Underpinnings of RAP

As noted above (Section 1.2) the modern concept of risk owes its origins to the long history of the emergence of rationality in western thought. Ancient philosophers claimed that reason enables human beings to strive for the good. They took it for granted that there is but one good, not considering the possibility that there may be many goods – both in an ethical and a material sense – among which people can and must choose. Medieval Christianity was concerned with the fact that human freedom includes the possibility to choose between good and evil (a problem ignored by RAP). It neglected the human ability to choose among goods. It was this ability, however, that became the focus of modern political philosophy with the publication of Hobbes' *Leviathan*.[10]

In the *Leviathan* Hobbes (1651/1968, Ch. XI) painted human beings as driven by "a perpetual and restless desire of power after power, that ceaseth only in death." *Felicity* was "continual success in obtaining those things which a man from time to time desireth" (Ch. VI). He took the existence of a wide variety of desires and aversions as a fact of human

[10] Our presentation of the philosophical origins of RAP has greatly profited from publications produced as part of the Foundations of Rational Choice Theory project, in particular Hollis and Sugden (1993), Heap, Hollis et al. (1992), and Sugden (1991).

life, and he considered that something can be said to be good for a person if it helps to realize the desires of that person. Hobbes rejected the idea of a good that could be apprehended by human reason. To say that a person can focus his or her desires on what reason apprehends as being good independently of these desires – a crucial claim of ancient and medieval ways of thinking – seemed nonsensical to Hobbes and his followers.

Reason, in this view, has nothing to say about the ends toward which human agency is oriented. Once ends are given, however, reason can help to find suitable means to attain them. In particular, reason makes it possible for human beings to understand that they can greatly improve their chances to realize their desires – especially the widespread desire to preserve one's life – by monopolizing violent means of action in the hands of a single authority. This authority, the *Leviathan*, will then be able to enforce contracts and prevent violence between individuals.

Given the existence of human desires, Hobbes argued that reason can help to establish rules useful for the attainment of desires.[11] The theory of the state which Hobbes developed from this premise is still highly influential. His arguments have been elaborated with sophisticated instruments such as the theory of games and by the application of economic concepts to political institutions (Buchanan and Tullock 1962; Olson 1965). As a result, contemporary political science has become one of the most important fields of application for RAP.

The view of human agency proposed by Hobbes was refined a century later in Hume's *Treatise of Human Nature*. He reformulated the dichotomy of means and ends with his famous statement that "reason is, and ought only to be the slave of the passions" (1740/1978, p 415). Hume did not intend people to behave in an uncontrolled "passionate" fashion, in the sense of "passion" that is now common. He saw "passion" as a faculty that is complementary to that of cold, calculating, manipulative reason; it expresses the warm, intuitive and also spontaneous part of our consciousness. Still, his view implies that through reasoned debate people can reach an agreement about means only, not about ends. If these are not shared from the outset, there is little chance of reaching a common decision.

Hume's two states of mind – reason and passion – are related to the concepts of beliefs and desires. The validity of a belief can be tested by comparing it to existing states of affairs in the world. If it does not correspond, the belief is judged untrue. Desires work the other way round. An inconsistency between one's desire and the world does not require that the individual reject the desire. Often, people attempt to change the

[11] This argument prefigures later forms of rule utilitarianism.

world so that the desire can be realized. According to this view, beliefs can and must be checked against reality, while there is no similar check for desires. There is a deep divide between facts and values, and therefore between "is-statements" and "ought-statements." Freedom then means both the ability to desire what one pleases and the ability to strive for these goods according to one's beliefs.

Hobbes and Hume both acknowledged that a person may hold desires that are incompatible under given circumstances. They did not clarify how a human agent could make a rational choice among such desires. That answer came from Jeremy Bentham (1789/1970), who claimed that what people actually desire is happiness, and that the contribution of an object to a person's happiness can be described as a quantitative variable. He named that variable *utility*.

Equipped with the concept of utility, the study of human agency could become an exercise in constrained maximization, amenable to description with the mathematical means of the calculus. Bentham's concept of utility prepared the ground for the development of RAP as a general theory for the study of human affairs. The immediate consequences, however, were of a more philosophical nature. Bentham treated utility much like a physical substance, like water, quantities of which could be added together, both for a single individual and, collectively, for a society as a whole. Technically, Bentham assumed a *cardinal* utility which allowed for interpersonal comparisons. The direct consequence of this remarkable assumption was a utilitarian approach to ethics. Assuming that everyone seeks to maximize utility, ethical judgment should be focused on the consequences of human actions and institutions for the utility enjoyed by human beings.

In this manner, the separation of a domain of facts and a domain of values emerged in modern philosophy. Similarly, Descartes proposed the separation of mind and matter in the development of modern science. Matter has been explored as a world amenable to factual knowledge. The position and momentum of physical bodies can be known by objective measurement, and from such knowledge future positions and momenta may be forecast. Values are not to be found in this world. They are crucial, however, for the world of mind. On the other hand, mental entities are notoriously difficult to describe with the kind of objectivity known from classical physics. Or, as the popular mid-19th century witticism went: "What's matter? Never mind. What is mind? No matter."

The separation of facts from values implied that the latter cannot be grasped by scientific reasoning. It took a long and painful development to acknowledge this consequence of the strict separation between "is-statements" and "ought-statements." The failure to do so is known in modern philosophy, after G. E. Moore, as the "naturalistic fallacy." Rationalist philosophers like Descartes and Spinoza certainly thought that rational arguments as rigorous as those of mathematics could lead to reliable knowledge about matters of ethics and even of religion. Hume did not share this confidence in the power of reason, but he was convinced that although there is no scope for rational deliberation about moral issues, human beings can orient their actions towards a set of common values apprehended by the faculty of passion. But gradually, a different insight gained ground in modern culture: *If values cannot be justified with rational means, then people can differ about questions of value to a very large extent indeed.*

Earlier, we pointed to Michelangelo's David as the exemplar of Renaissance man: free to make and pursue projects of his own will. The figure of David, however, conveys a supreme sense of accountability toward both his political community and his God. David certainly belonged to a moral community and felt able to explain with arguments acceptable to this community why the project he had chosen was just. But with the shift in emphasis from action oriented to projects to action oriented to purely subjective preferences another possibility gained prominence: the possi-bility of a human life geared to arbitrary goals, unjustified and unjustifiable. Michelangelo had presented the promise of self-fulfillment enshrined in the modern idea of freedom as the ability to choose between many goods. From this idea, it took politicians and philosophers a small step to the view that moral issues are not amenable to reasoned debate at all. The eventual intellectual outcome was nihilism. In the arts, Dostoevski represented this moral nihilism with the figure of Ivan Karamasov. Nietzsche may have been the first to express this possibility in philosophical terms, and building upon Nietzsche, Foucault (1971) provides a sophisticated case for nihilism with his way of linking truth to power.

If values, goals, desires, and preferences are presuppositions of rational action about which rationality has nothing to say, and if they are not shared by human beings as the result of some prestabilized harmony, then human beings are faced by a solitude unknown to ancient thinkers. It is the awareness – or is it an illusion? – of this solitude that forces the worldview of modern times to rely on a basically atomistic view of rationality. Every individual must deliberate for her or himself before a group of individuals can agree about rules for living together.

Surprisingly, perhaps, this type of social atomism opened the way to a rational understanding of social orders. If human beings exist independently of each other, but interact in social settings, they can stabilize their interactions with appropriate institutions, such as the state, the church, and the market. Basically, rational actors pursue their self-interests by engaging in relations of exchange with each other, they don't try to solve problems by talking to each other. These relations of exchange, however, may establish a remarkable degree of order.

There is a strong similarity between the atomism of RAP and that of classical physics. Newtonian physics described the world of matter in terms of atoms located in space and characterized by quantitative properties like position and momentum. Interaction between these atoms follows natural laws which can be described in mathematical terms. Along these lines, a comprehensive theory of physical systems could be formulated and used for the development of an amazing variety of technologies.

RAP offered the prospect of analyzing the world of human actions in a similar way: by studying rational actors conceived as atoms, which form complex patterns through their interactions. Eventually, RAP evolved into a comprehensive theoretical framework with the theory of economic equilibrium. Each atom can be characterized by an action space and a set of preferences. The interaction between these atoms unfolds according to social laws which again can be described in mathematical terms. This theory is not amenable to technological applications in the world of matter, but it has a tremendous practical impact when it comes to the shaping of social institutions in contemporary societies.

Development of the sciences of matter and of mind were similarly shaped by an atomistic philosophy.[12] It turned out that modern mathematics – especially by means of the calculus – could be used to express theories in an atomistic format. While classical physics was formulated from its origins in mathematical terms, RAP originated in political philosophy without mathematical tools. Mathematical modelling became possible, however, with Bentham's analysis of rational choice in terms of utility. The calculus, which had been developed in order to attain a mathematical description of mechanical systems, became the fundamental mathematical tool for the description of social reality, too. In physics, the calculus was used to design differential equations representing dynamic systems; in economics the calculus was used to represent patterns of constrained maximization. In both fields, the idea of an objective reality consisting of atomistic entities interacting in well-specified patterns and amenable to an unambiguous description seemed self-evident. Mechanistic think-ing provided the worldview in which the paradigms of classical physics and RAP could flourish.

[12] Freudenthal (1982) provides a remarkable analysis of the relation between atomism in social philosophy and in Newtonian physics.

2.2 Economics: RAP as a General Theory

2.2.1 Trusting the Invisible Hand

It is impossible to understand the relevance of RAP for contemporary culture without paying careful attention to its role in shaping our understanding of economic institutions. Few would doubt that the global economy is one of the most influential realities of today's world. And few would doubt that it is driven to a considerable extent by the interplay of a multitude of agents in self-interested pursuit. We all use notions of supply and demand to explain prices or price changes obtaining on all kinds of markets. And we have some rough idea that supply and demand are related to the actions of enterprises striving for profits, and of consumers and households trying to make the best of their resources. This idea usually goes along with the related notion that this interplay of rational actors operating in a market setting brings about some kind of equilibrium. A market economy seems to have a remarkable capacity to allocate milk and Coca-Cola, cars and TV sets, computers and plastics in such a way that, as a rule, supply and demand match each other quite well. Contemporary economic policy is informed by more or less sophisticated models relying on this same body of ideas.

Adam Smith was the first thinker to provide a comprehensive analysis of economic matters in RAP terms. He seized upon the concept of rational action developed in political philosophy and applied it in a highly original way to the reality of capitalist economies. Although the metaphor of the "invisible hand" plays only a marginal role in Smith's writings, it has since been used to convey the thrust of his overall argument. This argument provided the first comprehensive analysis of how a modern market economy enables rational agents to coordinate their actions efficiently. Smith's *Inquiry into the Nature and Causes of the Wealth of Nations* (1776/1976) showed how the atoms of the social universe could arrange themselves into the complex patterns of the social division of labor.

His achievement was threefold. First, he was able to collect a multitude of empirical observations, mainly anecdotal or historical in kind, about the global economy gradually emerging in the long wake of the Middle Ages. Second, he was able to give clear advice to governments trying to find out how best to deal with this new reality. At first sight, his advice was paradoxical: When it comes to economic matters, governments should not try to govern. By and large, an economy without government interference would produce outcomes which are far more desirable than the ones which could be achieved by government intervention. Smith was by no means dogmatic on this point; he saw several situations where government actually should intervene in economic processes, as when the economic use of unskilled labor deprived many people of the chance of personal self-realization. But his basic advice remained one of laissez-faire – leave things alone. His third achievement consisted in justifying this advice with a careful analysis of how markets operate. In other words, he offered good reasons to trust the invisible hand of the market.

It is the last point that is most relevant. Smith developed a remarkably sophisticated analysis of market dynamics, and without this analysis RAP would be much less relevant for the social sciences and for our understanding of the economy. Therefore, we now look at some crucial steps of his argument.

Smith claimed that consumers and producers alike have a shared idea of an adequate price for a product. In classical political economy this price has been called the natural price. In Smith's view, the quantity demanded at the natural price is given by historical contingencies. If supply matches demand, trade takes place at the natural price. If supply falls short of demand, however, two things happen. In the short run, market prices rise above natural prices so as to bring demand to a level which corresponds to supply. In the longer run, entrepreneurs react to these price increases by expanding capacity with additional investment. And they do so precisely because they want to maximize their utility, which in this case corresponds to their profits. As long as the market price lies above the natural price, profits are exceptionally high, and in their chase for profits entrepreneurs will invest in the corresponding line of production, thereby expanding capacity until demand can be satisfied again at normal prices. If supply exceeds demand, an opposite but symmetrical process occurs. If government, however, interferes with market forces, it runs a serious danger of destroying, or at least impairing, the adaptive potential of markets.

This analysis treats entrepreneurs as rational actors interacting in the specific institutional setting of a market economy. And it shows how the interplay of these actors leads them to satisfy a social task of considerable importance: satisfying the needs and desires which consumers express at given normal prices.

Although the preceding sketch does not exhaust Smith's overall argument, it is sufficient to show that Smith's use of RAP to explain the operation of market mechanisms was far from trivial. In certain respects, as with the distinction between natural price and market price, it was even more sophisticated than more recent versions of RAP. On the other hand, it lacked an explicit analysis of consumer behavior and was weak in its treatment of interdependent markets. Filling in the lacunae in Smith's argument turned out to be a daunting task, so much so, in fact, that it remains to a large extent unfinished business. This is all the more remarkable as a large number of scholars have devoted a tremendous amount of work to the problem, beginning with the Lausanne school of economics and sociology (Pareto 1927; Walras 1874-1877), continuing with eminent mathematicians (von Neumann 1935-36/1945; Wald 1936), and leading to modern general equilibrium analysis (Arrow and Debreu 1954; Debreu 1959).

If we deliberately postpone the treatment of uncertainty, the state of the art which has resulted from this long cumulative research effort may be characterized by three steps: rational decisions; competitive markets as based on rational decisions; and two main results of RAP with regard to economic matters.

Rational decisions can be codified with three statements:

- Rational actors can choose between different possible actions. Actions may differ in kind and in scale.
- Rational actors can order possible actions according to their preferences.
- Rational actors try to choose an action which is optimal according to their preferences.

These statements can be specified in mathematical language so as to yield an axiomatic basis for a formal theory. Research questions that can be addressed in this framework include the following: Under what conditions does a decision problem have a unique optimal solution? To what extent do people and organizations actually behave as rational actors? How can the theory be used to improve actual decision making processes? Can the theory be generalized so as to allow for changing preferences, ill-defined alternatives and the like?

The same RAP structure underlies competitive markets. Markets can be described by specifying two kind of rational actors: consumers and entrepreneurs.

- Consumers are rational actors faced by given prices. They sell resources (like labor services and the right to use existing capital stock, land and the like) and use the money so obtained to buy products. They prefer having certain combinations of products and services to others.
- Entrepreneurs are rational actors faced by given prices. They sell products to each other and to consumers and they buy products from each other and resources from consumers. They need resources and different products to produce their own products. They prefer having a larger profit to a smaller one.

These statements also can be expressed as mathematical formulae, permitting many research questions to be addressed: Under what conditions can the behavior of consumers and entrepreneurs be described by simple supply and demand curves? How can the theory be generalized so as to allow for money and credit, for foreign trade, for markets where single entrepreneurs can actively influence prices?

The two most fundamental research questions, however, are the following: Under what conditions will supply and demand brought about by rational actors match? And in what sense can this match be said to be a good thing? With regard to the latter question, it is important to bear in mind that within RAP any standard of desirability must somehow be geared to the preferences of actors. The answers to these two questions are given by two theorems which can be outlined as follows:

- With suitable assumptions about the structure of consumer preferences and about the production possibilities available to entrepreneurs, at least one array of prices exists at which supply and demand for all traded goods match. The corresponding situation is called an economic equilibrium.
- With suitable assumptions about the structure of consumer preferences and about the production possibilities available to entrepreneurs, it is impossible to find a situation where all actors are better off according to their own preferences than they would be in an economic equilibrium.

These theorems can be proved mathematically if the statements introduced in the first two steps are expressed as mathematical axioms.

Generations of researchers have developed RAP into an extremely powerful theory of human action. In particular, it has been used to develop a fascinating picture of how modern economies work. RAP combines models of human actors with models of the interaction between them. The former are kept simple in order to keep the intricacies of the latter manageable. In many respects, these models operate as metaphors to describe human actions and interactions. A metaphor is not literally true, but a vehicle for connecting seemingly dissimilar ideas. In a similar sense, RAP provides an important orientation for researchers and practitioners dealing with human affairs in general and economic phenomena in particular.

So far, we have deliberately avoided the theme of uncertainty. But clearly, in many markets uncertainties abound. Tourism and agriculture are heavily weather dependent, and the weather is notoriously hard to predict. Entrepreneurs investing in long term projects inevitably have to deal with many unknown and unknowable aspects of the future. And in the business of insurance uncertainty is not simply a difficulty one has to deal with, it is the basis of the business itself. Under these circumstances, it should come as no surprise that much effort has been devoted to the treatment of decisions under uncertainty in the framework of RAP. The resulting insights and research problems will be discussed at length in the present volume. In order to understand the general thrust of these efforts, however, it is crucial to have at least some understanding of how RAP deals with utility.

2.2.2 Adventures of Utility

The utilitarian tradition has had a long and rich, though sometimes tortured, history in economics (Meeks 1984). That history offers an especially apt illustration of the unshakeability of general theory (RAP), despite repeated assaults on and reformulations of a related special theory – utility theory in this case. Utility theory seeks to explain the purposive actions of individuals. In its classical formulation, based upon the work of John Stuart Mill and Jeremy Bentham, utility was equivalent to happiness. Operationally, happiness meant "pleasure" or "the absence of pain." Bentham (1789/1970) took this one step further. To him, pleasure was not simply the ethereal stuff for poetry and novel, but a quantifiable concept. It could be measured precisely.

Bentham thought that utility is the reason for preferences. People prefer, say, to live in nice landscapes than in ugly ones because the former contribute more utility to their overall level of happiness than the latter. In this conception, differences in utility are measurable, just like spatial distances from some point of origin. And, as with distances, one may choose different measurement units and different points of origin. But while it is arbitrary what level of utility we want to treat as a zero level and what amount of utility we want to use as measurement unit, all other properties of the utility function are fixed (see Figure 2.1).[13] A utility function which can be modified only in these two ways is called a cardinal utility function.

With cardinal utility functions, utility levels can be added to each other. It makes sense to ask how large the overall level of utility of a given person is, and it makes sense to ask how much different goods contribute to this utility level. In the 1870s authors such as Jevons (1871/1970) began to elaborate a mathematical apparatus for the treatment of utility functions in economics. The famous law of decreasing marginal utilities, for example, could be expressed in terms of derivatives of utility functions. It is empirically highly plausible that the satisfaction people derive from an additional unit

[13] We take some pains to represent utility curves because the analytical relations between ordinality, cardinality, and risk are fundamental to RAP and often misunderstood.

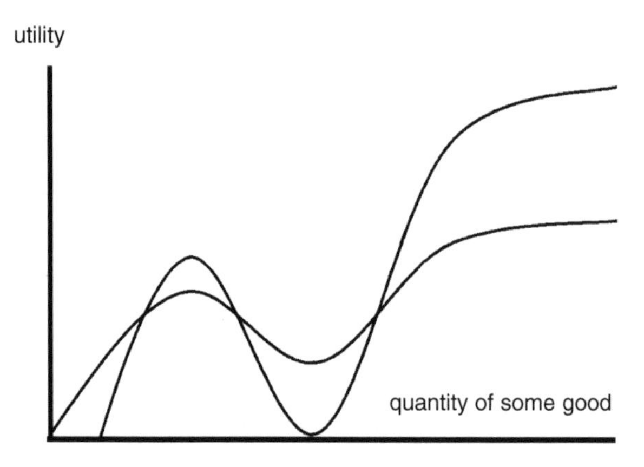

Figure 2.1: Two representations of the same cardinal utility function (differing only in definition of zero level and of unit on the utility axis)

of some good usually decreases with the number of units. A thirsty person will enjoy the first glass of water more than the third one, as the famous example goes.

If utility can be measured quantitatively, however, an interesting question arises. Suppose we take away a glass of water from a person who is not very thirsty and give it to a very thirsty person: Isn't it plausible that the first person has lost less utility than the second one has gained? This is the problem of interpersonal comparability of utility. One may claim that utility is a cardinal concept for any given person, but that it does not make sense to treat utility as a cardinal concept across persons. This claim, however, looks dangerously close to a thinly veiled strategy to protect the rich against requests for redistributing their wealth to the poor. The least one can say is that the notion of cardinal utility has a strong egalitarian connotation. As there is little reason to give different weights to the utilities of different people, equality in the distribution of wealth and income seems a desirable goal.

With the emergence of the "marginal revolution" in economics, the utility concept reached its heyday in economic thought (Meeks 1984). Yet, ironically, while utility was becoming more firmly embedded in *Homo Economicus*, the interpersonal measurement problem admitted of no easy solution. Indeed, key members of the marginalist school – e.g., Walras and Jevons – held that interpersonal comparisons of utility were impossible. Theory continued to develop in the absence of measures to test the theory, but utility theory in economics appeared to be ready for demise.

The day was saved – at least temporarily – by Edgeworth's (1881) bold design of a "mathematical psychics". He argued that measuring utility was an appropriate operational quest. The problem was that utility theorists had gone about it in the wrong way. They had been trying to make cardinal measurements of utility when utility was, in fact, an ordinal concept. Cardinal measures are numerical magnitudes that have an established zero point and where all unit values between numerical magnitudes are equal. For example, the increase in cardinal magnitude from 1 to 2 is equivalent to the increase from 9 to 10. Ordinal numbers represent rankings – 1st, 2nd, 3rd, 4th – where no measurable distance may be defined between ranks. Edgeworth suggested that efforts be devoted toward measuring utility in ordinal terms.

Pareto (1927) stepped up to that challenge and developed an ordinal system of measurement that emphasized, not the numerical measurement of utilities, but the rank preferences of different sets of goods. In Pareto's understanding, there is no such thing as a "level of utility" reached by an actor. All there really is are preferences by which actors rank different alternatives. Moreover, an ordinal utility concept implies that interpersonal utility comparisons are meaningless.

Notice, however, that Pareto's approach requires that preference structures satisfy a series of non-trivial assumptions: in particular, each actor must hold unique preferences over all possible alternatives, and these preferences must be transitive. For economic applications it is also usually assumed that when faced with two quantities of an economic good, an actor always prefers the larger one. Whether such a preference structure is less of a "metaphysical entity" than a cardinal utility function is open to debate.

Thanks to the notion of ordinal utility, RAP could rely directly on preference relations and treat utilities simply as shortcuts to represent certain preference structures. If, for a given actor, preferences satisfy the required axioms, they can be represented by ordinal utility functions (which associate each object of preferences with a real number representing its utility for the actor in question). An ordinal utility function can be represented with many different shapes. All that is required is that when an actor prefers alternative A to alternative B, A is assigned a larger utility than B.

Despite the Edgeworth-Pareto advancement, economics was not yet Camelot. Of concern to a next generation of economists, especially Hicks (1940), was whether utility should be measured at all. After all, embedded in the measurement of utility were a number of challengeable psychological hypotheses, not to mention philosophical assumptions. Invoking Occam's razor, many economists concluded that the measurement of utility – including its psychological baggage – was inessential to a theory of consumer demand. Instead, consumer demand could better be explained

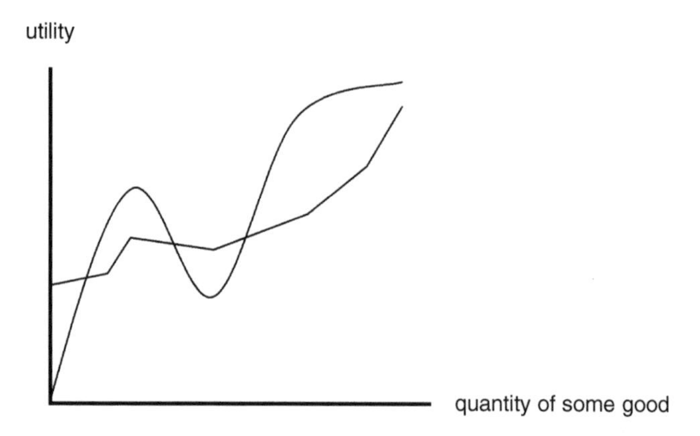

Figure 2.2: Two representations of the same ordinal utility function (increasing and decreasing segments coincide)

with a concept and analytical device purged of psychological speculation. So, the idea of ordinal utility was boiled down to the notion of indifference curves – the mathematics stayed the same, but the intuition geared to it changed. The idea behind indifference curves is to locate sets of goods regarded by some consumer as equivalent and to rank these sets according to this consumer's preferences.

An especially nice feature of indifference curves turned out to be their ability to reproduce the "law of diminishing marginal utility." The story of the thirsty person, for example, can be told in terms of indifference curves. If the person in question is indifferent, say, between the first glass of water and a sum of $10, it is quite likely that she will be indifferent between the third glass of water and a sum of money which is smaller than $10. The behavior that had been described in terms of diminishing marginal utility can equally well be described in terms of the shape of indifference curves.[14]

Despite the adoption of indifference curves, consumer demand theory was still not free of psychological underpinnings. Ordered preferences were inferred from axioms implying psychological proclivities, such as "consumers prefer more to less." The solution proposed by Samuelson (1950/1969) was to ignore consumer reasoning leading to consumption choices and to instead use the operative theoretical principle of "revealed

[14] The crucial feature is for indifference curves to be strictly convex. This means that any point on a linear segment between two points on an indifference curve should lie above the indifference curve. Thus, getting an apple and an orange is always preferable to getting two apples or two oranges.

preference." The doctrine of revealed preference is an operational procedure for purging psychological assumptions behind individual choices. Knowledge of a consumer's preferences will not be found in calculated utility or in "concealed indifference," but in the consumer's actual consumption behavior, specifically, behavior that manifestly reveals those preferences. In other words, we can work back to an understanding of a consumer's preferences about a product by observing the choices he or she made in the past about similar products.

To this point we have disregarded the problem of decision under uncertainty. In fact, utility theory has been tied time and again to problems of uncertainty. The basic idea has always been to use probabilities as weights for utilities. If one believes in cardinal utilities and in objective probabilities, it is quite plausible to handle a decision problem under uncertainty as follows. For each action under consideration, form a list of its possible outcomes, of the cardinal utilities of these outcomes, and of the objective probabilities of the same outcomes. The utility of an action is then equal to the sum of the utilities of each outcome weighted with the probabilities of these outcomes. This is called the expected utility of an action. The case of certainty then results simply as the degenerate case of a unique outcome with probability one. In this setting, a rational actor will strive to maximize expected utility.

Now consider a problem for which no objective probabilities are available. This may be due to simple lack of information, or to the fact that the decision under consideration is unique, or to other reasons. Then a rational actor can still have subjective probability judgments. These judgments are now quite similar to preferences. Just as people may prefer apples to oranges or the other way round, they may consider that apples are more likely to foster their health than oranges or the other way round. In such situations, one can still form an average of utilities weighted with probabilities. To stress the subjective character of these probabilities, the notion of subjective expected utility (SEU) is used for these cases.

Combinations of an action with its possible consequences are often called lotteries (Kreps 1988). Choice under uncertainty can then be discussed as a choice between lotteries. SEU theory claims that rational actors hold subjective probability judgments as well as ordinal utility judgments about the different outcomes and that they try to choose that lottery whose subjective expected utility is greatest.

At first sight, subjective expected utility theory may seem to break down when cardinal utilities are dropped. As we have seen, ordinal utilities may be stretched nearly at will, and this would mean that one and the same action had many expected utilities, subjective or otherwise. In the 1940s, however, von Neumann and Morgenstern turned this argument upside down (von Neumann and Morgenstern 1944). After all, as long as actions have only one possible consequence, the idea of a cardinal utility function had no clear empirical meaning. All that is needed in these cases are the shapes of the indifference curves with the corresponding ordinal utility function. When it comes to choices between lotteries, however, things are different. According to von Neumann and Morgenstern, it is only now that cardinal utility functions do have empirical meaning. They demonstrated that, for decisions under conditions of uncertainty, a cardinal utility function *could* be constructed from actual choices, as long as these choices satisfied certain postulates. The initial formulation, along with the steady stream of work it generated, elevated the expected utility model to "the operational definition used in empirical work related to tests of rationality" (Hogarth and Reder 1987, p 4).

Consider the choice between (1) investing a million dollars in a risky project which after a year may yield a rate of return of 15% if all goes well and a rate of return of 0% if things go wrong, or (2) investing the same million dollars for one year at a predefined rate of interest. And suppose that, for no particular reason, perhaps, the person faced with that choice firmly believes that success of the project is twice as likely as failure. Finally, suppose the person in question is indifferent between the two projects at a rate of interest of 5%. At a lower rate of interest, she prefers the lottery, at a higher one, she goes for the rate of interest.

Figure 2.3 shows the use of cardinal utility functions for such choices. Call the rates of return X1 and X2 (corresponding to 15% and 0%). As long as only choices under certainty are considered, the agent under consideration prefers X2 to X1. Accordingly, to describe such choices any ordinal utility function will do as long as it yields a higher utility for X2 than for X1. Take any such function to start with (this implicitly defines the zero level and the unit of the cardinal utility function at which we are aiming). Now take our agent's subjective probabilities and compute the agent's expected value X3 by weighting the two outcomes with the respective probabilities (this yields 10%). The utility of the lottery <X1, X2> corresponds to point C – this is the weighted average of utilities A and B, using the probabilities of X1 and X2 as weights. The 5% rate of interest is point X4 – this is the certainty equivalent of the lottery. Therefore, its utility must be equal to C. In other words, the utility of certainty equivalents is no more arbitrary.

Along these lines, we can construct a utility function for choices under uncertainty. In our case, it will yield a higher value for the utility of the expected value than for the lottery: If our agent were offered a rate of interest of 10%, she would prefer it to the lottery. Along these lines, a cardinal utility function can be construed for all relevant outcomes. The curvature of the utility function may be used as a measure of risk aversion. Only with a linear utility function do the utilities of lotteries coincide with those of their expected values. The actor is then said to be risk neutral.

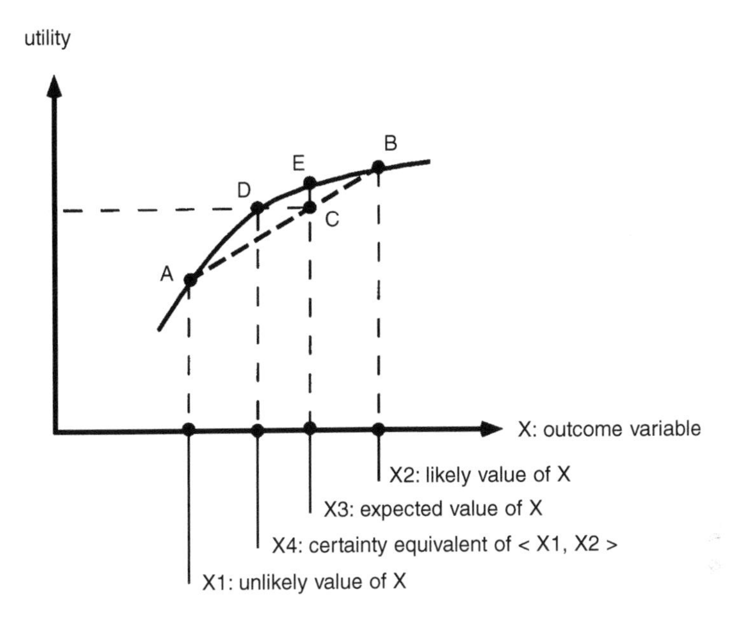

Figure 2.3: Cardinal utility function for choice under uncertainty

The issue of linearity is essential in another respect, too. SEU theory assumes that the utility of a lottery is linear in the probabilities of the different outcomes. Consider a lottery which yields $10 if a tossed coin shows heads, and nothing if the coin shows tails. According to SEU theory, the utility of this lottery – which is a cardinal magnitude because of the presence of risk – must be half the utility of $10. A large body of research about decision making under uncertainty revolves around empirical evidence refuting this claim (Machina 1987a, 1987b).[15] But of course there are many situations for which SEU theory is perfectly reasonable.

Paradoxically, it is precisely the redundancy of cardinal utility functions for the study of choice under conditions of certainty which makes this tool available for the generalization of RAP to conditions of uncertainty. And this in turn greatly facilitates the inclusion of various kinds of uncertainty into the study of market processes. Not only is RAP able to provide powerful metaphors and insights about the working of the economy, it can also do so while explicitly considering choices under conditions of uncertainty.

[15] The basic idea of the present book owes much to one of the authors (Jaeger) reading Machina's brilliant analysis of the paradoxes arising in RAP-based risk research. The history of these paradoxes – from Allais to 'versky' – reminded him of the tension between paradigm and anomalies as discussed by Kuhn (1962).

The tools used for this purpose can be used for the study of risk outside the realm of markets, too.

The crucial step turns out to be the distinction between actions and outcomes which may or may not happen as a result of actions. This does not necessarily imply a consequentialist view of human agency. Take the action of lying. One may want to judge it exclusively in terms of its effects, but one may also want to say that lying is bad intrinsically. The latter view can be expressed in the present framework by defining as one of the outcomes of lying the fact that the actor in question has lied. And one may then add that the actor in question prefers having said the truth to having lied.

If we now explicitly include problems of risk and uncertainty, RAP as a general theory may be characterized by the following four statements:

- Rational actors can choose between different possible actions, each of which may lead to one or several possible outcomes. Actions as well as outcomes may differ in kind and in scale.
- Rational actors assign (objective or subjective) probabilities to various outcomes.
- Rational actors can order actions as well as their outcomes according to their preferences. Preferences for actions involve some (positive, zero or negative) degree of risk aversion for specific choice situations.
- Rational actors try to choose an action that is optimal according to their own preferences.

2.3 Inside the Atom: RAP and Psychology

For the social sciences, RAP provides a powerful micro-macro interface. It does so in a bottom-up perspective: Given the preferences and probability judgments of individual agents, the overall social outcome follows as soon as their opportunities for action are given. Global markets, population dynamics, war and peace can be analyzed with this schema. The simplicity of this reductionist construction, however, conceals a subtle feedback mechanism – *for any individual, the opportunities for individual action, as well as the possible outcomes of her actions, depend on the actions of all other agents.* The intricacies of RAP-based social science, especially, but by no means only, of economics, are due mainly to the complexity implied by this feedback. Individual actors, on the other hand, are treated as mere black boxes, social atoms that require no further analysis.

This does not imply, however, that RAP excludes such analysis. It is of obvious interest to ask how preferences and probability judgments come about and how rational actors gather and process information about choice situations. Psychological research has much to say about these questions. Two basic approaches dominate the psychological literature in this area: habituation and value-expectancy (Jungermann, Pfister et al 1998).

The *habituation approach* is inspired by the basic belief that humans act either as instinctive, biologically driven creatures or as socially determined members of larger social units. In this view, actions are not the result of reasoning about potential consequences; instead, they originate from conditional learning on the individual level (as epitomized in its behavioristic branch) or from social pressure and the drive for conformity (as epitomized in its sociobiological branch) (Hamilton 1964). Behavioral conditioning is strongly dependent on the use of reinforcements. A drive for conformity, however, is spurred on by a universalist trait – seeking social acceptance.

Rational behavior – behavior selected because it was believed to optimize personal satisfaction – is certainly compatible with the habituation approach. Rationality is a vehicle for satisfying basic needs for sociobiological existence – positive rewards and social support. Single actions in the habituation approach may be opposed to one's personal utility, but they are part of a larger cluster of behavioral elements used to satisfy instinctive needs (whatever they may be) or to adapt to social norms.

The strong behaviorist flavor of the habituation approach has lost some of its influence within psychology, with the ascendancy of cognitive science (Dawes 1988; Dulaney 1968). However, advances in neurophysiology and evolutionary biology reemphasize the notion of basic needs as major drives of human behavior (Trivers 1985; Wilson 1978). While the grand claim of sociobiology to be able to explain social behavior entirely in biological terms is far from being scientifically credible, more specific insights deserve careful attention. In this vein, beyond the simplistic human instinct model (for example, McDougall 1920), the basic needs model based on Maslow (1954) or Rostow (1971), or the conditional learning paradigm (for example, Skinner 1971), yield new typologies of human needs and drives.

According to these typologies, human needs and drives are based on biological functions, but are at the same time socially and culturally shaped through socialization procedures (Eibl-Eibesfeldt 1974; 1979). However, the traditional object of psychology, i.e., the individual, turns out to be too narrow a focus to explain the variety of values and motives in different cultures and social milieus. The new theory claims that biology and anthropological factors define the limits of the range in which culture and social factors exercise their designing and modifying power. According to this view, it is the legitimate task of psychology to investigate the biologically or anthropologically given limits for human actions, while the investigation of the reasons for the variety within these limits should be left to the field of sociological or cultural inquiry (Rappaport 1971; Steward 1955).

The habituation approach, however, has come under attack from more recent cognitive research (Neisser 1967; Fischhoff, Goitein et al 1982). This research puts much greater emphasis on deliberative mental processes. Therefore, it provides fertile ground for the hypothesis that human beings act in response to a personal evaluation of the expected consequences of their and other people's actions. Personal values operate as yardsticks for evaluating consequences. Subjective probabilities operate as yardsticks for estimating the likelihood of consequences. This *value-expectancy approach* is almost identical to rational choice theory within the rational actor paradigm (Miller, Galanter et al 1960; Simon 1987). The focus is slightly different from the economic viewpoint, however. Here, human action is believed to be driven by the subjective expectation of the most likely consequences of one's choices. These consequences may be "real," "perceived," "irrational," or "pure products of fantasy." The emphasis, then, is on the cognitive processes by which human beings perceive possibilities for action and likelihood of outcomes (Kaminski 1983).

Both approaches provide evidence for their claims. Many personal decisions are driven by the motive to increase one's own utility, while others are non-reflective routine actions by individuals in correspondence to social norms or individual habits. The debate has therefore shifted from the fruitless question of who is right, to the more interesting problem of which approach is more suitable for gaining new insights into human behavior (the descriptive-analytic question) and which approach is ethically more appropriate (the normative question).

On the descriptive-analytic level of debate, both sides can claim success. The proponents of habituation theory have collected massive evidence indicating that, under the right circumstances, humans are capable of learning or imitating behavior that is opposed to their own interests and values. The infamous Milgram experiments, however repre-hensible, illustrate this point (Milgram 1963). The proponents of value-expectancy approaches were also able to demonstrate that, under different circumstances, the learning approaches failed completely and were replaced by considerations about expected consequences in the test subjects (Trabasso and Bower 1964). Thus, the debate can be partially resolved by specifying the circumstances under which one or the other approach is a better predictor for human behavior. Humans seem to be driven by biological drives and by social routines as well as by deliberate cost-benefit analyses.

On the normative question, it is difficult to defend habituation claims. Depicting humans as prisoners of biological drives and social needs ignores some of the unique features of *Homo sapiens sapiens*, namely the capability for complex reasoning. Critics of the habituation approach labeled its proponents "cynics" or "reactionaries," because they stress the importance of inherited biological factors over cultural learning (Vayda and Rappaport 1968). This essentially reduces human motivations to basic needs, including social support. For a long time, the habituation approach was a "theory *non grata*," dismissed as mechanistic, reactionary, and neo-Darwinistic. The simplistic "nature versus nurture" conflict epitomizes the ethically laden debate, in which one side was clearly enjoying an advantage, not because of better evidence, but because of a more appealing ethical justification.

The temporary victory of the value-expectancy approach in psychology on the basis of ethical considerations created an awkward situation for the whole discipline. As with the rational actor paradigm (RAP) in general, values and goals are excluded from analysis in the value-expectancy approach. Instead, they are taken for granted. Scientific inquiry into human decision making is reduced to the investigation of the mental processes in which personal values are applied to perceived consequences and to how people perceive consequences. The absence of explanations for the genesis of individual values, as well as the rejection of integrating subconscious elements into human motivations (other than mystically influencing value formation), were the two major flaws that led to a growing dissatisfaction with the value-expectancy approach among various psychological scientific communities (Gardner and Stern 1995; Graumann and Kruse 1990; Seamon 1987; Stern 1992). The omission of the question "From where do the values come?" resulted in a confusing alienation of cognitive psychology from other major traditions, most notably the psychoanalytic, motivational, and developmental psychology schools.

These traditions emphasize the importance of emotions, effects, motivations, archetypes, subconsciously triggered associations, as well as biological drives. They have developed their own schools of thought, and they have been a thorn in the side of the value-expectancy camp, serving as a reminder of its shortcomings and problems. In fact, these traditions include elements of both mentioned approaches to human decision making. On one hand, they emphasize (early) socialization as a major drive for value formation as well as value deformation. They also investigate the role of imitation, imprinting, and social reinforcements in the development of personal motives or values. On the other hand, although they integrate value-expectations in their framework of human motives, their focus on deviant behavior often leads to an emphasis on potential distortions of personal perceptions of consequences, and misallocations of values, rather than in "normal" processes of personal utility optimization.

A second line of criticism of the value expectancy approach came from the opposite side of the political spectrum within psychology, the proponents of a critical psychology aligned with ideas of the New Left (Holzkamp 1983; Markart 1990). Although the traditional left had violently opposed the old habituation tradition, they also suffered the consequences of the "value black hole" within the value-expectancy approach. Critical theorists missed the inclusion of positive, emancipatory values. Leaving the value orientations to individual preferences was regarded as a static viewpoint, one which did not incorporate the eschatological vision of a better society promoted by Marxist and humanist perspectives.

However, the New Left did not want to give up the notion of rationality as a means to accomplish predefined goals. RAP provided at least a basis for the possibility of engineered (in contrast to random) social change, but it provided no clue for specifying the goals of human actions. By different-iating concepts of rationality, such as instrumental, social, and cultural, critically minded researchers tried to incorporate the perspective of rational actors within the framework of a rationally derived set of social values, visions, and goals.

The continuing problem with this approach has been the inability of theorists to find a logically coherent and intersubjectively valid deduction of values and goals for all individuals and society as a whole. Traditional Marxists continued to believe in "hidden" laws of historical development from which they believed universal values could be deduced. Neomarxist and critical analysts focused on the structural circumstances under which individuals can develop and express their own preferences in the absence of class pressures and political power. According to this view, socially binding orientations should be the result of social discourse, in which individuals are given the opportunity to reflect their own interests as well as the common good of the whole group in an atmosphere of openness, sincerity, and equity. The discourse should be the place where individuals develop and specify both the goals and the means to accomplish these goals in an open debate, ideally through a consensual decision process (Renn 1998b; Webler 1995).

While the traditional Marxist approach has lost much of its attractiveness due to serious doubts about the validity of its historical model, the discourse model has gained momentum within and outside psychology. However, as Luhmann has pointed out, discourse theory has three major flaws (Luhmann 1971; 1997). First, it presumes that individuals can engage in an open and non-strategic discourse even if they belong to different social or cultural systems. Second, the theory does not provide a real explanation or a normative guideline for the genesis of social values; it only places constraints for designing the communicative circumstances under which such values are negotiated (rather than developed). Third, it narrows the field of possible procedures to articulate socially binding values to a single instrument, i.e., the process of discourse. Although valuable as a procedure for legitimizing decisions, discourses are not the panacea for providing value orientations in modern societies, which are characterized by value diversity and pluralism.

What then is the present status of the rational actor paradigm with regard to psychology? The older habituation approach fitted well with RAP, and the newer value-expectation approach fits even better. In this respect, RAP is quite relevant to cognitive psychological research. It also provides a research framework that seems to be internally consistent and empirically productive.

The criticism of both the habituation and the value-expectation approach, however, points to significant tensions between important psychological research and RAP. Four difficulties arise. First, preferences may be much less stable than RAP presupposes. In actual decision making, preference change may be a normal and influential process. Second, judgments of likelihood may differ in important respects from the mathematical structure of probability theory (Gigerenzer, in press). Third, the set of possible actions may not be as well defined as presupposed by RAP. And fourth, because of limited time, memory, etc., human beings may be unable to perform the systematic comparisons between alternatives required by RAP. We will meet these difficulties time and again in the following chapter as they will turn out to be major reasons for the search for a superior alternative to RAP. Here it must suffice to note that psychologists have not yet reached a consensus on any such alternative.

Finally, we should recognize that RAP assigns a very comfortable intellectual position to psychologists. If social reality is to be explained in terms of individual decisions, then psychology is part of the very foundations of any scientific inquiry into human affairs. By sharing the individualism which is so pervasive in contemporary culture, RAP gives prominent place to psychology as the study of human individuals. On the other hand, RAP can build on psychological theories which are very different from the simple framework of preferences, probabilities, and optimal choices as long as these theories can be seen as explaining the elements needed for this framework. In this sense, RAP is less a general theory within psychology than a more general theory of the social sciences which gives a unique place to psychology.

Peaceful coexistence between psychology and RAP thus seems not only the rule of past research, but also a plausible perspective for the future. Things are different with sociology, which is faced much more directly by RAP's claim that social atomism is the proper orientation for understanding human societies.

2.4 Sociology and RAP: The Fundamental Tension

Sociology owes its birth to three midwives: first, a tumultuous epoch demanding explanation; second, the prevailing intellectual apparatus, RAP via classical economics, for supplying an explanation; and third, the need to distance itself from psychology – a new and competitive discipline. Sociology emerged in the midst of the industrial revolution, by scholarly consensus one of the greatest periods of economic, political, and social upheaval in history. So vast, so pervasive, so cataclysmic and so obvious were the social changes produced by this revolution, that virtually all thoughtful observers recognized the need for explanation – the need for a coherent account of the apparent resulting social incoherence. This age of change was athirst for an explanation of itself.

The principal social science and the dominant theoretical paradigm of the time, of course, was classical, *laissez-faire* economics: the economics of Adam Smith, Thomas Malthus, and David Ricardo, but also its further development by James Mill, John Stuart Mill and the utilitarians, among others. As a founding perspective in social science, classical economics, and the RAP foundation that buttressed it, enjoyed the advantages that typically accrue to pioneer developers, whether of land or of intellectual landscapes; they establish first claim in the staking out of new terrain. Thus, as sociology emerged as a new discipline it had to confront the already laid claims of classical economics. Resting as it did on assumptions of the isolated, rationally calculative actor, the RAP of classical economics was rejected by sociology's founders for failing to recognize the importance of macrosociological and historical forces, on the one hand, and of non-rational, individual human forces on the other.

Classical economics – due to its RAP foundations – was seen as woefully inadequate to account for the tumult this era of change had showered on the Western world. Indeed, it is an irony of history that amidst an epoch of such social turmoil, where destitution was the lot of many and where for most people daily life was a brute struggle for existence, a rational social order could be discerned at all. In one important sense sociology arose in reaction to RAP – as a corrective to a blurred lens that could only see cohesion and advancement in the cataclysmic social upheaval of the epoch of industrialization.

2.4.1 The Macro-Micro Link and the Problem of Consciousness

All sciences are laden with presuppositions, unchallenged first principles, and orienting perspectives of theoretical method. Disciplined inquiry is driven by theoretical method. The idea of "theoretical method" used here reflects a concern with the development of cumulative research that is principled, systematic, coherent, and fruitful (see, for example, Freese 1980). Theoretical method may be considered as part of epistemology, but distinguished from it by its emphasis on procedures for generating knowledge claims. In contrast, epistemology emphasizes the bases, the logic, and the soundness of knowledge claims. Theoretical method is a framework for guiding inquiry that, appropriately applied, results in theoretical propositions corroborated by empirical regularities – social theory is only useful to the extent it can explain features of the social world. RAP clearly relies on a theoretical method that proceeds from the micro to the macro, while sociology typically proceeds from macro to micro.

The ideas of the Enlightenment both shaped the foundations of the human sciences and prompted a seemingly unavoidable dualism, one of foundational importance in conducting inquiry about human action. We can refer to this dualism, after Tarnas (1990), as "two temperaments" or "general approaches" to human existence. Translated into the orienting perspectives of the emergent human sciences, these temperaments represented a clash of first principles over theoretical method: Was the individual prior to society? Or was society prior to the individual? We can think of the first of these as an "individualistic" perspective and the second as a "sociolistic" or "sociologic" perspective. A corollary of the respective perspectives was the appropriate theoretical method for proceeding with inquiry: a micro, bottom-up approach for the individualistic and a macro, top-down approach for the sociolistic – or methodological individualism, in the first instance, and methodological holism in the second.[16]

The macroscopic orientation of sociology established the boundary between sociology and both RAP and psychology. RAP and psychology started with the little, sociology started with the big. On the substantive axis the thickest boundary was established over the idea of consciousness – the more generalized awareness of both social actor and social observer. For early sociology the idea of consciousness, as a more comprehensive concept than the narrowly rational thought of RAP was a more promising keystone conceptualization. Here, it was thought, lay the key to unlocking the secrets of *la condition humaine.*

[16] It is important to recognize that this corollary between temperament and theoretical method is not always matched by operational method; that is, by whether qualitative, interpretive analyses versus quantitative, scientific analyses are brought to bear on specific research problems.

Marx, in some ways, inadvertently laid the groundwork for the focus on consciousness.[17] The view that the more conscious manifestations of humans concerning their social lives were largely rationalizations held an important place in his thinking. The idea of "false consciousness" led to consideration of consciousness more generally. Distinguished intellectual historian H. Stuart Hughes, in his modern classic *Consciousness and Society: The Reorientation of European Social Thought* (1958, p 3), summarized the point by saying "...for the great social thinkers of the next generation [i.e., after Marx] the crucial concern was with the irrational, virtually unchanging nature of human sentiments – what Freud usually referred to as 'drives' and what Pareto awkwardly termed 'residues'..." To understand the body social, therefore, required something more than simply a focus on rational thought and action, the bedrock of RAP. It required a focus on the more fundamental conceptualization of human awareness – consciousness.

The premised requirement is consistent with the following reasoning. Question: Is it reasonable to suppose that social actors are at every moment in the strategic mode of thought presumed by RAP? Answer: No, that doesn't seem reasonable. Question: Does that then mean that when not in RAP mode they are not thinking, not aware at all? Answer: No, that is equally unreasonable. To be aware, they must be engaged in some other form of thought. Question: Is that kind of thought, then, unimportant? An equally resounding "No," said the founders of sociology. What, then, is that other form of thought? The founders of sociology answered: "Consciousness," at once a more basic and vastly more comprehensive idea than rationality; indeed, it was the basic stuff of social existence.

Consciousness was therefore a more comprehensive, engulfing conceptualization of human capacities, subsuming both rational and nonrational qualities within a single orbit. The nonrational was more than a weak sister in shaping the body social; it was an active partner via consciousness. So central was the idea to early sociologists that Hughes (1958) characterizes their era as an age of consciousness. He writes:

[17] It was doubtless inadvertent because Marx's principle premise, turning Hegel on his head, was that materialism was the first order of things. The reintroduction of consciousness by those coming after Marx represented something of a resurrection of Hegel. But it was more than that, for it was a manifestation of an influence of the German intellectual tradition on conceptual substance, as well as theoretical method. Elements of German idealism – that ideas are primary – shaped the eagerness by sociology's founders to incorporate consciousness into social thought. Hughes (1958) writes: "In the dominant Anglo-French tradition the primacy of sense perception and the validity of empirical procedures were taken for granted as naturally as the supremacy of the 'idea' was accepted in German" (p 184).... In Germany there had lingered on from the early nineteenth century a nondogmatic religiosity, an emphasis on the spiritual and a distrust of the material world. " While not necessarily distrusting the material world, the founders of sociology did find comfort in the emphasis on the spiritual values of mankind."

> The problem of consciousness early established itself as crucial.... rationalists and empiricists alike[18] agreed on an identity of view between actor and observer in the social process, and on assuming this common attitude to be postulated by scientific investigation or utilitarian ethics. All other standpoints, it had been argued, could be dismissed or discounted as intrusions of irrelevant emotion. Now rather suddenly a number of thinkers independently began to wonder whether these emotional involvements, far from being merely extraneous, might not be a central element in the story. (p 15)

It, indeed, became a central element in the sociological story. It was also the pivotal concept reorienting the way social thinkers approached the world. Again, Hughes provides the fitting description: "The study of society they gradually came to see as a vastly more complicated matter than one of merely fitting observed data into a structure of human thought that was presumed to be universal." (p 15) ... "The result was an enormous heightening of intellectual self-awareness." (p 16) ... It inevitably led to "the new definition of man as something more (or less) than a logically calculating animal" (p 17). Once the theoretical potential of this new orientation enlivened their curiosity, "the major intellectual innovators...were profoundly interested in the problem of irrational motivation in human conduct. They were obsessed, almost intoxicated, with a rediscovery of the nonlogical, the uncivilized, the inexplicable" (p 36). The intoxication spelled a sobering reassessment of the rational actor paradigm (RAP) – beginning with its very core.

> ...the more imaginative thinkers came to the conclusion that the "former conceptions of a rational reality" were insufficient, and that human thought would have to make "concessions" to a reality that could no longer be conceived as an orderly system. In this process of concession and adaption, the "activity of human consciousness" for the first time became of paramount importance. For consciousness seemed to offer the only link between man and the world of society and history... (p 428)

The reintroduction of consciousness into social inquiry was a first-order substantive boundary establishing conceptualization. But, it had second-

[18] With this description Hughes is combining both RAP based thinkers, such as the classical economists, with positivist thinkers such as Comte and Spencer.

order boundary setting effects, too, especially in demarcating how RAP based social science, on the one hand, and the new sociologic approach, on the other hand, addressed the issue of "unintended consequences."

In fact, one pivotal point separating classical economics from early sociology was in the treatment of unintended consequences. RAP's view of social order, as explicated by classical economics, is explicitly based upon a notion of unintended consequences. Social order is not intended; it emerges as the outcome of independent, individual actions. RAP is atomistic in orientation. It assumes that individuals will pursue their own self interests, using rational calculation to assess alternative choices, mindful of competitors doing the same, but basically unmindful of the influence of cultural rules, social context, history, institutions, or the state. These myriad individual, "anarchic" actions are collected together by the market which, acting like an invisible hand, coordinates "the private interests and passions of individuals" in a way "which is most agreeable to the interest of the whole society" (Smith 1776/1976, pp 594-595). In answer to the question of "By what mechanism does society hang together?" the RAP answer, via classical economics, basically was: *the market.* The market converts selfish motives into social harmony – into social order.

Probably no better, and certainly no more famous summation expresses RAP's framework for understanding the relationships between individual actions and social cohesion than that provided by Adam Smith himself, in the classic treatise launching modern economic theory, *The Wealth of Nations*, published in 1776:

> It is not from the benevolence of the butcher, the brewer, or the baker that we expect our dinner, but from their regard to their self-interest. We address ourselves, not to their humanity, but to their self-love, and never talk to them of our necessities, but of their advantage. (1776/1976, p 14)

Beginning with Adam Smith (at the incipient stages of the industrial revolution), through the years of the industrial revolution, economists strove to account for social cohesion emanating from the unintended consequences of individual decision making. Sociology's founding fathers were driven by the same quest as the economists, avatars of RAP.

For classical economists the answer was to be found in the unintended consequences of rational choice and action. For the upstart sociologists the key lay in the unintended consequence of consciousness. Consequences could be traced to nonrational, nonreflective, contingent choice and action – shaped to no small degree by institutional and social forces. Understanding the body social required study of the body social directly rather than of its elementary parts. Despite their common quest, the sociological route departed sharply from the RAP foundations of economics, especially from the assumption that human choice and behavior rested on rational dispositions alone. Thus, to the founders of sociology, the "invisible hand" was a sociological sleight of hand.

Even accepting that competition and the market accounted for the smooth coordination resulting from selfish motives, to Karl Marx that ordering was not the harmony proclaimed by classic economics; rather, it was a cacophonous score of disparities. Capitalists, on the one hand, enjoyed a relentless increase in riches, while the workers were just as relentlessly impoverished. Beneath the harmony apparent to classical economics was a persistent and unavoidable disharmony – inevitable class conflict. In *The Eighteenth Brumaire of Louis Bonaparte* Marx provided one of the most explicit statements summarizing one of the founding perspec-tives of sociology (sociology's own version of unintended consequences), a perspective keenly influenced by and mindful of history and macrosocial context:

> Men make their own history, but they do not make it just as they please; they do not make it under circumstances chosen by themselves, but under circumstances directly encountered, given, and transmitted from the past. (1852/1963, p 15)

Modern history, said Marx, was shaped by the laws of capitalism and it was these laws that he was devoted to unlocking, especially capitalism's inherent contradictions that ensured recurring crises and an eventual breakdown.

Max Weber was similarly intrigued by capitalism as an institution. But rather than seeking to understand the laws of capitalism, Weber was concerned with the puzzle of capitalism: Why did capitalism come into existence in the first place? The answer to the puzzle was provided by Weber in what was to become a classic work in Western thought: *The Protestant Ethic and the Spirit of Capitalism* (Weber 1904/05/1993). Capitalism came into existence, not by the "natural" emergence and growth of self-interested entrepreneurial actors, as RAP would have it, but on the coattails of an entirely different institution – religion. The Puritan version of Protestantism was the exemplary shaper of capitalist values. Puritans, in their drive to seek salvation and avoid damnation, demonstrated worthiness for salvation by acting frugally and industriously in order to produce tangible evidence of salvation in the form of capital accumulation. Hard work and prosperity were in the name of God, rather than secular profits.

In the process of laying out the institutional antecedents to capitalism, Protestantism, especially the Puritan version, Weber also reaffirmed the central role that consciousness – much more comprehensively than the means-end rationality emphasized by RAP – would play in the founding of sociological analysis. Unlike classical economists who claimed that capitalism was driven by selfish, self-interested, rational actors seeking to make the most profits, Weber showed that they were motivated by religious fervor – nonrational forces, in RAP's view.[19] Early capitalists may have been acting as if rational, but in their heads it was a consciousness – the hope of eternal salvation – that was driving them to save and accumulate. Thus, capitalism itself – a new institution – emerged as the unintended consequence of organized religion – an old institution.

2.4.2 RAP Knocks on Sociology's Door

For a long time, sociology developed as an antithesis to RAP. Not until an initial foray in the 1950s by George Homans (1950), followed in the early 1960s with the appearance of a more precise theoretical statement (Homans, 1961),[20] and a comprehensive theoretical formulation by Peter Blau (1964) could one discover any trace of RAP in sociology. The reappearance of social actors who were also rational appeared in the context of a variety of versions of what was termed by these theorists as "exchange theory." The basic idea is that social actors engage in a wide variety of exchanges (sentiments, psychic rewards, etc.), not just monetary exchanges. These exchanges are the presumed outcome of the rational assessment of the costs and rewards of exchanges.

Homans and Blau "connect themselves with the positivist traditions of behaviorist psychology and economics and make exchange eminently rational" (Collins 1988, p 338). Homans, for example, built his theoretical structure from the ground up, from psychological principles, particularly the behaviorist psychology tradition of B. F. Skinner, that he believed underlie all social relationships. But in his major theoretical statement, *Social Behavior: Its Elementary Forms* (1961), Homans went further, translating those psychological principles into the language and interpretation of RAP based economics. "Homans thus reunited two branches of the utilitarian tradition (which derived originally from Adam Smith and Jeremy Bentham), which emphasized that society is built out of the commonsense, rational behavior of individuals, especially as they exchange rewards" (Collins 1988, p 343).

[19] Weber's conclusion was deepened by his comparative studies of religion (see Weber 1920–1921/1988; Weber, Mills et al 1958), where he shows that Buddhism and Hinduism lacked the Puritan proscription encouraging an entrepreneurial life of "rational" business activity.

[20] Homans also issued a challenge to the predominately macro orientation of sociology, thereby launching the well-known, continuing micro/macro debate.

Building in part on the work of Homans, Blau set out on a more comprehensive quest: not only to account for interpersonal processes, the emergence of power, the emergence of social norms, but also to specify the principles of macro-level exchanges – such as between organizations. In *Exchange and Power in Social Life* (1964) he attempted to specify how the macrostructures of society were the logical end of a sequence of microstructures, and that the microstructures of social life could be traced to the rational calculus of exchange.

Despite the work of Homans, Blau and a scattered few other theorists, such as Emerson (1972), the attempt to breathe sociological life into RAP was less than a resounding success. Exchange theory generated pockets of interest and some empirical work, mostly experimental, but it never achieved the pervasive presence of the structural-functionalist sociology it sought to supersede. A second wave of RAP revival, a decade or two later, would, however, prove more successful. The second wave, termed "rational choice",[21] incorporated many features of exchange theory, which it intended to supersede, but within a larger theoretical frame. The wave, it appears, was initially stimulated by economist Gary Becker who, in *The Economic Approach to Human Behavior* (1976; see also Becker 1996), sought to re-introduce economics into sociology.[22] There followed a sustained effort by such theorists as Jon Elster (1979; 1983; 1989a), Michael Hechter (1983; 1987), and, of special importance, by James Coleman (1990).[23]

A central thrust of modern RAP theory takes it back to one of the central concepts in classical sociology: norms. But rather than take norms as a "given" of social structure, as did Durkheim, Parsons, and others, RAP sociologists ask how norms arise in the first place. Modern RAP theorists, especially game theorists, have attempted to show how norms emerge from rational calculations (Axelrod 1984; Coleman 1990). Indeed, it has been argued that here is where the richest theoretical challenges for RAP lie; "the future of rational choice lies in the analyses of norms and institutions" (Levi, Cook et al 1990, p 1).

The contemporary effort to revive RAP in sociology has enjoyed considerable short-term success, in the form of lively theoretical debate, the beginning accumulation of empirical findings, and growing institutional

---

[21] It has also been called by some "rational action" – a version of the terminology we have favored throughout our discussions – and called, by those applying rational choice to policy, "public choice" theory.

[22] In large part for this effort he was eventually awarded the Nobel Prize in economics.

[23] Distinguished as a sociologist for decades, known around the world, and a former president of the American Sociological Association, Coleman was a visible avatar of the renewed sociological enthusiasm for RAP. In addition to the contribution his stature and visibility provided, Coleman also contributed to the institutionalization of RAP in sociology by co-founding and editing a new journal, *Rationality and Society.*

recognition. It is, therefore, no longer difficult to detect RAP's presence in sociology, in the United States, Europe, and elsewhere.

Nevertheless, despite its increased visibility in sociology, rational choice is still a minority perspective in the field. Perhaps as a consequence of this status it has felt the need to attack such hoary, fundamental, core theoretical problems of the discipline as: What is the basis of social order? What is the relationship between the individual, culture, and social institutions? What are the forces of social change? One important consequence of this focus on traditional sociological questions is that RAP oriented sociologists have not turned their intellectual curiosity toward risk (but see Short, 1997). They have virtually ignored the topic of risk. Thus, in comparison to their counterparts in psychology, economics, engineering, and decision sciences, RAP sociologists have contributed little to the topic of risk or to theoretical debates about risk.

Despite the recent upsurge in the popularity of RAP in the field, the much longer history of sociology has been generally unaccommodating to RAP approaches. Sociology, from its beginnings, has favoured approaches to understanding the social world that provide for, if not irrational motivations, at least unconscious ones with their systemic unintended consequences that are best uncovered with a macro-orientation to analysis. It is in this sense that Parsons (1937) reaffirmed a dichotomy between explanations based upon norms and circumscribed choices and those based upon strategic, rational decision processes. And until recent times social scientists have oriented toward theory along one or the other side of this dichotomy. It is in this context – that is, comparing RAP's orientation of self-calculation to the normative orientation of Durkheim and Parsons – that Harvard economist James Duesenberry's (1960, p 234) famous aphorism that "economics is all about how people make decisions while sociology is about why we don't have any decisions to make" is so insightful.

3. RISKY DECISIONS OF A SINGLE AGENT

3.1 The Monarch and Risk: a Love-Hate Relationship

Monarchs, real or metaphorical, have a variety of propensities in common. They have their preferences and they often like to get involved in dangerous activities that challenge their power. They either meet these challenges in victorious splendor, or exhibit weaknesses and strengths during the activity resulting in a new arrangement by which they need to share power with others, or they fail completely and lose their throne. The rational actor paradigm, the king of social sciences, has faced many such potentially dangerous challenges in recent years. One of the most challenging topics for the rational actor paradigm is risk.

How to deal with uncertainty when anticipating consequences follows as a natural extension of the traditional elements of any rational decision. But it also implies a potential threat, since the possibility that uncertainty might not be manageable by strictly rational means would constitute a serious challenge to the monarch. Of what use would a theory be that proposes a complicated sequence of actions and deliberations, if such a sequence would lead to predictions that are unrelated to observed outcomes?

This threat would spell disaster for the normative as well as the descriptive branch of RAP. First, it would be difficult to accept a normative procedure that yields individual or collective outcomes that are no better than intuition or competing procedures. Second, it would be hard to argue that individuals and social institutions persistently engage in actions that result in outcomes that are contrary to their expectations. Empirical evidence (see below) offers only weak support for individual decision making in accordance with the rational choice model. Even if we adopt a moderate version of RAP, which assumes that the rational procedure is the best analytical description of individual behavior without claiming that individuals actually pursue such an approach consciously, we would be faced with theoretical crisis. The result would be a theory that does not describe real thought processes and, at the same time, does not predict actual outcomes. Such a theory would attract few supporters.

In response to this threat, different schools of the rational actor paradigm have modified this body of thought. Uncertainty became, for example, a major element of decision analysis and game theory. The structural problems of social interferences and external effects were adopted as major study objects by welfare theories, resource and environmental economics, the economic theory of politics, functional schools of sociology, conflict theory, and others. The distributional effects of risk taking received increased attention from analytical and contract-based philosophy, welfare economics, and social systems theory. The absence of positive values in traditional economics has inspired a new interest in business ethics and, within this field, the handling of risky choices as an ethical dilemma. This list is illustrative only, in recognition that the monarch has taken the challenge and responded with a series of activities to cope with the crisis. An indication for the seriousness of this crisis is the fact that three recent Nobel prizes in economics were awarded to researchers (Simon, Becker, and Selten) who have dedicated their work to issues of uncertainty and decision making related to basic features of RAP.

In essence, risk is closely connected with RAP, but is a threat as well to its validity as a general theory in both its normative and descriptive implications. The discipline of physics faced a similar situation with the study of light. For a long time, the phenomena of optics could be studied as a natural extension of classical physics. Then came the discoveries that the speed of light could not be surpassed, and that light at times behaved much like a set of particles, at times more like a set of waves. These discoveries led physicists to reframe classical theory as an important, but still special case. Risk may have a similar quality for the rational actor paradigm in the social sciences.

It is too early to decide whether the monarch will evolve victoriously from this challenge, will suffer losses but no defeat and share power with new princes, or will fail the risk test and be replaced by one of the newly evolving theoretical approaches. As noted above, one key objective of this book is to conduct a thorough and critical review of the various attempts to cope with the challenge of risk and uncertainty within the framework of RAP.

In the Introduction, we offered a working definition of risk. In Chapter 2 we discussed RAP as a general theory with major impacts across the whole spectrum of the social sciences, and we provided a review of disciplines and perspectives that investigators have used in RAP-based analyses of risk. The remainder of Part I will be devoted to a detailed description and critical analysis of the dominant risk perspectives within the rational actor paradigm. We will explain the underlying rationale, demonstrate the links to the rational actor model, and develop the prospects, limitations, and problems with each approach. This will enable us, in the second part of the book, to consider possible alternatives to RAP-based risk studies and to discuss the question of whether the risk debate is actually fostering a far-reaching paradigm shift in the social sciences.

3.2 Insurance and Portfolio Investment

How do professional decision-makers, such as managers and entrepreneurs, deal with uncertainty and risk? The main risks to be considered are financial losses. The relevant concept of a loss, however, is far from trivial. If an investor considers a rate of return of 7% as normal, but expects only a 5% return from a specific investment, then this investment implies an opportunity cost of 2% of the invested sum, not a net gain of 5%.

3.2.1 Investment Decisions and Expected Values

It is common wisdom that entrepreneurs seek to maximize their profits. Such a claim is well defined as long as these actors are gifted with perfect foresight of the costs and benefits that any possible investment project will generate in the future.[24] As Kreps (1992) remarks in his textbook on microeconomics: "Played by the sort of individuals who populate the economies of this book, a chess match would be a boring affair; the players would come on stage, look at each other, and agree to whatever (forced) outcome the game happens to have." (p 772)

In real life, however, the future is largely unknown. Familiar ideas of profit maximization need to be modified if they are to be applied to such situations. If uncertainty concerning an investment project is represented by a series of probabilities, such a modification is straightforward. Consider a project which can be expected to yield benefits of $5 million costing $4 million. Suppose the manager considers that there is a 10% probability of serious production problems which would increase costs from $4 million to $6 million. He or she concludes that there is a 10% possibility for a $1 million net loss and a 90% possibility of a $1 million net gain. This situation illustrates one paradigmatic case for the use of the word "risk". The risk of production problems can be said to be the product of expected losses multiplied by the probability of the event in question. The expected value of the total project then is computed to be $800,000 (90% of $1 million gain minus 10% of $1 million loss). Note that in this case it is not total costs that are considered, but net losses. Will investors be indifferent between such a project and an opportunity to get $800,000 by investing $4 million? Only if they are risk neutral in the sense introduced in Section 2.2.2.

The word risk – like "time" and many other useful words – is used in a series of different ways which are related to each other not by one common trait, but by a complex pattern of partial similarities. The definition of risk proposed in Chapter 1 is broad enough to cover most elements of the relevant pattern. This should not lead to neglect of the many ambiguities which can arise in actual uses of the concept.

---

[24] This includes knowledge of the discount rates needed to compare costs and benefits arising at different points in time.

The importance of polythetic concepts has been emphasized in much philosophical work originating with Wittgenstein (1953). With regard to risk, Rayner and Cantor (1987) argue that risk is a polythetic concept, and that it may be severely misleading to reduce its meaning to the product of costs and probabilities. On the other hand, referring to this product is one important way of using the word risk. As our example shows, however, even in such cases there may be more than one reasonable way of defining costs.

Now consider a risk-neutral investor faced with various investment projects, each characterized by probability distributions of well-defined costs and benefits. It would be quite rational for her to maximize the expected value of the profits of the various investment projects. However, some investment projects may involve risks of such magnitude that, if realized, would drive her into bankruptcy. Suppose for the sake of illustration that the risk of bankruptcy occurs with an average probability of 10% in the projects with the highest expected value of probable profits. Under these circumstances, attempts to maximize the expected value of probable profits will lead to a steady elimination of one tenth of the firms. Competition then will favor firms that base their investment decisions on more than the principle of maximizing expected value. An instructive example is given by firms that invest part of their money in insurance contracts to avoid bankruptcy even if these contracts do not maximize the expected value of probable profits.

3.2.2 Insurance

Suppose a hustler in the rear seat of the bus offers the following opportunity to his fellow riders. Pay $100 and roll a die. If the result is a six, win $480. Otherwise, the gambler gets nothing. Few people would consider this a wise bet. The reason, of course, is that the expected value of this lottery situation, (1/6) x $480 + (5/6) x 0 = $80, is less than the $100 investment. On first glance, the mathematics shows this to be a ridiculous offer. But the business of insurance is founded on similar relations.

Consider a population of firms of similar size, each of which faces a risk of experiencing a loss so great that it would drive the firm into bankruptcy. The loss, L, is expected to occur at some probability, p. An insurance firm might offer an insurance contract to handle this risk. From the point of view of the insurance firm the premium, X, needs to cover two kinds of costs. The first cost has to do with covering promised payments. If N contracts are sold, a loss L is to be expected pN times a year. The second cost is associated with the costs of doing business. Each premium has to cover 1/N of operation costs C, including salaries, rent, and the like. Finally, it is often not enough to merely cover costs; the premium is expected to yield some positive rate of return, r. The premium then will satisfy formula (1):

$$X = (p \cdot L + \quad) (1 + r) \quad \frac{C}{N} \qquad (1)$$

Oddly enough, this is precisely the case of a game, like the hustler example above, where the cost of playing is substantially greater than the expected value. Nevertheless, firms are happy to buy insurance against the risk of bankruptcy. The reason, of course, is that they want to avoid the impact of bankruptcy.

To apply this approach to the case of a manager buying an insurance contract, we must reformulate the manager's maximizing behavior so that it refers not to profits, but to a cardinal utility function of these profits (see Section 2.2.2). More precisely, we are concerned with the risk that a firm sees its equity wiped out by huge losses. From the point of view of a manager assessing an investment project, this risk cannot be balanced by the chance of making the same profit with the same probability. The risk that the existence of the firm will come to an end is of an entirely different type of risk than the chance of making additional profit. If an excellent investment project includes the risk of making a moderate loss with some given probability, then the project will be far from excellent if the possible loss is increased by a huge factor, even if its probability is decreased by the same factor. When the survival of the firm is at stake, risk can no longer be described as the product of probability and expected monetary losses. A more appropriate description in these cases can be attempted in terms of cardinal utilities.

Clearly, the same analysis holds for the insurance firm itself. Competition will eliminate insurance firms selling insurance for risks so high that they exceed their financial possibilities. And, by the same reasoning that led us to consider insurances for firms in the first place, we come to the possibility of reinsurance for insurance suppliers. Eventually, the limits of possible insurance are set by the finite size of the total capital market at a given moment in time.

This argument sheds light on the problem of limitation of the size of firms. This problem "arises because of the argument that under perfect competition in all markets there would be no limit to the size of the firm. The restriction is found in the unwillingness of the entrepreneur to borrow so much as to risk the wiping-out of his equity and in the imperfection of the capital market, due to the unwillingness of individuals to lend more than a limited amount to any one firm, again because of risk-feelings [...] The whole subject is wrapped up in the abstruse mysteries of capital theory and the coordinating function of the entrepreneur" (Arrow 1970, p 6). The study of firm size (see Granovetter 1984) seems to require an analysis of economic risks that goes well beyond received ideas of optimal firm size based on U-shaped cost curves.

The "abstruse mysteries of capital theory" are quite relevant to assessment of RAP (Harcourt 1972). Sociologists thinking about rational action tend to assume that RAP is the appropriate approach for the study of economic reality and ask themselves whether it is also appropriate for the study of other phenomena. However, the problems facing RAP within the economic realm itself are serious enough to warrant careful re-examination. Perhaps a more comprehensive approach is warranted, even for studying economic decision making.

3.2.3 Portfolio Investment

A better understanding of utility functions for insurance contracts can be gained by looking at contracts as items of an investment portfolio. Investment portfolios rely on creating a diverse set of options, some independent, and some correlated negatively or positively.

To illustrate this point, imagine a situation that offers pairs of lottery tickets for a $5 price. Each ticket gives a 50% chance of winning $10 and a 50% chance of winning nothing. The pairs of tickets come in three versions. In version A, the two lotteries are independent of each other. In version B, they are linked by a strictly positive correlation. In version C, they are linked by a strictly negative correlation. Version A has four possible outcomes (0, 10, 10, 20); version B has two extreme outcomes (0, 20); and version C has two equal outcomes (10, 10). In all cases the expected value is the same, but the distribution of possible outcomes is different. Clearly, each has its own advantages and disadvantages. A manager who tries to reduce the risk of bankruptcy from a given investment project by buying an insurance contract is forming a portfolio with two negatively correlated assets. In portfolio analysis, this distribution is often measured by the variance, which is then called the volatility or the risk of the portfolio. (This marks a specific usage of the word "risk" not to be found in other fields of study.)

An optimal strategy reduces risk by creating a portfolio with items that are either uncorrelated or negatively correlated. Risk that cannot be eliminated by such means is called systematic or market risk. If we compare portfolios which display only systematic risk, increasing profits can be achieved by accepting higher systematic risk. To design an optimal, profit maximizing portfolio, then, requires the specification of the level of risk which the investor is willing to take. The specification of acceptable risk in portfolio investment cannot be the result of an optimization process, it is the precondition for subsequent optimization to make sense.

The fact that in portfolio investment profits can be increased by increasing the risk of a portfolio is sometimes seen as pointing to the very essence of profit making. As Arrow (1970, p 6) wrote, "The more recondite phenomena associated with the occurrence of risk in the minds of many economists include the existence of profits." In this view, profits include two elements. One corresponds to interest on the capital invested and must be explained along with an analysis of the rate of interest. The other consists of "pure profits" and is a reward for risk taking.

The preceding argument clearly supports the view that profits are an incentive for risk taking. However, this does not imply that risks taken for this incentive are always reasonable. Risks that must be described as highly unreasonable may be taken precisely because high rates of return offer a strong incentive. This may be the case with arms races as well as with environmental disruptions. A somewhat paradoxical situation arises where the pursuit of profits contributes to the risk of financial crises that threaten this very pursuit (Feldstein 1991; Strange 1986).

RAP is indispensable for an analysis of financial markets. The strategies of risk diversification elaborated in the framework of portfolio theory are vital for the management of financial risks by banks and monetary authorities. However, the crucial role of trust relations in financial markets points to a dimension of decision making that is difficult to understand in terms of the atomistic rationality presupposed by RAP. It would be wrong, however, to construct a contrast between "economic" and "sociological" man, with the former oriented toward self-interest and competition, while the latter engages in relations of trust and cooperation. RAP also offers extremely valuable contributions to the understanding of such relations, especially in bargaining situations (see Section 4.1).

3.3 Decision Analysis

Profit maximization gives a quantitative criterion for selecting alternatives. But clearly RAP is not limited to cases where decision alternatives can be compared in quantitative terms. As long as they can be compared at all, one can always try to capture a decision-maker's preferences for various options with a utility function (Edwards 1954; Keeney and Raiffa 1976; von Winterfeldt and Edwards 1986; Jungermann, Pfister et al 1998; Pinkau and Renn 1998). Where this endeavor seems promising, the task of decision analysis can be undertaken.

The task is to provide tools to assist decision-makers by reducing the decision as far as possible to an exercise of logic. Through interviews, surveys, or informal discussions, analysts elicit values from the decision-maker that they feel are relevant to the choice under consideration. The goal is to produce a logical hierarchy of decision options, according to the utility the options offered the decision-maker. The decision-maker is then asked to assign weights to each decision option on each value. Probabilities are also assigned to each option. This amounts to conducting a risk assessment. Knowledge about consequences and effects is used along with personal judgments to arrive at scores. Theoretically, the decision-maker's preference for any decision option is reflected in the sum of the scores calculated. The normative assumption implicit in decision theory, as "imported" from economics, is that the decision-maker should choose the decision option that results in the highest utility. Two important techniques for performing decision analysis are Multi-Attribute Utility (MAU) analysis and value tree analysis (see section 3.3.1).

In general, decision analysis approaches can be captured in a sequence of steps, as given in Table 3.1. In decision analysis, risk assessments are estimations of negative or positive consequences associated with each option (magnitudes) and the assignment of probabilities to each outcome. The desirability of outcomes is specified as a cardinally measured utility providing a hierarchical order from beneficial to adverse effects.

Provided the total range of consequential effects (from best to worst) is taken into account, and all value dimensions are non-redundant and independent, the decision-maker's utility function aptly considers variations in probabilities of outcomes. This step-wise model has proven to be an excellent normative guideline for rational decision making (Fischhoff, Lichtenstein et al 1981). However, the model is still limited to individual decision-makers.

This poses a serious problem for collective decision making. With democratic decision making – involving groups with different criteria and values – this simple model fails, because rationally derived means to summarize values or to aggregate weighting between individuals are missing. The most widespread criterion for comparing different alternatives involving more than one actor is the Pareto criterion (Pareto 1927). According to this criterion, alternative A is superior to alternative B if and only if in the situation given by alternative A at least one actor is better and no actor is worse off than in the situation given by alternative B. Out of a set of alternatives, an alternative is Pareto optimal if no alternative in the set is superior to it. Unfortunately, in most cases many alternatives will be Pareto optimal and the criterion does not allow for selection among these. Kaldor (1939/1969), Hicks (1940), Samuelson (1950/1969) and others tried to find a more powerful criterion with the compensation principle. According to this criterion, alternative A can be said to be superior to alternative B even if in both cases some actors are better and other ones are worse off. The trick is to check whether the former could in principle compensate the latter so that a Pareto ranking would result. If this principle

Generic Steps of the Decision Analysis Approach	Example: Municipal Solid Waste Disposal
Structure the problem and specify goals for the choice situation ahead	Priorities: reduce waste generation, encourage voluntary re-use and recycling, mandate recycling, incineration and landfills.
Extract appropriate value-dimensions from stated goals	Equity of risk exposure, compensation, cost effectiveness, minimize impacts
Define criteria and indicators for each value dimension (or attribute)	Meta-criteria: health risks, environmental risks, social and cultural risks.
Assign relative weights to each value dimension	Health risk = 40%; environmental risk = 35%; cost = 25%
Describe alternatives or options that seem to meet the criteria	Option A: Regional recycling centers and an expanded landfill. Option B: a new landfill in community W.
Measure how well each decision option performs on the criteria	Geological borings, probabilistic risk assessments, collect statistical data, elicit citizen's responses
Assess probabilities for each possible outcome	Option A: health risk = 11; eco risk = 21; cost = 82. Option B: health risk = 34; eco risk = 75; cost = 20
Sum each decision option's utility on each criterion multiplied by the weight of each value dimension	Option A = 32; Option B = 45
Perform a sensitivity analysis to incorporate changes in criteria composition, outcome generation, assignment of weights and probability assessment.	Option A (28, 32, 56); Option B (16, 45, 47)

Table 3.1: Steps of a decision analytic approach

works, a social utility function can be constructed which ranks alternatives in a meaningful way which is stronger than the Pareto criterion. Unfortunately, the compensation principle is often arbitrary, depending on the details of the compensation scheme.

With a powerful theoretical argument, Arrow (1951) has shown that attempts to design a social utility function on the basis of given individual preferences are doomed to failure: there simply are too many possibilities to avoid arbitrariness. In practice, all attempts to construct social utility functions on that basis are either too abstract or are vulnerable to strategic maneuvering (Green and Shapiro 1994; Shrader-Frechette 1985). Neither Pareto's optimality rule nor the Kaldor/Hicks criterion can provide a rational justification if conflicts are unresolved. In particular, the necessary amount of compensation to be paid to those deprivileged by a specific outcome cannot be determined on a rational basis (Kunreuther 1995). Recently, game theoretical models have been proposed to describe more precisely the behavior of groups in conflict, but they cannot provide generally applicable instruments for public policy makers to resolve conflicts about how to make decisions when values and evidence are disputed.

3.3.1 Decision Theory and RAP

In the evolution of RAP methods to cope with uncertain outcomes, decision analysis is particularly close to the axiomatic presumptions that define RAP. It overcomes some of the deficiencies of other quantitative approaches by enhancing the technical concept of rationality with a multicriteria model of choice. In contrast to the technical understanding of risk implicit in many quantitative and economic approaches, decision analysis does not claim to yield objective results independent of the decision-maker's views or preferences. At the heart of that model lie the subjective expected utilities of an individual decision-maker. Based on microeconomic rationality, decision analysis relies on the subjective judgments of an individual decision-maker to arrive at risk assessments of alternative decision options (Phillips 1979; Schoemaker 1982).

As Table 3.2 shows, the "individual" hypothesized in RAP is manifested in decision analysis as the decision-maker.[25] Correspondingly, "alternatives" are the decision options. Preferences in decision analysis are not wholly subjective, but are interpretations made by the decision-maker. The decision-maker may draw upon results of technical analyses, upon

[25] Decisions made by more than one actor are typically outside the domain of decision theory. Decision analytic techniques have been used for group decision making; however, there is no agreement on how to overcome the problem of aggregation. Two approaches predominate. One approach is to have a group of individuals consensually agree on each entry into the

Presuppositions of RAP	Implications in Decision Analysis
A set of alternatives knowable to an actor exist	Alternatives are a set of decision options
The future consequences of those alternatives can be assessed	Consequences are known through statistical studies, but also through judgments. Expectations from the decision maker (or some other source such as an advisory panel) are used.
Alternatives are evaluated against a set of unchanging preferences	Depending on the decision, preferences may be private or public. In either case, it is the decision maker's estimation and interpretation of those that matters
The best alternative is the one most preferred.	The best decision gives the decision maker the highest utility.

Table 3.2: Presuppositions of RAP and the corresponding implications in decision analysis

his own intuition, judgment, or speculation, or upon the input of others when forming his preferences. He may also have different motivations for building his specific set of preferences (e.g., he wants to make his brother happy).

Analytic studies of decision making are either normative or descriptive. Normative studies seek to discover the appropriate procedure for selecting the "best" decision option from a group of alternatives. Descriptive studies focus on explaining or predicting activities or outcomes of decision processes. Both types of decision analysis can be combined by adopting a decision making process that the affected and responsible individuals or institutions regard as appropriate (descriptive analysis), and then by optimizing the outcomes of each step according to the values, preferences, and aggregation rules of those same parties (normative analysis).

Because risk only has meaning with respect to the values and expectations of the decision-maker, decision analytic approaches reject

---

decision analysis. Consensus-building becomes the major challenge in this case. The alternative approach is to have each individual go through the decision analysis exercise, then aggregate results according to the democratic criterion of one-person, one-vote. Obviously, problems remain with both of these solutions.

the claim that risk is simply an objective property of a system or situation. Risk assessment can include as many dimensions as the decision-maker desires (Jungermann, Pfister et al 1998). It is not necessary, for example, to limit end points related to adverse health effects or environmental damage. The decision-maker must define risk in terms he or she regards as appropriate. Risk assessment based on decision analysis may include concerns for distributional consequences of released hazards by adding criteria such as equity and justice onto the list of evaluative criteria. The distributional effects are assessed and evaluated according to the preference structure of the decision-maker.

Probabilities and expectations in risk assessments are deliberately derived from subjective judgments. These may be informed by measured data, but they are, foremost, mental constructs (Winkler 1968). As constructs, they include the strength of personal beliefs about the occurrence of specific outcomes. An advantage of this approach is that it permits assessments to address the problems of human interaction with risk sources, thereby accounting for subjective probabilities of human failure (a major problem with the method of probabilistic risk assessment, described in Section 3.4.2). Of course, beliefs about risk are also based on perceived regularities of past behavior of the risk source under consideration. Subjective judgments for calculating probabilities ought to be informed by the best statistical evidence available. However, it is up to each individual participating in the decision analysis to decide how much his or her subjective judgment is based on statistical risk assessments.

The decision analytical concept regards risk assessment as a structural tool to improve decisions (Merkhofer 1987). Methods and techniques applied do not aim at an objective, clinical measurement of a physical entity, but at a consistent and logically sound model of a person's or institution's knowledge and preference structure. Risk aversion, proneness, or indifference are key factors that shape an individual's utility function. The assessment procedure varies from task to task; it can be close to the technical approach if the decision-maker believes that this is appropriate, or it can be very intuitive. Some analyses rely only on the elicitation of subjective beliefs, because quantitative data are unavailable or not appreciated by the decision-maker. Illustrations for the former approach can be found in Keeney (1992), a case study of the latter type is described in Borcherding and Rohrmann (1990).

Multi-Attribute Utility analysis (MAU) attempts to present individual or collective dimensions of benefits and risks on a quantitative basis (Borcherding and Rohrmann 1990). In line with decision theory, MAU relies on subjective probabilities to inform risk assessments. The evaluative dimensions are derived from the overall value system of the decision-maker. They should cover all the concerns an individual has with the decision outcomes. Representatives of the decision making group(s) are interviewed applying the usual survey methods. During the interviews the major concerns such as health effects, environmental damage, benefit for oneself, equal distribution of risks and benefits, aesthetic quality are ordered in a consistent and logical structure.

In a second step, possible outcomes of different decision options are collected and probabilities assigned to each outcome. Utility functions are constructed first for each dimension separately, taking into account different attitudes towards risk (prone, adverse, indifferent); then compound utility functions are constructed by determining preference probabilities (tradeoffs) between various dimensions. In the last step an aggregation rule for combining the different partial judgments is applied in order to determine the best solution. If independence of dimensions can be assured and if the utility functions follow a monotonic sequence (either ascending or descending), taking the sum across the evaluations of the prospected outcomes multiplied by their preference probabilities is the appropriate method to summarize the relative positions of each option on a one-dimensional scale (Edwards 1977). Multi-attribute decision making has proved to be a powerful tool as a decision aid for a single or a homogeneous collective decision-maker. If conflicting values play a major role, different techniques of decision aiding have been proposed, in particular value tree analysis.

Value tree analysis: A value tree identifies and organizes the values of an individual or group with respect to possible decision options. In the process of structuring a value tree, analysts try to elicit the salient values and their attributes of individuals or groups (Keeney 1992). On the basis of these questions a tree is constructed with general values at the top and specific, measurable "attributes" at the bottom. For each group involved in the decision process (for example, regulator, industry, intervenor groups, environmentalist, sample of exposed population) a separate tree is constructed. Each alternative under consideration is measured on each of the attributes by assessing subjective probability distributions, constructing utility functions or measuring relative performance using a numerical scale (such as 0 = no utility, and 100 = highest utility), and then computing expected utilities. In group sessions, attribute importance weights or scaling factors are elicited from group representatives to determine tradeoffs between attributes and the group specific expected utility across all attributes. Each individual value tree can be used to construct a joint value tree for all groups or to level out compromises or compensation payments according to the degree of violation that each group is likely to suffer from each option assessed. Thus, value trees cannot dissolve conflicts or guarantee an optimal solution, but they provide a useful tool in facilitating resolution of conflict by structuring the negotiation process between stakeholder groups (Keeney, Renn et al 1984; von Winterfeldt and Edwards 1986; von Winterfeldt 1987).

3.3.2 Critiques of Decision Analysis

The Subjective Expected Utility (SEU) model of individual choice has long stood as a core feature of the Rational Actor Paradigm. It presumes that individuals, when faced with a pending decision, envision a number of action alternatives (Edwards 1954; Jungermann, Pfister et al 1998). Following this, they compute the expected utilities of each possible action alternative. This is done by first estimating the likely outcomes of each action alternative, and then assigning probability estimates of how likely it is that each outcome will occur. The model assumes that people do this in strict independence from each other, drawing upon their personal values and own estimates of the likelihood that certain outcomes will occur. Once the SEU for each alternative is computed, an optimization rule is applied. The rule is simply this: The action alternative that produces the highest subjected expected utility is preferred. In this way the "best" decision alternative is identified and can then be implemented.

On the basis of SEU, decision analysts have developed a series of procedures to improve actual decision making. Often they rely on multi-attribute utility functions enriched with stochastic reasoning. This implies that the overall utility associated with each decision alternative can be reconstructed as a weighted average of utilities relating to different attributes of each alternative. While a judicious use of these procedures is no doubt often very helpful, there are also several problems involved. For example, claims that decision analytic procedures always produce outcomes decision-makers prefer are suspect. Once he agrees to participate, the decision-maker may feel obliged to agree with the results, even if, deep in his heart, he prefers another choice (Pinkau and Renn 1998). By committing himself to the decision analytic process, the decision-maker may find it politically impossible to reject the result.

This problem is further aggravated by the blind faith that some people have in numerical analysis and in computer models. They often do not recognize that the validity of the figures incorporated in decision analysis depends on the ability of the decision-maker to transfer her preferences and expectations into numbers. In defense of decision analysis, however, it has to be said that independent scaling techniques and thorough sensitivity analysis help to assure a valid assessment of preferences and probabilities.

The difficulties of substantiating values and the inability of many variants of decision analysis to cope with multiple actors and conflicting values set strong limits for this application of RAP. Its scope mainly covers situations in which options and values are well defined and all subjective input is a product of a consistent and homogeneous value structure.

The argument that decision analytic approaches are the best way to reach the "right" decision would indeed be compelling if analysts could be certain that the methods involved guarantee a true representation of individuals' preferences and expectations (Camerer 1992; Harless and Camerer 1994). The problem is that this simply is not the case. In spite of the highly sophisticated methods of assessing the personal judgments in decision analysis, many studies demonstrate that only a low percentage of decisions follow the rules of normative decision analysis. In other words, people often fail to select the decision option with the highest expected value. It is also yet to be proven that numerical values and mathematical operations serve as valid representations of verbal reasoning and emotional commitments. As long as sufficient proof for these assumptions is lacking, we cannot know whether actual decision processes are violating the rules of rationality or if the technical skills to understand subjective rationality are underdeveloped.

We simply do not know whether people make decisions in a way that contradicts their own preference structure or if the techniques employed to represent those preferences and synthesize them according to logical procedures are unable to cope with complex value and attitude structures.

The problem of how well an analyst is able to measure the true preferences and beliefs of people touches another dimension even more relevant for a critical assessment of this approach – manipulation and bias. Risk assessments in decision analysis rely heavily on the subjective expectations of the decision-maker for data. Verifying the authenticity and accuracy of these data is much more difficult than in quantitative, economic, or toxicological risk assessments, which, while not wholly "objective," do acknowledge that the data and methods informing conclusions can be examined and held to intersubjective conventions.

Decision analysis presupposes that the value system of the analyst does not interfere with the decision making process, another dubious presumption. Since most decision-makers have difficulty understanding the procedure and the calculations involved, they are at the mercy of the analyst for guidance. The analyst may suggest a categorization scheme or criterion that does not reflect the authentic preferences and expectations of the decision-maker. The selection of scales, the way of presenting data, and the procedures to elicit probabilities and preferences provide a wide range of influential tools for the analyst to introduce his or her own preferences and beliefs into the decision making process. This points to the paramount importance of informing decision-makers of the limitations and conditions inherent in each assessment technique. Decision analysis is a powerful tool to improve decisions and to cope with risks, but it has limitations. Consulting with more than one analytical team is one way a decision-maker can protect herself from manipulation or bias. Comparing the results of different teams may help to gain an appreciation for where the assessment helps to clarify one's own judgments and where it starts to reflect the ideas and thoughts of the analysts.

But it is not only the analyst who is capable of manipulating procedures. Decision-makers can influence the outcome of the analytical process by exaggerating or misrepresenting their values and utilities to an analyst (von Winterfeldt and Edwards 1986). They might be motivated to do this in order to "sell" the outcome of an analysis as a scientifically proven or formally correct choice. Like all applications of RAP, decision analysis does not question the type and quality of values that decision-makers feed into the process. They can manipulate the procedure by adding values and relative weights to the desirable outcome that they would like to emerge as the winning option for strategic reasons. Analysts can discourage this behavior by eliciting value dimension weightings before asking for assessments on outcomes. If the number of decision criteria are high, and no few number dominate, it is more difficult for the results to be reliably manipulated.

A further criticism of the decision analytical approach points to oversimplistic assumptions of decision making processes (Beach and Lipshits 1993). This is especially cogent when the decision process being analyzed relates to public choices. According to the critique which is gaining in popularity among social scientists, decision making is characterized by strategic reasoning, power plays, interests, and institutional responses. Decomposing political decisions into outcomes, probabilities, assessments, and preferences naively implies that the decision-maker is willing and able to let reason dictate the decision, an honorable, but hardly realistic assumption in today's political world.

The German sociologist Volker Ronge has provided one of the most thoroughgoing critiques of this approach. He points out that political decision making is by no means governed by a coherent "logic." Nor is it in the interest of most political decision-makers to make their reasoning and goals known to the opposition and the public at large. Ronge noted that decision-makers respond frequently to the exerted power of influential social groups rather than setting priorities for selecting policy options (Ronge 1980).

In this circumstance, decision analysis often is reduced to a tool decision-makers use to push through their agendas. If the results of the analysis serve that need, the scientific legitimacy of the analysis is lauded, otherwise the analysis is ignored or discredited. The decision analyst's main role is to provide external legitimation that will rationalize preformulated decisions. This works because legitimation by science is an effective means for justifying decisions.

This line of argument reflects a general skepticism toward science in policy and decision making, skepticism often associated with a purist attitude that treats politics as a dirty game, too messy for scientists. It carries the message that scientists either refrain from politics (since the analyst loses his innocence) or agree to be "hired gun" for one side or the other in the debate. However, one need not deny that science can be marshaled in the service of one side of a political debate in order to recognize that this polarization is grossly exaggerated. Policy makers *do* pursue self-interests and *are* exposed to various social pressures. But this does not mean they are indifferent toward public interests (for example risk reduction). Nor does it mean that decisions are made without recognizable reasoning. Reactions by the media, the threat of withdrawal of public support, the risk of inciting protest movements, and critical examination by opposition groups require policy makers to justify decisions. One way they can do this is by providing evidence that the decision was rational. Moreover, if decisions have been made on the basis of poor or biased predictions of future outcomes, politicians will have a hard time explaining their reasoning after the real outcomes become visible.

In addition to these general criticisms, several more technical problems have been identified in the literature. The following examples may be noted (Merkhofer 1984):

- subdivision into probabilities and utilities is often difficult to carry out.
- elicitation of preferences is affected by the social desirability of each value dimension.
- assigning relative weights to each dimension makes sense only if the assessment is calibrated towards the "worst" or "best" case.
- calibration procedures are difficult to explain and likely to be misinterpreted by the clients.
- utility functions presuppose certain mathematical properties that may not reflect the utility structure of the decision-maker.
- aggregation of attributes across several dimensions must rely on mathematical procedures that might not correspond to the intuitive aggregating pattern of the decision-maker.
- decision analytical models provide unambiguous data only if a single decision-maker is presupposed. As soon as conflicting values and assessments arise, there are as many risk assessments as groups involved. Either a benevolent dictator is required to find the best solution for all parties involved (not a very attractive choice), or a negotiation

process has to be initiated in order to find a reasonable compromise. It is far from obvious that even single individuals do not entertain contradictory impulses and beliefs.

Thomas Dietz and Paul Stern have been in the vanguard of a handful of scholars explicitly aiming to revamp the SEU model of individual choice with a model that is more consistent with what has recently been learned about the evolution of human cognition, the existence of psychological heuristics, and the importance of social influences (Dietz and Stern 1995).

In their most concise statement on this problem, Dietz and Stern make four separate, but connected criticisms of the SEU model. First, with a glance to the literature on human and social evolution, they note that the relatively complex array of mathematics the SEU model presumes of people is not consistent with what we know about human cognition. People simply are not very adept at multiplying probabilities by expected utilities of different action alternatives, at least not without external aids. Human cognition, they point out, has developed along another track. What humans are good at is pattern recognition, classification, and the application of rules of thumb. In short, people are "skilled taxonomists" rather than "powerful calculating machines." They tend not to be as cognitively thorough-going as the SEU model assumes. Rather than systematically considering each action alternative, people are more likely to categorize similar action alternatives and make judgments about whole sets according to rules of thumb.

Second, Dietz and Stern note a very significant literature on psychological heuristics and their importance in decision making. Work by Daniel Kahneman, Amos Tversky, and others since the late 1970s reveals the presence of certain heuristic devices (Tversky and Kahneman 1981; Kahneman, Slovic et al 1982). We will discuss these heuristics below (section 3.6). Because it is empirically derived, Kahneman's and Tversky's research into psychological heuristics presents a strong argument in favor of a revisionist model of individual choice.

The third argument Dietz and Stern make for their new model of individual choice also has to do with how humans simplify complex decision making. Drawing on Schwartz's model of norm activation, Dietz and Stern assert that individuals consider the worthiness of different possible outcomes until one triggers a moral response. Norm activation presumes, for each individual, compelling norms. These supersede the consideration of other values and thus "drive" decisions. According to Dietz and Stern, these decision activating norms may not be single values, but rather value clusters. Again, following up on the idea of classification and simplification, they propose that people consider possible decision outcomes, not in terms of a myriad of single values, but of sets of values called clusters. They propose three value clusters: egoistic, altruistic, and biospheric.

Fourth, they draw upon the social movement literature to bring in the ideas of framing and social influence. Again, they turn to the experience of everyday life and note that one of the fundamental premises of SEU is invalid. People do not make choices in social isolation. Instead, they actually seek out the advice of their friends, spouse, children, and peers. They make themselves available to persuasion by media, authority figures, and others. Indeed, even if they do not seek out others' advice, people cannot escape social influence. Any search for information, grounding, or perspective is likely to encounter the motivations of other individuals. People frame descriptions and advice in strategic ways. As an example, Dietz and Stern point to results from contingent valuation studies that show that by merely rewording a survey question, the results can be significantly changed (Guagnano, Stern et al 1994). As an attempt to extend the social choice mechanism of markets to cope with decisions about non-market goods, contingent valuation represents a novel use of RAP. Dietz and Stern resist this temptation on the grounds that it employs a model of individual choice that is not accurate or appropriate to this situation.

Although an alternative model of individual choice seems to be emerging, it is yet incomplete. It is nevertheless reasonable to expect that the SEU model will not be fully replaced. It is still a powerful tool to use in interpreting why people do what they do, albeit only one way of looking at the world. The small group of scholars working on alternative models of individual choice provide us with additional perspectives. In the end, we may well find that one model provides a clearer picture of some kinds of decisions, while another one works equally well for others. Whatever the results, revising the model of individual choice lies at the core of the emergence of a paradigm to succeed RAP, a theme to which we will return in Part II.

3.4 Quantitative Risk Assessments

In practical risk management, estimation of the probability with which uncertain events may happen is a fundamental challenge. For this purpose, various kinds of quantitative risk assessments have been developed. Researchers identify and quantify relationships between potential risk agents (such as dioxin or ionizing radiation) and physical harm observed in humans or other living organisms. Assessments of harm may be based on toxicological studies (animal experiments) or epidemiological studies (comparison of a population exposed to a risk agent with a population not exposed to the risk agent). Modeling is used to isolate a causal agent from among several intervening variables (Lave 1987; Renn 1985).

Well calibrated expected values of risks associated with the operation of complex technological systems require statistical data on relative frequencies of failure. Such data often are inadequate or not available. In such cases, fault tree or event tree analyses – methods to assess failure probabilities for a system by aggregating the failure probabilities of each component – are used (Hauptmanns et al 1987; Kolluru 1995; Lowrance 1976; Morgan 1990). All probabilities within a logical tree can be aggregated to model the overall failure rate of the system. A probabilistic risk assess-ment (PRA) provides an average estimate of how many undesirable events one can expect over time as a result of human error or technological failure.

The normative implication is obvious – since physical harm is perceived as an undesirable effect (for individuals and society as a whole), technical risk analyses can be used to reveal, avoid, or modify the causal agents associated with unwanted effects. They can also be used to mitigate consequences, if causes are yet unknown, remote from human intervention, or too complex to modify (Renn 1998a). Their instrumental functions in society are, then, oriented to risk redistribution and sharing, through mitigation of consequences, standard setting, and improvements in the reliability and safety of technological systems.

This simple perspective on risk has drawn much criticism (Hoos 1980; Douglas 1985; Mazur 1985; Beck 1986/1992; Freudenburg 1988; Clarke 1989; Adams 1995; Margolis 1996). First, what people perceive as an undesirable effect depends on their values and preferences. Second, the interactions between human activities and consequences are more complex and unique than the average probabilities used in most risk analyses are able to capture. Third, the institutional structure of managing and controlling risks is prone to organizational failures and deficits which may increase actual risk (Freudenburg 1992). The interaction between organizational malfunctions and risk is usually excluded from these risk analyses. Fourth, statistical treatments typically assume equal weight for magnitude and probabilities. The implication of indifference between high consequence-low probability and low consequence-high probability events with identical expected values is contrary to distinct preferences found in empirical research (Fischhoff et al 1982; Renn 1990; Slovic 1987).

From a normative perspective, this practice of risk minimization implies a clear distinction between experts and laypersons. Moreover, risk reduction or mitigation is based on the assumption that risk should be reduced in proportion to the expected or modeled harm to humans or ecosystems (Morgan 1990). These assumptions are highly contested: actions taken to cope with risk are not confined to the single goal of risk minimization, but include other objectives such as equity, fairness, flexibility, or resilience (Nowotny and Eisikovic 1990; Short 1984). The inclusion of these comple-mentary objectives requires participation by interest groups and the affected public.

3.4.1 The Actuarial Approach to Risk

The actuarial approach averages harmful events over time and space, using relative frequencies (observed or modeled) as a means to specify probabilities. The base unit for risk (the expected value) is based on the relative frequency of an event averaged over time. Undesirable events are confined to physical harm to humans or ecosystems, which can be objectively observed or measured by appropriate scientific methods. Thus, for example, fatalities in car accidents for the coming years can be forecast. The expected value can be extrapolated from statistical data about fatal accidents in previous years (Hohenemser, Kates et al 1983). This approach to risk relies on two conditions. First, enough statistical data must be available to make meaningful predictions. Second, the causal agents responsible for the negative effects must remain stable over the predicted time period (Häfele, Renn et al 1990).

Actuarial risk analysis provides only aggregate data over large segments of the population and for long time duration. Each individual, however, may face different degrees of risk depending on the variance of the probability distribution (Shrader-Frechette 1991). A person who is exposed to a larger risk than the average person may legitimately object to a risk policy based on aggregate calculations. The extent to which a person is exposed to a specific risk also rests on lifestyle factors and may be assessed by anecdotal knowledge, both of which are largely unknown to those who perform risk analyses. Brian Wynne (1991) documented the failure of risk experts to recognize the extensive knowledge of local farmers about sheep farming and the physical environment when they conducted a risk analysis of the Chernobyl fallout in the United Kingdom. Finally, some critics argue that the dominance of science in risk policy making provides too much power to an elite that is neither qualified nor politically legitimated to impose risks or risk management policies on the general population (Jasanoff 1982).

These criticisms can be summarized in three propositions:

- Actuarial analysis relies on relative frequencies as a means to estimate probabilities. This approach excludes unexpected events and tends to blur differences in space, populations, and time.
- Actuarial analysis confines undesired events to physical harm to humans and ecosystems, ignoring social and political costs. This one-dimensional concept of risk stands in opposition to the multi-dimensionality of risks, as experienced by individuals and social collectivities.
- Actuarial analysis assumes a mirror relationship between observations and reality, thus failing to recognize that both the causes of harm and

the magnitude of consequences are mediated through social experience and interaction, and depending on social definitions of what constitutes sufficient cause and undesirable effects.

In our opinion, all of these critical remarks are well taken. Actuarial risk analysis is a narrow framework that should not be the sole criterion for risk identification, evaluation, and management. Actuarial risk analysis rests on many "transscientific" assumptions (Weinberg 1972), including selection rules for identifying undesirable effects, the choice of a probability concept, equal weighting of probability and magnitude, and many others (Pinkau and Renn 1998). These conventions in risk analyses can be defended through RAP-based reasoning, but they represent poorly what individuals and society experience as risk (Jasanoff 1986; Kasperson et al 1988; Kunreuther and Slovic 1996). Complementary risk analyses by the social sciences are necessary to capture the areas of risk experience that are either ignored or dismissed by technical approaches to risk analysis, such as the actuarial approach.

Clearly, however, actuarial risk analyses are necessary and relevant to broader concepts of risk. They serve a major purpose, for people do get hurt or killed in accidents, in natural disasters, and by pollution. Our position, consistent with our definition of risk, is in agreement with Short (1989a), who insists that risk cannot be confined to perceptions and social constructions alone, but that objective outcomes are an integral part of the social processing of risk (1989b). Actuarial risk analyses help decision-makers to estimate objective outcomes such as physical harm. They provide important knowledge about actual damage that is theoretically or empirically linked with action possibilities.

The narrowness of this approach constitutes both its weakness and its strength. Abstracting a single variable from the context of risk-taking makes the concept of risk one-dimensional but also universal (Covello 1991). Confining undesirable consequences to physical harm excludes other consequences that people might also regard as undesirable. Physical harm may be one of the only consequences that (nearly) all social groups and cultures agree is undesirable. The evaluation of consequences differs considerably among groups when undesirable effects include impacts on values, inequities, or their social interests. These additional effects may or may not be more relevant than physical harm to different actors. Because they must rely on subjective preferences, and since physical harm appears to be universally accepted as a negative effect, there is agreement that physical harm should be avoided.

3.4.2 Probabilistic Risk Assessment

Probabilistic Risk Assessment (PRA) is an entrenched methodology, with widespread and growing use, for assessing the overall risk of technological systems (Lowrance 1976; Rowe 1977; Hauptmanns, Herttrich et al 1987; Kolluru 1995). A child of the aerospace and nuclear age, it originated as a means for predicting the effects of failures of small components on the operation of complex aerospace systems. Its use was eventually broadened to assessing the risks of complex commercial nuclear systems, the paradigmatic risk of our age. It lives on as a tool of risk assessment and safety management of complex technological systems – nuclear and others.

In the early years of the age of commercial nuclear development, power reactors were relatively small, simple, and easy to manage. Because of their size and simplicity these early systems were well-defined, fairly easy to understand and manage, and from an engineering point of view, "deterministic." System functioning could be traced along a chain of clearly discernible cause and effect sequences. Nevertheless, the possibility that a serious accident could occur was recognized in the beginning stages of the development of commercial nuclear power. Unknown, however, were the *probabilities* of such an accident or its consequences. To address this contingency, containment structures (designed to "contain" radioactive releases) were included in reactor designs. With some exceptions (especially in the former Soviet Bloc countries), they have been a requisite design feature ever since. Thus, *system* safety, largely a function of the small scale of the system itself and of the ability to contain accidental radiation releases, was not a serious concern in the early history of nuclear power.

Accompanying the rapid growth of nuclear power in its early commercial days, however, was growth of another sort: growth in the size of reactors – indeed, staggering growth. The rapid scaling up of reactor size was driven by the economic need to compete with conventional power plants. An unavoidable concomitant of rapid reactor vessel growth was growth in the complexity of the systems the vessels would serve. This transformed the character of the entire system, from deterministic to stochastic. As a consequence, system safety, taken for granted with small reactors, now became a major concern. A new assessment tool was needed. It was provided by British physicist F. R. Farmer (1967a, 1967b), with the introduction of PRA as a tool for assessing reactor safety (Gesellschaft für Reaktorsicherheit 1979, 1989; U.S. Atomic Energy Commission 1975; see also Weinberg 1981). As a risk assessment tool, PRA has been the centerpiece of the safety of nuclear power ever since.

As a methodology, PRA combines Cartesian reductionism with Bayesian statistical techniques adapted to the practical problems of engineering design and reliability (Renn 1985). Overall system failure (the obverse of reliability) is assessed by reducing the system to a collection of its operating components: valves, pipes, switches, claddings, rods, and – in the case of nuclear operators – even people. Because actuarial data on failure are almost never available, estimates of a system failure are typically based upon fault tree and event tree methods (IAEA 1995). Fault tree analysis assumes that while probabilities of system failures are not known, probabilities of component failure are known. Fault tree analysis always begins by identifying an undesirable event, such as a loss of coolant that keeps the reactor from overheating. The goal is then to estimate the probability of this occurring. Initiating events leading up to a loss of coolant accident (LOCA), for example, are then diagrammed in a sequence of event boxes. The sequence is extended to lower and lower levels of complexity until reaching so-called primary faults: valves, switches, etc. The probability of failure of each "event box" is then identified or calculated.

For many components, identification is possible because actuarially based probabilities are available, based upon similar applications. Other components, for which no actuarial data exist, pose a special problem. To address this problem, practitioners substitute "synthetic probabilities" for the unavailable empirical probabilities (Häfele et al. 1990; Renn 1992b). Synthetic probabilities are often based on the reliability of a given component in another system (a specific type of valve in, say, an aerospace application) or, more often, on the reliability of a similar component in another system. Another problematic feature is the treatment of human actions in the system (IEC 1993). Traditionally, humans were either ignored or treated as an afterthought. Human Reliability Assessment (HRA), a set of procedures designed to assess the impacts of key human actions, have not – where used – been applied with the same rigor as hardware assessments, but have been applied *post hoc*. Thus, while human actions are one of the least analyzed elements of such systems, they have, in fact, been treated as if they were well understood.

Once the tedious process of assigning probabilities to each of the systems' constituent components is accomplished, the overall reliability of the system is determined by aggregating the separate failure probabilities of each of the components. Typically, the aggregation is performed via Boolean algebra. (See U.S. Atomic Energy Commission 1975 for a general review of the PRA methodology.)

PRA embeds normative assumptions, both in its methodological approach and in its policy application. Methodologically, its *raison d'être* rests on two key assumptions: that complex technological systems are accessible to detailed human comprehension and that its approach of "componential reductionism" is the best way to yield that comprehension (Marcus 1988). The possibility that the large technological systems humans create are beyond detailed comprehension is not entertained. Also disregarded is the possibility that other approaches – such as holism – might lead to better understanding of risks associated with technological systems. Moreover, the singular focus on technological systems and hardware components renders irrelevant the human systems co-determining overall system safety. Important human considerations such as motivational contexts, formal organization, and organizational culture are ignored (Reiss 1992).

As a policy or management tool, PRA endorses the normative claims that qualitative differences between different types of risk – such as between low probability-high consequence risks versus high probability-low consequence risks – are irrelevant. PRA reduces risks due to system failures to the common metric of probability. Since there are no absolute standards for determining an acceptability level – (whether 1 x 10^{10} or perhaps 1 x 10^4 is acceptable), determining acceptable risk is, in large part, a function of engineering practices or imposed standards (Priddat 1996).

3.4.3 Toxicological and Epidemiological Approaches to Risk

Toxicology and epidemiology manifest the essential features of RAP in the way they conceptualize and approach issues of risk (Lave 1987; National Research Council 1991; WHO 1977). Both of these approaches begin with an uncontested assumption, and their association with RAP is very clear: health is an essential component to human happiness. The corresponding conclusion is that rational individual actors seek to improve their overall happiness, partly by preserving their personal health. Rational actors will then make choices about risk-related behavior so as to minimize the overall negative impacts, including health impacts among others, on happiness. It follows that the normative force of toxicological and epidemiological characterizations of risk is realized in regulations and standards meant to protect human health by reducing exposure to risk agents that have been shown to have damaging effects.

We group toxicology and epidemiology together because both attempt to identify causal relationships between a risk agent and a victim. The difference is that toxicology uses the results of animal experiments, while epidemiology looks at diseased populations in comparison to a control or unaffected population to characterize the causal mechanism.

Toxicological approaches to handling choices about risk revolve around portraying cause-effect relations on a dose-response curve. Uncertainty enters into the process at four levels. First, there are uncertainties associated with natural variability among the population being subjected to the dose of a risk agent (Crawford and Wilson 1996). For example, if 1000 mice are injected with differing doses of a toxin suspected to produce cancer, differing proportions of the mice will die from the toxicity of the doses. Plotting the incidence of cancer against delivered doses will not produce a perfectly smooth dose-response curve, however, because some individuals will be more cancer sensitive to the toxin than others. This type of uncertainty is handled using standard statistical techniques; e.g., confidence intervals on dose response curves (Kolluru 1995).

Second, there are uncertainties associated with extrapolating to points at the low or high ends of the curve (Meijers, Swaen et al 1997). In order to generate results with a reasonable number of sacrificed animals, toxicological experiments must deliver doses that are several orders of magnitude higher than the range of interest. To extrapolate to the far bottom of the curve, assumptions must be made about the shape of the curve near the origin. One may choose among a variety of curves, such as a linear, low dose-high response, or threshold function. Without a firm scientific basis for resolving this issue, this kind of uncertainty can be quite intractable. Selecting a shape for the curve is politically relevant because toxicological information is often used as the basis for setting regulatory standards for exposure. The decision is often made through a political process that incorporates scientific educated guesses by scientists, stakeholder group concerns about the consequences each model is likely to hold for their interests, and the mandate of the regulatory agency to protect human health without undue burdens on the risk generators (Doull, Rozmann et al 1996). For instance, the US EPA promotes the use of a linear model based on the fact that it is more conservative than the multi-stage or other models, but this is currently being challenged as being too conservative and thereby too costly for business and consumers.

Third, there are uncertainties associated with translating the results of toxicological studies across species. Even when the biochemical mechanisms for harm are known, it is impossible to be sure that one species will experience the same effect as another. An agent that causes liver cancer in rats may produce a completely different kind of cancer in humans, or it may have no effect at all. And rarely are the results of animal data tested on human populations. These uncertainties are not addressed within the realm of toxicology itself. In policy making about risk standards, the customary argument has been to use a factor of 100 for interspecies translations. For example, if the no adverse effects level for a chemical in mice is 100 micrograms per unit, then the human standard might be set at 1 microgram per unit. This somewhat arbitrary factor did not come about through a formal process, but has emerged as a compromise between stakeholders, especially government and industry.

Fourth, uncertainties enter into toxicological assessments of risk when individuals are exposed to more than one risk agent at a time and the agents have a synergistic effect (National Research Council 1988). Because the number of synergistic combinations is potentially infinite, there is little that toxicology can do to frame this uncertainty. For the most part, this issue is not addressed in standard settings that rely on toxicological results.

Epidemiological approaches to estimating risk rely on computing odds ratios. An odds ratio expresses the increased likelihood that an exposed population will experience deleterious consequences (usually cancer or death) over an unexposed (or control) population (Arms 1990). Like actuarial analysis, epidemiology relies on data acquired from already experienced outcomes. In retroactive studies, the population being studied has experienced the harm already (e.g., an elevated rate of disease was noticed among a certain population). This segment may be defined geographically, in which case a specific risk source may be the cause, or it may be defined according to other characteristics, such as exposed infant children in low income families (as was the case with lead poisoning). In proactive studies, either a mechanism for harm is already postulated (e.g., toxicological studies suggested that plastic-lined water pipes may cause cancer in humans; therefore an epidemiological study of a town with these water pipes is organized) or an earlier study has identified an at-risk population (e.g., people who work within high voltage electromagnetic fields). Proactive studies follow an at-risk population over several years to track the onset of disease with the pattern of exposure. The *Framingham Heart Study* is an example of a proactive epidemiological study.[26] Cancer registries, medical records, personal interviews about past behavior and health problems, and death certificates are some of the key data sources in epidemiological studies (WHO 1977).

Uncertainties enter into epidemiology in several places. First, there is the problem of identifying the affected and control population. Epidemiology is best known for retroactive studies of geographically-defined populations; but genetics, behavior, workplace, consumption patterns could all be defining characteristics for a study population. Each risk agent's causality with the disease can only be assessed by comparing the affected population against a properly matched control population. The challenge is to accurately define the affected population and to identify control populations that are identical to the affected population in all regards except for exposure.

Second, intervening variables add uncertainties to epidemiological studies. These uncertainties can be reduced with appropriate experimental

[26] Documented at www.nhlbi.nih.gov/about/framingham/bib-menu.htm.

designs. Control populations must be carefully selected so that all intervening variables associated with personal behavior and exposure to the suspected risk agents are nullified. When computing the estimated dose acquired by members of the at-risk population from a risk source, the experimental design must ensure that intervening effects of secondary risk agents are kept constant. For instance, if water pipe linings are suspected of delivering a carcinogen, then the experiment must first ascertain that no other sources of water contamination exist (compared to the control population). However, since the causal mechanism is not yet known, only suspected, uncertainties associated with intervening variables can never be completely eliminated, even in the best experimental designs.

Third, epidemiologists must cope with time-related uncertainties. In retroactive case-control studies, time presents the greatest difficulty. For most forms of cancer there is a latency period (often years) between the exposure to the risk agent and the development of the disease into diagnosable symptoms (e.g., a tumor). This period can be as long as thirty years for some forms of cancer. The more time that passes between the onset of exposure and a diagnosable symptom, the more the opportunity for other causes to intervene. The more time that passes between the onset of the exposure and the study, the more tenuous the reliance on personal interviews or other kinds of data records becomes. In addition, there is the problem of population migration or population retirement (i.e., subjects in either the case or control populations move away or die).

Toxicological and epidemiological approaches are similar to the actuarial approach in their reliance on historical empirical data, differing primarily in the way they quantify undesirable effects. Rather than relying on statistical data about existing impacts in order to chart out trends, epidemiology and toxicology seek to identify causal relations as the basis for determining expected levels of harm. Projections of different levels of harm are based on empirical data about the population and the risk agent, dispersion and transference models, and causal models that depict how the risk agent induces harm. Toxicology and epidemiology often use actuarial data to help develop these models or to measure the vulnerability of different population segments to given risk agents.

Toxicology and epidemiology adopt a definition of risk that is meant to be objective, not socially constructed. There, risk is a property of physical systems, regardless of whether a person is aware of his / her exposure, or the likelihood of suffering harm. Viewing risk as isomorphic with physical reality in this way glosses over the fact that risk is also a political issue. By thus blinding themselves to socially relevant features of populations, however, toxicology and epidemiology are liberated from concern for distributional aspects of harm. For example, epidemiological findings concerning the dangerous effect of lead paint on children of low income families carries the same scientific legitimacy as results about risks to higher status groups or classes. But while this may provide a potential resource to disadvantaged groups, their ability to take advantage of such a resource may be structurally limited.

Toxicology and epidemiology are considered rational means to uncover causal relations between risk sources and health consequences. The underlying assumption is that these results should be the basis for a strategy to manage or eliminate the risk. Typically, results of these studies are used for standard setting and regulation. Toxicology and epidemiology have become essential elements of policy making and regulatory institutions. Before a new chemical can come onto the market, either as a food additive, a farm pesticide, an industrial ingredient, or another kind of consumer good, studies must be done to estimate the potential for harm related to intended usage. Toxicology provides a basis for making informed guesses about the likelihood that a chemical will carry deleterious side effects. Epidemiology provides a means to follow up on the impact these risk agents have on people over the years that people remain exposed to them.

3.5 Natural Hazards

The experience of natural hazards has been one of the most influential challenges for human societies from the beginning of cultural evolution (White 1974; Burton, Kates et al 1978; Drabek 1986). As humans strive to find meaning behind natural disasters and to explain the reasons why some people have fallen victims of natural events while others have been saved, societies have offered culturally constructed concepts to explain and interpret such sudden disruptions of social functions and to justify and direct societal efforts to cope with natural disasters. Sociologists of religion and culture have analyzed the structure and patterns of these cultural explanations as they appear and reappear in different social settings and times (Douglas 1966; Nelkin and Gilmam 1988; Wiedemann 1993).

One of the pervasive patterns is the attempt to explain the obvious randomness of victimization as a mythological link between social behavior and metaphysical punishment, avoiding the trap of linking personal behavior to experiencing disaster (which was too easy to falsify). Another attempt is to extend the realm of scientific reasoning to natural disasters by using probability concepts that give the impression of calculability of natural events and their outcomes (Banse 1996; Treml 1990). In line with this argument, any social action that is not taken to prevent the negative consequences of a natural disaster is perceived as a social failure of the institutions which are responsible for dealing with such disasters. A third alternative is the tolerance of fate as an inevitable part of human life that escapes any scientific explanation and leads to avoidance responses, i.e., behavior that respects natural forces and aims at arranging a lifestyle of co-existence with the hazardous environment (Turner 1995).

While fate implies that means of coping with natural disasters other than avoidance are not available, and the metaphysical explanation includes at least an element of fatalism (the punishment will come regardless of the preparedness of society to combat natural disaster) – the scientific attempt to protect societies against natural disasters implies that human actions can indeed alter the consequences of natural disasters. Although most modern ethical philosophers still make a distinction between responsibility for actions and responsibility for inaction (Spaemann 1980), the prevailing secular view of natural disasters extends the responsibility and accountability of social institutions to preventing or at least mitigating harmful consequences of natural events, regardless of whether they are caused, promoted or intensified by human actions. Natural disasters may be caused by forces beyond the reach of human intervention; avoiding or mitigating their consequences are, however, legitimate tasks of social systems, for which they are held accountable (White 1961).

In terms of RAP, natural disasters represent constraints of human actions that can hardly be avoided or influenced. Recent studies on natural disasters stress the factor, however, that many so-called natural hazards are really consequences of human actions leading to natural responses (Susam, O'Kefe et al 1983; Emel and Peet 1989). This observation creates even more incentive for societies to scrutinize their own actions and accept a sense of social responsibility for natural events that were originally blamed on God or Nature. In response to this attribution of responsibility and accountability to human society for natural disasters, risk management agencies have evolved with the mandate to prevent potential human-induced causes for natural disasters, to prevent negative consequences from such disasters and to mitigate unavoidable negative impacts.

In pursuing this task, natural risk management agencies first developed models of prediction. Without some reliability in prediction, rational action is limited to avoidance (such as leaving flood areas undeveloped or removing settlements from earthquake-prone regions) or to resilience (building flood-proof or earthquake-proof housing). Both strategies are suboptimal in the rational sense, since avoidance is associated with high opportunity costs and attempts to guarantee resilience may lead to overprotection (Petak and Atkinson 1982). In order to balance costs of protection and expected damage costs, agencies need a risk estimate that includes the calculation of a probability distribution over expected damage. Once such a distribution is established, optimal protection may be expected when the costs of expected damages (probabilities multiplied by the damage compensation costs) are equal to the investment costs of operating a system of protection.

With respect to floods, the most frequently used benchmark is to protect a region against a flood that is expected once in a hundred years (Fattorelli, Borga et al 1995). It is assumed that floods beyond that magnitude are still possible, but so infrequent that the cost of protection would exceed the cost of damage compensation. Similarly, houses in earthquake-prone areas are required to withstand an earthquake of a specific magnitude, again representing a threshold based on probability estimates. The rationale is the same in all such cases: protection is rationally justified for events that appear frequently enough so that expected damage losses outweigh the investment costs for protection. In addition, building infrastructure for mitigating the effects of natural disasters once they have occurred (such as building hospitals or emergency vehicles) is also a rational means to cope with the consequences of disaster. Since maintaining an infrastructure for mitigation is usually less expensive than protection, the break-even point between damage costs and investment costs is not identical with the break-even point of protection. Therefore, mitigational strategies can include natural events with a lower probability, and hence a higher magnitude, than protective strategies. For example, a region might decide to have sufficient hospital beds available for earthquake victims for more severe earthquakes than the reference scenario used for establishing the building code.

Other types of hazard require other strategies. Protecting somebody against hurricanes is almost impossible. The most rational strategy here is to warn people in advance so they can avoid the hazard (Petersen and Jensen 1995). The advantage of this strategy is that the avoidance response is only necessary when the natural disaster strikes, but at all other times the region at risk will be inhabitable. Potential destruction of agricultural land or threats of famine may require storage of food and other supplies as an emergency support for the victims of such events. Whatever the strategy might be, the basic rationale for defining the scope and intensity of the strategy is to balance costs.

The behavior of people in natural disaster-prone areas was one of the first observed "natural laboratory" deviations from the utility theory of RAP. After the March, 1964, Good Friday earthquake in Alaska, researchers observed that most businesses and homeowners did not have earthquake insurance, even though it was highly subsidized by the federal government – and, therefore, a rational purchase. Similar observations were made after tropical storm Agnes in the summer of 1972 (Kunreuther, Ginsberg et al 1978). These deviations from the expected utility approach of RAP were a challenge to analysts of protective behavior against natural hazards

(Drabek 1986; Quarantelli 1988; see also case studies in Wright and Rossi 1981).

First, most people resisted buying insurance against natural hazards, even when such disasters were fairly probable. Second, protective measures against floods or earthquakes were difficult to enforce and often met with substantial opposition by the people affected. Third, victims of natural disasters showed hardly any inclination to improve their protection or to avoid the hazard in the future. Only days after a disaster had occurred, people were busy rebuilding their homes at the same location and with the same level of protection. Fourth, many people were reluctant to comply with mitigational efforts such as evacuation or sheltering. Rather, they stayed in their premises and faced the negative impacts of the disaster. Fifth, in cases where the disaster was beyond the prespecified level of protection (such as a 500 year flood), people were eager to deflect blame to risk management institutions for providing inadequate protection or mitigation.

These observations were quite puzzling to most observers, even more so to RAP theorists. One response was "blaming the victims," asserting that people act irrationally and that their imprudent behavior should not affect the normative guidelines of what ought to be done. However, the consistent patterns of violations in rationality made it difficult to sustain the claim that the whole world acts irrationally except for the few analysts of natural disasters. Another more successful strategy was the attempt to discover a hidden rationality in the alleged irrational responses of the affected population (Jungermann 1986). Each of the observations could be explained in a revised quasi-rational framework.

First, the reluctance to buy insurance could be reinterpreted as a somewhat rational strategy, based on the observation that society would hand out emergency relief funds to victims of natural disaster, thus providing the benefit of insurance without the costs. Furthermore, prospect theory explained the apparent contradiction between the reluctance to buy flood insurance and the willingness to buy other, seemingly less necessary, kinds of insurance (Kahneman and Tversky 1979; see Section 3.6). Spending a little money on insurance premiums against extremely improbable events, such as a 100 year flood, was in line with the notion of risk aversion. People like to protect themselves against catastrophes even if they are highly unlikely. The regret of losing small amounts of money provides less disutility than the prospect of losing one's livelihood. It is the potential threat that provides the disutility and not its likelihood. The insurance people do not buy is characterized by high premiums for relatively likely but usually limited damage (Kunreuther, Ginsberg et al 1978).

Second, the reasons that people oppose protective measures although it would enhance their overall utility were twofold: interviews demonstrated that many people opposed regulation and governmental supervision rather than taking protective measures themselves. Freedom of action, rather than adequate protection, was at stake. In addition, many respondents showed the "invulnerability syndrome." Their subjective perception of the threat was much lower than the official estimate (Weinstein 1980; Covello, Slovic et al 1988). Rational choice is always based on the subjective probability and magnitude estimates of the decision-maker; for a person unconvinced of the pending threat, protective measures are not warranted, and therefore a waste of money.

Third, the experience that people rebuild their homes at the same location where they were struck by a natural disaster has been most difficult to explain. Again, some analysts claimed that this was not irrational at all (Kates 1976). Given that it was rational in the first place to build a home (with adequate protection according to the expectations of the owner) in the high risk area, then it is equally rational to rebuild the identical home if a worse-than-expected disaster accounted for the destruction. In these terms, such higher-than-expected disasters were deliberately accepted in exchange for lower protection (or insurance) costs. However, that line of argument was not very convincing because the level of protection in most of such houses was less than needed to protect against a reasonably probable event, and often insurance would not even cover the premises, indicating that any cost-benefit balancing would require another structure or a different location. So it was suboptimal from a rational point of view to build the house there in the first place. The "invulnerability syndrome" was also not a very convincing explanation, since these people had just experienced a major disaster (Mileti, Drabek et al 1975). Two other hypotheses were brought forth to explain this phenomenon. The first hypothesis originated from risk perception studies which demonstrated that many people misunderstood probabilities as signs of expected occurrences (Quarantelli 1988). But the expectation of occurrence within some broad time span gives no hint of when within that span a disaster will occur. Victims believed that once a 100 year flood took place, they were safe for approximately another hundred years. Additional investments in protective measures or insurance would be inefficient given the belief that the next strike was a hundred years away. According to the second hypothesis, relocating their houses or adding protective measures would decrease the property value of these premises (Kunreuther, Ginsberg et al 1978). Houses removed from the oceanfront (in particular if all the neighbors stay at the oceanfront) or houses that show obvious signs of protection (thus indicating this house is in an endangered area) will lose market value. One study, for example, demonstrated that people who had decided to live all their lives in their present homes were more inclined to invest in protective measures than were residents of homes they regarded as only a temporary place to live.

Fourth, the reluctance of people to engage in evacuations or shelters can also find rational explanations (Perry, Greene et al 1980; Covello, Slovic et al 1988; Nigg 1995). One hypothesis refers again to the "invulnerability syndrome" that influences people's calculation of potential consequences. Another hypothesis refers to the intrinsic value of property; many people have a strong emotional relationship to their property. Leaving such a property unattended and open to the forces of disaster (or even worse, looting), causes more pain and disutility than risking their lives. Since they might be fearful of looting, they balance the probability of being harmed against the probability of being robbed. If they feel rather secure in their home, that balance will be in favor of staying.

Lastly, the reason individuals blame institutions for not "overprotecting" them, i.e., providing them more protection than is rationally required, can be explained by the fact that the costs of overprotection are shared by many individuals (taxpayers) of which each one contributes only a marginal part. From the perspective of the individual (and taking into account risk aversion as predicted by prospect theory) this marginal amount for extra protection is worth the cost, in particular after the effect (Perry 1983). They may have opposed the relevant taxes or premiums before the event, but still would blame institutions if the protection turned out to be inadequate for a worse-than-expected disaster. Again, prospect theory provides a framework for understanding this preference reversal after the fact. Before the effect, premiums or taxes for protection are losses; after the effect, the missed opportunity of overprotection would have led to a gain. The utility curves for gains and losses are different – most people are risk averse when focusing on losses, risk prone when focusing on gains, even when the same alternatives are considered. So the illusion of marginality in conjunction with preference reversal can be used as an explanation for this behavior.

This line of argument is an interesting demonstration of how rational actor theories are applied in a major research area dealing with uncertain outcomes. Researchers, agency managers, and emergency planners start with a theoretical proposition of what a rational response to natural disaster should be. These responses are tested against reality and basically refuted. One possibility would be to change the model; another would be to make a clear distinction between normative and descriptive models and accept the fact that most people violate the norms of rationality. But the overwhelming response to that challenge has been to reinterpret the behavior of the alleged irrational people and find some way to construct a rational explanation (see critical remarks in Watts 1983) – and to preserve RAP. All these explanations have a certain degree of elegance and ingenuity. However, they leave the impression that RAP is beyond any doubts. It cannot be contested. In fact, one may argue that any type of behavior is necessarily based on some kind of rational foundation so that the task of the researcher is to reconstruct this rationality. Such a model can never be falsified or tested, since along these lines any expression of behavior can always be interpreted as a manifestation of subjective utility. In essence, it leads us to the truism that every person has some reason for doing what he or she does. The task of the researcher, then, is to unpack the rationality behind the action (Green and Shapiro 1994).

The use of tautologies in explaining people's responses to natural hazards has made the field of natural hazard researchers more skeptical about the value of rational actor theories (Susam, O'Kefe et al 1983). In addition, the belief of advocates of RAP that they can predict probabilities of natural disasters and thus prepare a rational protection strategy has been under increased scrutiny in recent years. The uncertainties are higher than expected and the frequency of natural disasters has increased over time, making reevaluation necessary. Nobody can really tell whether the increased frequency in natural disasters is a random variation of the old trend or the beginning of a new trend (say, due to global warming). This second order uncertainty transforms a rational response based on balancing costs into a risk in itself. Society may overprotect its members by assuming a new trend or underprotect by assuming random variation. Rational actor theory provides no solution to this problem.

Moreover, rational actor theories have recently been challenged by the resurgence of the non-rational responses to natural hazards – the transcendental and the fate concept. While transcendental concepts have little normative value in a secular scientific environment, they are used as descriptive explanation for people's behavior. Some explanatory schemes based on emotional or associative models of behavior show promise in explaining deviations from rational actor models. Fate concepts have been used for explaining actual behavior, but have also fertilized normative concepts as they lead to an emphasis on resilience and avoidance of vulnerabilities regardless of the technical means of coping with predefined disasters. Avoiding high risks may actually turn out to be the most "rational" strategy when the probabilities and the magnitudes are in flux and genuine uncertainty prevails. The natural hazard field is a telling example of the accomplishments of the rational actor paradigm, but also of its implicit limitations.

3.6 Cognitive Psychology

Technical analyses of risk are usually based upon an objective definition of risk or on an intersubjective rationale for estimating and evaluating risks. In contrast, the *psychometric paradigm* draws on cognitive psychology to conceptualize risks as subjective expressions of individual fears or expectations about unwanted consequences of actions or events. Paul Slovic, one of the principal contributors to the psychometric paradigm, has characterized the paradigm in the following way:

> It elicits current preferences; it allows consideration of many aspects of risks and benefits besides dollars and body counts; and it permits data to be gathered for a large number of activities and technologies, allowing the use of statistical methods to disentangle multiple influences on the results.... Another distinguishing feature of our studies has been the use of a variety of psychometric scaling methods to produce quantitative measures of perceived risk, perceived benefit, and other aspects of perceptions (e.g., estimated fatalities resulting from an activity). (Slovic 1992, pp 118f)

The psychometric school of risk analysis expands the realm of subjective judgment about the nature and magnitude of risks in four ways.

First, it focuses on personal preferences for probabilities and attempts to explain why individuals do not base their risk judgments on expected values, as decision analysis would suggest (Lopes 1983; Luce and Weber 1986; Pollatsek and Tversky 1970). An interesting result of these investigations was the discovery of consistent patterns of probabilistic reasoning that are well suited for most everyday situations. The theory to develop from these findings, prospect theory, emphasizes the propensity for people to be risk prone when focusing on gains and risk averse when focusing on losses (Kahneman and Tversky 1979). Furthermore, many people balance their risk-taking behavior by pursuing an optimal risk strategy which does not maximize their benefits, but assures both a satisfactory payoff and the avoidance of major disasters (Jungermann et al 1998; Simon 1976; Tversky 1972). As mentioned earlier (Section 3.2.3) portfolio theory is one example of a no-regret-policy sacrificing potentially greater payoff for a desired level of security. This example and many others suggest that deviations from the rule of maximizing one's straightforward utility are less a product of ignorance or irrationality than an indication of one or several intervening contextual variables. From the point of view of the orthodox utility theorist, these deviations are interpreted as irrationality on the part of the actor. Yet, such variables often make perfect sense when seen in the light of the particular context and the individual decision-maker's values (Brehmer 1987; Lee 1981). Furthermore, deviations from the maximization rule may still be considered as manifestations of the rational actor paradigm, since actors assess and evaluate likely consequences systematically, but use a different decision rule. This decision rule is more complex than the rational choice literature would suggest, and comes close to concepts of bounded rationality. Paul Slovic emphasizes this point:

> More generally, psychometric research demonstrates that, whereas experts define risk in a narrow, technical way, the public has a richer, more complex view that incorporates value-laden considerations such as equity, catastrophic potential, and controllability. The issue is not whether these are legitimate, rational considerations, but how to integrate them into risk analyses and policy decisions. (Slovic 1992, p 150)

Second, more specific studies on the perception of probabilities in decision making have identified several biases in people's ability to draw inferences from probabilistic information (Festinger 1957; Tversky and Kahneman 1974; Ross 1977; Kahneman and Tversky 1979). These biases include:

- *Availability:* Events that come to mind immediately (due to memory or imagination) are rated as more probable than events that are less mentally available.
- *Anchoring effect:* Probabilities are adjusted to the information available or the perceived significance of the information.
- *Representativeness:* Singular events experienced personally or associated with properties of an event are regarded as more typical than information based on frequencies.
- *Avoidance of cognitive dissonance:* Information that challenges perceived probabilities that are already part of a belief system will either be ignored or downplayed.

These are clear violations of what one would expect of a "rational" decision-maker within the RAP framework. However, because many laboratory situations provide insufficient contextual information to provide enough cues on which to base judgments, their implications might have been overstated in the literature (Nisbett and Ross 1980; Fischhoff, Lichtenstein et al 1981; Lopes 1983). Relying on predominantly numerical information and being unfamiliar with the subject, many subjects in these experiments revert to "rules of thumb" in drawing inferences. In many real life situations, experience of, and familiarity with, the context provides additional information to calibrate individual judgments, particularly for nontrivial decisions (Heimer 1988). Moreover, with complex decisions and removed from the time constraints of the laboratory, individuals often seek appropriate information and other aids to their decision making. At the same time, biases do exist in human judgment and the research on errors in drawing inferences demonstrates that individual perceptions cannot substitute for rational reasoning, but augment its scope of applications.

Third, the importance of contextual variables for shaping individual risk estimations and evaluations has been documented in many studies on risk perception (Jungermann and Slovic 1993a, 1993b; Renn 1998a; Slovic 1987). Psychometric methods of scaling risks on different dimensions have been employed to explore these qualitative characteristics of risks. The following contextual variables of risk have been found to affect the perceived seriousness of risks (Covello 1983; Gould et al 1988; Jungermann and Slovic 1993b; Renn 1983; Slovic et al 1981; Vlek and Stallen 1981):

- *The expected number of fatalities or losses:* Although the perceived average number of fatalities correlates with the perceived riskiness of a technology or activity, the relationship is weak and generally explains less than 20 per cent of the statistical variance. The major disagreement between technical risk analysis and risk perception is not on the number of affected persons, but on the importance of this information for evaluating the seriousness of risk (O'Riordan 1982).
- *The catastrophic potential:* Most people show distinctive preferences among choices with identical expected values (average risk). Low probability-high consequence risks are usually perceived as more threatening than more probable risks with low or medium consequences.
- *Situational characteristics:* Surveys and experiments have revealed that perception of risks is influenced by properties of the risk source or situation. Among the most influential factors are: the perception of dread with respect to the possible consequences; the conviction of having personal control over the magnitude or probability of the risk; the familiarity with the risk; the perception of equitable sharing of both benefits and risks; and the potential to blame a person or institution responsible for the creation of a risky situation. In addition, equity issues play a major role in risk perception. The more risks are seen as unfair for the exposed population, the more they are judged as severe and unacceptable (Kasperson and Kasperson 1983; Short 1984).
- *The beliefs associated with the cause of risk:* The perception of risk is often part of an attitude a person holds about the cause of the risk, e.g., a technology, human activity, or natural event. Attitudes encompass a series of beliefs about the nature, consequences, history, and justifiability of a risk cause (Thomas, Maurer et al 1980; Otway and Thomas 1982; Drottz-Sjöberg 1991). Due to the tendency to avoid cognitive dissonance, i.e., emotional stress caused by conflicting beliefs (Fe-

stinger 1957), most people are inclined to perceive risks as especially serious and threatening if related beliefs contain negative connotations and vice versa. A person who believes that industry policies are guided by greed and profit is more likely to think that the risks of industrial pollution are only the "tip of an iceberg". On the other hand, a person who believes that industry provides consumers with goods and services they need and value is likely to link pollutants with unpleasant, but essentially manageable, by-products of industrial production. Often, risk perception is a *product* of these underlying beliefs rather than the cause (Clarke 1989; Short and Clarke 1992a). In a cross-cultural comparison of risk perceptions, environmental versus technology-related beliefs of the respondents were better predictors for perceived seriousness of risk than national differences or other explanatory variables (Rohrmann 1990).

Fourth, studies based on psychometric scales have revealed different classes of meanings of risk depending on the context in which the term is used. Whereas in the technical sciences the term risk usually denotes the probability of adverse effects independently of context, the everyday use of risk has different connotations. With respect to technological risk the following semantic images are of special relevance (Renn 1989, 1998b):

- *Risk as a pending danger (Damocles' Sword)* : Risks are seen as a random threat that can trigger a disaster without prior notice and without sufficient time to cope with the hazard involved. This image is linked to artificial risk sources having large catastrophic potential. The magnitude of the probability is not considered. Rather its randomness evokes fear and avoidance responses. Natural disasters, in contrast, are perceived as regularly occurring and thus predictable or related to a special pattern of occurrence (causal, temporal, or magic). The image of pending danger is therefore particularly prevalent in the perception of large-scale technologies.
- *Slow killers (Pandora's Box):* Risk is seen as an invisible threat to one's health or well-being. Effects are usually delayed and affect only few people at one time. Knowledge about these risks is based on information by others rather than on personal experience. These risks pose a major demand for trustworthiness in those institutions that provide information and manage the hazard. If trust is lost, people demand immediate actions and assign blame to these institutions even if risks are very small. Typical examples of this risk class are food additives, pesticides, and chemicals in drinking water.
- *Cost-benefit ratio (Athena's Scale)* : Risks are perceived as a balancing of gains and losses. This concept of risk comes closest to the technical

understanding of risk. However, this image is only used in people's perceptions of monetary gains and losses. Typical examples are betting and gambling, both of which require sophisticated probabilistic reasoning. People are normally able to perform such probabilistic reasoning, but only in the context of gambling, lotteries, financial investment, and insurance. Laboratory experiments show that people orient their judgment about lotteries more towards the variance of losses and gains than towards expected value (Pollatsek and Tversky 1970).

- *Avocational thrill (Hercules' Theme)* : Often, risks are actively explored and desired (Machlis and Rosa 1990). These risks include all activities for which personal skills are necessary to master the dangerous situation. The thrill is derived from the enjoyment of having control over one's environment or oneself. Such risks are always voluntary and allow personal control over the degree of riskiness.

This list of factors further demonstrates that the intuitive understanding of risk is a multidimensional concept and cannot be reduced to the standard definition of risk as the product of probabilities and consequences (Allen 1987; Vlek 1996). Risk perceptions also differ considerably among social and cultural groups. However, it appears to be a common characteristic of public risk perception in almost all countries in which studies have been performed that most people perceive risk as a multi-dimensional phenomenon and integrate their beliefs related to the nature of the risk, the cause of the risk, the associated benefits, and the circumstances of risk-taking into a consistent belief system (Kleinhesselink and Rosa 1991, 1994; Renn 1998a).

The psychometric perspective on risk includes all undesirable effects that people associate with a specific cause. Whether these cause-effect relationships reflect reality or not, is irrelevant (Slovic 1992, p 119). Individuals respond according to their perception of risk and not according to an objective risk level or the scientific assessment of risk. Scientific assessments are part of the individual response to risk only to the degree that they are integrated in the individual perceptions. Furthermore, relative frequencies or other (scientific) forms of defining probabilities are substituted by the strength of belief that people have about the likelihood of any undesirable effect to occur.

While the psychometric paradigm has deepened the understanding of risk considerably, it suffers from the same micro-orientation as does RAP. The focus on the individual and her or his subjective estimates leads to serious difficulties (Wynne 1984; Mazur 1987; Plough and Krimsky 1987; Dake 1991; Sjöberg 1996; Marris, Langford et al 1997). The broadness of the dimensions that people use to make judgments and the reliance on intuitive heuristics and anecdotal knowledge make it hard, if not impossible, to aggregate individual preferences and to find a common denominator for comparing individual risk perceptions. More specifically, the psycho-metric paradigm faces the following conceptual problems:

- It assumes that an individual will act on his or her subjective estimates of consequences and probabilities. It is not certain, however, if an individual actor would not actively pursue a strategy to verify or validate his or her perceptions before acting upon them (Sjöberg 1997). And the paradigm provides no evidence that people act in accordance with their perceptions.
- It creates a world without interactive effects. If perceptions differ between and among individuals, how can one predict and evaluate consequences if other people's responses will have an effect on the consequences of one's own actions? (Luhmann 1990)
- It does not provide a strategy for designing collective decisions and actions. If aggregation of individual preferences is not possible, there is no meaningful concept of collective decision making (Bonss 1996).

Psychometric research into risk perception has revealed that contextual factors shape individual risk estimations and evaluations (Renn 1998a; Slovic 1987). The identification of these factors, such as voluntariness, personal ability to influence risks, familiarity with the hazard, and the catastrophic potential provides useful information about the elements that individuals process for constructing their interpretation of risks.

While psychometric research is relatively close to problems of practical policy making, other psychological research centres more on theoretical issues. A large body of literature documents paradoxes of decision making under uncertainty which are hard to digest for RAP (Machina 1987a, 1987b). There can be little doubt that most people prefer lotteries based on objective probabilities (as when a coin is tossed to select one of two outcomes) to lotteries based on subjective probabilities (as when it is simply unknown which of two outcomes will happen). Moreover, most people prefer a sure gain of $100 to a 30% chance of winning $300, while they prefer a 30% chance of losing $300 to a sure loss of $100.

The most influential attempt to explain at least a large part of such paradoxes is *prospect theory* (Kahneman and Tversky 1979). It generalizes the basic structure of SEU-theory in two steps. First, probabilities are transformed into decision weights, with the result that small probabilities are amplified, moderate and high probabilities attenuated. Second, outcomes are characterized as gains or losses with the help of a reference point. The utility function is S-shaped: for gains it displays risk-aversion (seeking sure gains) for gains, for losses risk-seeking choices (avoiding sure losses).

Prospect theory represents a skilfull generalization of SEU-theory. Such generalizations are not incompatible with RAP, but they stretch its conceptual network quite far. And it is a small step to ask whether at least some reference points could be set by social norms – a step which might lead beyond the social atomism of RAP.

For practical purposes, psychometric and other psychological studies can help to create a more comprehensive set of decision options and provide additional knowledge and normative criteria to evaluate them (Fischhoff 1996). They often fail to explain, however, why individuals select certain characteristics of risks and ignore others. Furthermore, they tend to focus on a world of isolated individuals whose actions do not interfere with each other and who do not need to take collective actions.

3.7 Problems of Policy Making

As approaches for objectively quantifying risk, actuarial analysis, probabilistic risk assessment, and epidemiology/toxicology all assume that risks can be assigned a value of harmfulness *independent of the social, economic, political, or cultural context.* There is an objective worth (which is still conceived only as a negative value in these contexts) for each specific risk. Reasoning that society and individuals seek to minimize risk exposure, quantitative risk assessment approaches advocate risk management actions accordingly. Just as economic actors seek to maximize their personal utility, rational actors are supposed to seek to minimize their exposure to risk. An objective understanding of risk is clearly relevant to making collective choices about the tolerability of new risk sources and about actions to reduce existing risk burdens. But these assessment techniques go beyond factual information about the seriousness of a risk. They lend themselves to normative judgments about rationally defining how safe is "safe enough." Although normative judgments are not directly related to each of the three assessment approaches, as normative advice can never be deducted from factual knowledge alone, the type of information collected and the rationale underlying each of the three approaches provide almost perfect matches for specific normative procedures for determining the acceptability of risks. In this sense, actuarial analysis is closely related to risk-risk comparisons, probabilistic assessments to safety and performance standards and epidemiology/toxicology to health standards. The special class of carcinogenic risks assumed to have no threshold of negative impacts provides the borderline to include normative procedures that balance benefits and costs. These normative procedures are the focus of this section.

3.7.1 Comparing risks

Actuarial analyses compare risks on the basis of frequencies. There is no causality proven in the statistical accumulation of outcomes matching predefined independent conditions, although causality is expected. Since actuarial analysis relies on extrapolation of past observations, it does not need to specify the causal relationships between an initiating event and its negative consequences. Risk analysis produces frequency distributions of the past to construct a trend for the future. How events and outcomes are causally connected is not part of the analysis. This is partially a strength of this approach (since causality requires a much better understanding of any system under observation), but a shortcoming, as well, since no one can be sure that causal relationships do not change over time. Therefore, actuarial analysis can accurately assess only well known risks.

Actuarial risk assessments produce expected values for negative impacts. These numbers have a variety of functions for risk management. Traditional insurance applications need expected values to calculate premiums or to define the minimum number of insurees to make risk sharing feasible. In terms of risk evaluation, expected values can serve as illu-strations of how different risks compare with each other keeping constant either the probability or the magnitude of the outcomes. These risk comparisons can also be used as criteria to defend definitions of "safe enough" by demonstrating that the suggested level of risk tolerance is consistent with the present existing tolerance level of other factually accepted risks. This way of comparing risks is linked to the past behavior of individuals or societies. Risks that they have accepted in the past without coercion and with adequate knowledge can be used as yardsticks for evaluating new risks. For instance, if actuarial data show the risk of dying in a train crash during a trip from Zurich to Paris is higher than the risk of dying on an amusement park ride, then people who do not hesitate to travel from Zurich to Paris because of the risk should, it follows, not fear to take the amusement park ride. The fact that people tolerate the risk of train crashes, which is revealed by their unhesitating use of trains, suggests the risk is within the tolerance level.

One reason it is difficult to use this argument to justify tolerability of new risks is that a person's overall risk burden may increase beyond the tolerance level through the accumulation of small "tolerable" risks. An alternative solution is that each person has a portfolio of risks that he or she deems acceptable. If new risks are imposed or taken by such an individual, the new risks should either replace a risk inside that portfolio or at least only marginally increase the portfolio (being indifferent between the two portfolios). A person or society should be indifferent to a risk portfolio that with the same benefits is no riskier than the risk portfolio they have already tolerated in the past (Gethmann 1993). This normative claim has been partially supported by the empirical observation that many individuals increase their risk seeking behavior when the risks in the portfolio are decreased. For example, when some car manufacturers introduced anti-locking brakes in their cars, drivers drove more aggressively, resulting in a similar number of accidents as occurred prior to the installation of the safety enhancing feature.

Similar in reasoning to the revealed preference approach is the expressed preference approach. Instead of looking at what people have done in the past to reconstruct preferences, surveys and questionnaires are used to ask people directly. Consider, for example, a man who calls for a reduction of highway deaths so that this risk is lower than that of taking a train from Zurich to Paris. If this person responds that he has never taken the train to Paris, but would gladly do so if he had the slightest reason, then his call for highway risk reduction would be inconsistent with his tolerable risk level.

Often, risks are seen as justified when the risk level does not exceed the level of the *natural background* risk. This comparison assumes that natural risk levels, i.e., those that humans are exposed to by being part of the natural environment (such as natural radiation, afflatoxins, pathogens), are "natural" benchmarks for judging the acceptability of human created risks. After all, the human race has been able to adapt to these natural risks over its evolutionary history. Although comparisons with natural background are popular in the literature, it is rarely used as an abstract decision rule. One reason may be that progress is often seen as consisting precisely in lowering risks with which we were faced by our natural background.

Two variations are found in the literature relating to the relation between "natural" and "anthropogenic" risks: comparisons with ranges of natural background and replacement of natural risks through newly established artificial risks. In the first case, technical risks are compared to a range of natural risks to which most individuals demonstrate indifference in choice situations. For example, when somebody moves from Boston to Denver, natural radiation background level increases by a factor of 10. If the moving person is indifferent about this increase in natural radiation level, he or she should also be indifferent about moving into a neighborhood with a nuclear power plant, since the increase in radiation from the plant is far less (assuming no major accidents) than the factor of 10 in the move from Boston to Denver. Critics of this argument, however, point out that the increased natural radiation level may indeed introduce a disutility to the moving person (assuming the person knows about the increased level of radiation) which is compensated for by other benefits of moving to Denver (such as better air quality). Conversely, constructing a nuclear power plant in a vicinity may add a higher disutility to some individual than the indirect benefits related to the operation of such a station.

The second argument is closely related to the moderate revealed preference approach. In case an artificial risk could substitute for a natural risk with a higher potential for harm, the replacement should be preferred, assuming all other conditions are kept constant. There may be few examples in which this reasoning applies to an actual choice situation. However, if the information about the risk sources is complete and benefit levels are equal, a risk-risk comparison can be justified as a normative yardstick if and only if the choice includes the replacement of a larger risk by a smaller risk, all other factors being equal.

Even such an obvious rule of rational choice is problematic. Risks are always expressions of uncertainty. Newer risks are often based on more uncertain and less reliable data, resulting in larger confidence intervals and a higher degree of ignorance over the true extent of possible variations. Thus, it may be prudent to choose a higher risk (expressed as the expected value of a probability distribution over potential losses) with narrow confidence intervals than a smaller risk with large error bounds. An alternative is to base one's judgment about the acceptability of risks on the 75% percentile or another upper boundary line to be on the safe side. As soon as one considers the varying degree of uncertainty as an important decision criterion, there is no single rational solution to selecting the least risk alternative.

Moreover, risk calculations do not specify the distribution of potential losses with respect to time, space, and social groups. Risks with high catastrophic potential (such as plane accidents) receive the same numerical value as risks with continuous losses (such as car accidents) if the expected values of the two risks result in the same number. As numerous experiments have proven, humans demonstrate preferences for continuous rather than catastrophic risks and are willing to tolerate more losses on average if the worst possible accident involves only a limited number of victims than the other way around. Prospect theory provides an interesting psychological explanation for risk averse behavior. Economists have also tried to find a rational explanation for these preferences, claiming that aversion to catastrophic events expresses a rational balancing of increased social costs for coping with catastrophes versus the more manageable costs of losing an individual time after time. Preferences are not limited to variations in time. People show also preferences for spaces and social groups. If only one region is affected, or a special social group (such as Native Americans, for example), society may put more or less importance to such an inequitable distribution of potential risk outcomes.

In sum, risk comparisons provide a rather weak and problematic basis for judging the acceptability of risks. They may be useful in situations where a well known risk source promises to replace an existing higher risk without imposing new inequities to the people affected. In addition, old and new risks should have similar confidence intervals and distribution functions of losses over time. Caution about the application of risk comparisons for normative decisions does not compromise their use as illustrations for judging the seriousness of a given risk. They can also provide a perspective for prioritizing policies since they are easy to construct and they constitute a rough indicator for the magnitude of potential harm.

3.7.2 Establishing Standards of Acceptable Risk

If risk-risk comparisons provide only limited solutions to the question "How safe is safe enough?", the answer may be closer to a normative application of probabilistic risk assessments. These assessments produce values for the relative frequency of technological failures and magnitudes of the potential consequences. They rely on causal models, which may incorporate existing actuarial data from past failures, created actuarial data from accelerated testing, or Bayesian probabilities of failure rates. Results of PRA studies are normally used to suggest specific alterations to technological system designs, to suggest safety standards, or to suggest emergency response plans. They serve as evaluative aids in the design of technological systems and risk management plans. The technological system may be intended to replace or reduce existing risks (as in upgrading a facility through additions of pollution control or safety equipment) or they may introduce brand new risks (through innovation and development).

One way that PRA is used to help arrive at an answer to the question of "safe enough" is to provide a quantitative basis for risk comparison among technological systems and risk management plans. Just as actuarial analyses provided a baseline for comparisons invoking revealed and expressed preference approaches to risk tolerance, PRA can provide that baseline when actuarial data are missing or incomplete. One strength of PRA is that it can model hypothetical risk scenarios, which should allow some risks to be prevented. Another is that it can be used to conduct sensitivity analyses in order to answer "what if" questions. Most important, however, is the application of PRA for setting *quantitative safety criteria*. In many countries, such as The Netherlands, regulatory agencies demand that all technical systems should not exceed a risk of 10^{-6} per year for any accidental fatality linked to a specific industrial plant. In Switzerland, all cantons (states of the Swiss Union) are obliged to superimpose a standard of acceptability on quantitative estimates of risks. All risks below the line are acceptable, all risks above the line need to be reduced until they fall below it. In order to make the necessary comparisons between the risk acceptability line and the risk of a specific plant or facility, PRA analysis is required.

Determining quantitative safety criteria or setting acceptability thresholds remains a subjective and usually a political decision. Probabilistic risk analysis provides the data base for comparing actual risks with a preset threshold, but the decision about the value of the threshold is independent of the numerical analysis of risks. One can employ the aforementioned methods of revealed or expressed preferences to determine a level of tolerable risk. These methods are not very convincing, as argued earlier in this section. Preferences of what people personally feel is tolerable vary considerably. Under present conditions, neither economic theory nor political practice provide any rational means to aggregate numerous and varying preferences about risk acceptability.

Negotiations, mediation, and bargaining (including economic compensation) may constitute more practical possibilities to derive thresholds on which all affected persons can agree, at least in principle. However, trading economic benefits for potential loss of life or health has been laden with conceptual and empirical problems. Utility functions for loss of life are normally non-linear and often non-monotonic, making it hard to extrapolate over larger ranges of probabilities. Compensating each individual would be practically impossible, as individuals will not reveal their true preferences, but ask for more than they really need. They will also communicate with each other about their demands for compensation and then ask for equal treatment among each other (thus demanding the highest amount that the most risk averse person would regard as just compensation). In addition, many people are unwilling to trade any risk for even large benefits. Finally, averaging risks over time, space, and social groups may hide equity concerns that often govern the risk evaluation by the affected people.

Pollution risks are a class of risks characterized by this political problem. Many countries have responded to the problem of determining thresholds of acceptability by setting standards according to the present state of *best available control technology*. Rather than specifying the desired outcome of regulation, they reinforce the process of lowering the risk continuously by demanding that all available control technology that is technically proven and economically feasible needs to be installed. Some countries, such as Germany, demand retrofitting existing facilities if new pollution abatement technology has been developed and tested. Although such a policy of pressing technological advancement has been effective in reducing risks, and in sidestepping discussions of tolerable levels of risk, it is unsatisfactory for a variety of reasons. First, the availability of pollution abatement technologies is not proportional to the seriousness of risks that they promise to reduce. Some marginal risks may be reduced at high cost, while some substantial risks may be imposed on the public, only because no abatement technology is at hand. From an economic point of view such inefficiency is hardly justifiable and certainly a violation of basic RAP principles. Second, pollution control is defined as a technological problem. The reliance on control technologies perpetuates a system that continues to fight techno-logical errors by adding new technologies to the old. Interest in looking at the overall benefits that such a system provides seems lacking. Instead, the focus is on curing symptoms with the same type of medicine that made the patient sick in the first place. In response to this criticism, most countries have developed dual modes of regulation: strict standards for clearly intolerable risk levels (regardless of whether a technology to reduce the risk exists or not), and the application of the best available control technology for all risks below the threshold of intolerability. However, such a move towards dual regulation brings us back to the question of how to determine what is clearly intolerable.

Many regulatory systems rely on dose-effect data to make such judgment. Rather than focusing on emission data, they concentrate on the effects of energy or substances in terms of potential damage or sickness. Toxicological and epidemiological approaches to risk assessment expose and characterize existing causal relations between risks and consequences. Once the nature of these relations is documented, collective choices are taken to block or reduce the exposure and/or to compensate the victims. When the risk agent has a threshold for harm, risk managers have an easy task. They strive to achieve a level of exposure that is lower than the *NOAEL (no observed adverse effects level)*. Because of uncertain-ties in pharmacokinetic models and because of uncertainties with inter-species transferability and variance within populations, the NOAEL itself is not used. Rather, a number lower by some specified order of magnitude (say ten or a hundred times) than the NOAEL is adopted. Setting standards according to a cautious interpretation of the NOAEL is usually uncontested unless other risk distribution factors play a major role. Exposure of a specific class of people to a potential pollutant or toxic substance may then induce problems of acceptability even if the level of exposure is below any measurable effect.

When the risk agent has no threshold of safety (as with carcinogens) the approach usually taken is *ALARA (as low as reasonably achievable)*. The "reasonable" in ALARA implies the need for some method for weighing competing choices against each other in order to set priorities. ALARA is a similar approach to "best available control technologies". The direction of change is clear – towards further reduction of risks. However, as was the case for control technology, the level to which the risks need to be reduced remains unspecified. The term "reasonable" is only a euphemism for some hidden rationality that is not there. We are still faced with the original dilemma, i.e., that society has to make choices about what risks to accept and what risks to reject, but lacks the means to make a collectively binding decision that is convincing to all people affected by the decision.

Similar arguments can be articulated for the *precautionary principle* that has been included in many environmental bodies of law (for example in Germany). According to the precautionary principle, lack of scientific certainty is not a sufficient reason to delay an environmental policy if the delay might result in serious or irreversible harm. Such an approach may lead to a costly reduction of harmless pollutants and a tolerance for strong negative effects if reduction methods are not available. However, the precautionary principle has been justified on the grounds that risk minimization is warranted when consequences are highly uncertain and effects are either not known or only suspected.

Economists argue, however, that a general rule for risk minimization leads to inefficiencies and the neglect of priority setting. Even if uncertainties are high, it is more prudent to estimate or guess the probabilities of harmful effects rather than treating all potential risks as equal. Prioritizing may involve subjective judgment, but this is still superior to the assumption that all uncertain outcomes have the same probability and magnitude of negative outcomes. Furthermore, since the budget for safety is not unlimited in any economy, the precautionary principle demands a setting of priorities that is often not publicly admitted. Allocating money for risk reduction from a tight budget forces a "reasonable" judgment about the magnitude of effects associated with each reduction effort. The precautionary principle in itself cannot provide a decision rule for making such a judgment.

The interpretation of what constitutes a "reasonable" reduction necessitates a balancing act between benefits and risks. Looking at the risks in isolation does not provide any clear rationale for setting acceptability thresholds. In economics, the technique for balancing advantages and disadvantages of an activity which cannot be left to the market is *cost-benefit analysis (CBA).*

Cost-benefit analysis applied to decisions under risk is a way of measuring the preferability of one risk exposure scenario over another. The task may be to choose among a variety of plans to manage existing risk or it may be to decide on the tolerability of new risks. CBA rests on the assumption that a balance sheet for each alternative can be made by translating all kinds of costs and benefits into a single measurement unit, usually dollars. In two parallel columns the expected costs associated with alternative schemes are tabulated in parallel to the expected benefits.

A third column contains the cost-benefit ratio. The preferred solution is the one with the smallest cost-benefit ratio. Costs include not only the costs of implementing risk reduction measures, but also the costs associated with the remaining level of risk exposure. One scenario investigated must be "do nothing." This scenario then serves as a baseline for evaluation.

Despite CBA's apparent orderliness, the procedure of comparing risks with benefits often raises more questions than it provides satisfactory answers. Cost-benefit analysis reaches its conceptual limits when applied to stochastic outcomes. Utility functions (and indifference curves between public and private goods) for uncertain outcomes display irregular shapes and often discontinuities. Willingness-to-pay studies are sometimes conducted to assess the costs of public goods, goods for which a market does not exist. Key examples are environmental services, such as clean air, water, etc.

When the questions are framed as potential losses versus potential gains, even if the cases presented are exact mirror images of the same choice situation, they produce conflicting results. Furthermore, varying degrees of risk aversion among individuals may well make it impossible to find an aggregate level of utility that could be assigned to a collectivity of affected people. Finally, chains of probabilistic events can increase the overall uncertainty of the final outputs to such a degree that any balancing between a large set of unlikely events against some likely or even certain benefits makes no sense. For this reason, many economists, as well as policy analysts, believe that cost-benefit analysis is an inappropriate tool for rationalizing decisions about acceptable risks.

Less ambitious than cost-benefit analysis is the application of *cost-effectiveness* to decisions about risks. Cost-effectiveness is a principle for reducing existing risks by investing in safety only as long as the return is positive. When there are several investment possibilities with positive returns, priorities are set by first investing in the one with the highest rate of return. The guiding principle is to reduce risk at the rate of marginal increase of safety for each unit of money spent for this purpose. It may also work backwards by defining a level of *de minimis* risk and then allocate funds so as to first achieve the greatest risk reduction possible. Risk reduction continues until all known risks are below the *de minimis* level.

Cost-effectiveness is not relevant to decisions about the acceptability of new risks; it is only relevant to improved efficiency. By spending money wisely, risk reduction is achieved faster and at lower cost than otherwise. One can hardly argue against an efficient use of money for enhancing the overall safety effect. However, applying cost-effectiveness assumes that the activity itself is legitimate and the risk level acceptable in comparison to the overall benefits. It is another tool to reduce risk levels, given the desire to spend money on risk reduction. But it does not offer a satisfactory answer to the question of how safe is safe enough, since it avoids this question.

Optimal budget concept is a subset of cost-effectiveness approaches. It begins with the assumption that there is a defined, finite sum of money available to spend on risk management and an excessive number of risks to manage. The money is, therefore, neither sufficient to permit 100% risk reduction in all cases, nor may this be possible due to technological or other constraints. For these reasons, priority setting is necessary. The principle for rank-ordering risks is to allocate money so that the greatest reduction of risk for the fixed amount of money is achieved first, followed by the second greatest reduction, and so on, in declining returns. Optimizing a fixed budget for safety means that the maximum number of losses (expressed in whatever terms) is prevented on the basis that each dollar is invested in its most productive (safety enhancing) use. The optimal budget approach has certain advantages:

- It treats all humans equally. Regardless of who is affected, a person saved is a person saved.
- It treats all risk sources equally. Regardless of fashion or special interests, all activities undergo the same level of scrutiny and decisions to invest in safety are made on the basis of maximum return.
- It does not run into the fallacy of equating human lives with some monetary equivalent. The purpose is to spend the money wisely, not to legitimate lives lost by money lost.
- It helps to increase the efficiency of safety measures.

The first two advantages can also be seen as problems. Society may want to place more weight on saving special individuals than others. Social groups, and particularly nations, have a strong preference to save members of their own at a much higher expense than members of other groups or foreign nations. Any amount of money invested in intensive care stations in the United States or Western Europe would probably violate the optimal budget principle if people from other nations such as India or Brazil were serviced by them. Such differential preferences may also exist for different risk contexts. Society may be willing to spend more money on saving people from risks imposed on them than from those to which they voluntarily expose themselves. People evaluate context variables and assign different degrees of perceived risk, for example between imposed and voluntary risks, to situations with identical expected values for loss of life. Empirical studies demonstrate that the sum invested to save one life ranges from $10,000 to $10 million (Thaler 1991, pp 139ff). This may serve as an illustration of the power of revealed preferences towards differences in risk contexts. But even if these problems could be solved (for example, by constructing classes of similar risks to which the optimal budget concept is applied), the basic question of how safe is safe enough still remains unanswered. Like the simple cost-effectiveness method, the optimal budget concept does not provide a rationale for justifying existing or new risks. It only helps society to reduce risks more efficiently.

This discussion leads us to the conclusion that none of the methods discussed in this section is able to solve the problem of judging the level of acceptable risk. Some difficulties are based on inherent inconsistencies or faulty reasoning within these methods, but most are either caused or aggravated by the fact that they need to incorporate uncertain outcomes. It is risk itself that limits the effectiveness of these methods. The inability of these RAP-based methods to cope with the normative implications of the otherwise well-defined analytical models has contributed to the perception of crisis in RAP-based risk management. It has also sustained risk as a major focus of societal interest and conflict.

Two major difficulties for attempts to define thresholds of socially acceptable risk can be distinguished. First, it is usually impossible to define such thresholds on the basis of purely technical information. Such information is necessary, but not sufficient to judge whether some risk is acceptable or not. It must be complemented by preferences, evaluations, and judgements of likelihood which all have a strong subjective component. RAP has no difficulty in taking this into account. Quite the opposite, it is a basic strength of RAP to provide means for representing and analyzing subjective preferences as well as subjective probabilities. Individual differences in preferences and judgements of likelihood can be accommodated in a RAP framework, as can differences in risk aversion (which are a special case of differing preferences). In this sense, modern portfolio theory is much more sophisticated than actuarial risk analysis, but any portfolio manager is still able to use results from the latter.

Nevertheless, they share a major feature: thresholds of acceptable risk are a matter of subjective preferences which cannot be decided on the basis of technical computations, optimization procedures or any form of "objective" expert knowledge. How safe is safe enough, then, is a question each actor must decide for himself. How can these decisions be coordinated? In many cases, markets – for insurance and other forms of risk-sharing, as with the stock market – provide a possible solution. Research is then warranted to study how these markets operate, which risks they avoid and which ones they accept or even generate. We will return to these issues below (Chapter 7).

For many risks, however, markets are not available as a mechanism to coordinate individual decisions about acceptable risks. A market solution may be inappropriate, insufficient or impossible because of externalities, because the relevant risks range too far in the future, because what is at risk is a public good. In such cases, the second difficulty arises: a political decision is required to aggregate a variety of different utility functions into a social welfare function for the risk problem to be solved.

Focusing on homogeneous group decisions and "win-win" situations will not do to tackle the thorny problems involved in such aggregation. In these cases, defining thresholds of socially acceptable risk can become a formidable task. What can RAP contribute when it comes to reconciling conflicting values about risks, to analyzing and improving risky decisions of interacting agents?

4. RISKY DECISIONS OF INTERACTING AGENTS

4.1 Risk in the Theory of Games

Faced by risky prospects, an isolated actor can try to assess the probability of different outcomes. On this basis she can strive to maximize her expected utility from possible outcomes, taking into account her own negative or positive attitude toward risk. Faced with a second actor, however, things become more complex. The subjective probabilities of *ego* and *alter* are now interlocked. The probability *ego* can assign to different consequences of her actions depends on the actions of *alter*; these depend on *alter*'s probabilities, etc. Game theory developed as a means to resolve this infinite regress.

It adds the realism of the strategic actions of others to the calculus of the individual rational actor. At the core of the theory of games is the concept of a consequence – a positive payoff or an adverse effect. Monetary costs are the most popular measure, but adverse effects may include anything that can be ordered by a preference relation. If risk aversion is to be represented, payoffs must be represented in terms of cardinal utility functions (see Section 2.2.2).

In the theory of games, uncertainty is usually represented by probabilities. These are subjective probabilities, as they refer to the guesses *ego* makes about the actions of *alter* and then to the probabilities which *ego* assigns to her own actions when using mixed strategies. The theory of games relies on a peculiarly strong concept of rationality in order to achieve consistency between *ego*'s and *alter*'s probabilities.

4.1.1 Probabilities in Games

We can combine both of the foregoing points – the complexity of choice in the presence of a strategic opponent and the use of probabilities – in an example. Consider the game represented in Table 4.1. Each field shows the payoffs for a combination of strategies, the first number indicating the payoff for the row player. Payoffs are expressed in cardinal utilities reflecting each actor's risk aversion. Player A, attempting to maximize her expected value for these utilities, must decide between strategy A1 or A2. To do so, she assigns probabilities to Player B's strategies. These assignments are essentially guesses based on assumptions about Player B's motivations and intentions. Player B, it follows, faces the symmetrical problem.

Neither strategy A1 nor A2 can produce a known return for Player A, since the outcome necessarily depends on Player B's choice. A symmetrical argument applies for the strategies facing Player B. Player B may adopt strategy B2 on the reasoning that Player A will adopt strategy A2, since strategy A2 presents only positive gains for Player A. Then

		Player B	
		Strategy B 1	Strategy B 2
Player A	Strategy A 1	0 / 10	10 / 5
	Strategy A 1	10 / 0	5 / 10

Table 4.1: A simple game

again, Player B may worry that Player A has already predicted that he, Player B, would take this action. If Player A feels confident that B will take strategy B2, then A would be wise to adopt strategy A1, since it provides a better return for A. Second guessing makes the game even more unpredictable.

Thus, the question "What is rational for me to do?" cannot be answered without answering the question: "What belief is rational for me to hold about your decision?" But if I believe you to be rational, I must expect that you will choose whichever strategy is rational for you. Thus, the question "What belief is rational for me to hold about your decision?" cannot be answered without answering the question "What is rational for you to do?" But, since your decision problem is exactly symmetrical with mine, we are no nearer to an unequivocal answer to the original question, "What is rational for me to do?" Thus, the attempt to answer that question leads to an infinite regress (Sugden 1993, pp 74ff).

If player A begins with no idea about the choice of Player B, she may assign a probability of 0.5 to each strategy of Player B. Strategy A1 then has an expected value of 0.5 x 0 + 0.5 x 10 = 5, strategy A2 of 0.5 x 10 + 0.5 x 5 = 7.5. If, however, Player B should adopt the same reasoning, strategy B2 would clearly have the higher expected outcome. Therefore, assigning equal probabilities to the strategies of Player B is inconsistent with RAP.

This leads to an extremely important consequence. If RAP is to be used for the study of interactive choices, the idea of a rational actor must be extended to include the belief that other actors are rational actors, too.[27] In order to be consistent, the corresponding beliefs of interacting actors must be iterated: Player A believes that Player B believes that Player A believes that Player B is rational, and so on. As it relies on such strong assumptions about rationality, RAP may well be said to imply a notion of absolute rationality.[28]

If the outcomes of my actions depend on your actions as well as mine, there is no way to assign probabilities to the actions of each of us without substantial additional assumptions. It was one of the brilliant ideas of von Neumann and Morgenstern (1944) to show that a probabilistic treatment of interactive choice is possible with the introduction of mixed strategies. An actor is said to pursue a mixed strategy if she combines several strategies according to some random distribution. An illustration is given by a poker player who bluffs in one third of the cases where he has a given hand.

It is then possible to define *equilibria* for mixed strategies by asking whether an actor can improve her payoff by changing her strategy while the other actor sticks to his strategy. A Nash equilibrium is defined as a situation in which no actor can improve the outcome of a given combination of strategies by unilateral action. More precisely, in game theory a Nash equilibrium is defined as a strategy profile such that each strategy is an optimal answer to the other strategies belonging to the same profile (Nash 1950). With mixed strategies, it can be shown that in many cases (in particular all one-shot zero sum games) at least one Nash equilibrium exists. If this equilibrium is unique, rational actors may then use the corresponding probabilities both for their own strategies and for those of their opponent.[29] Isolated rational actors strive to achieve optima; interacting rational actors can achieve equilibria in which each actor has reached an optimum, given the actions of the other ones.

In zero-sum games, where a player can improve her payoffs only if the other player gets a lower payoff, a Nash equilibrium can be achieved by following the well known minimax rule. The minimax rule requires that a rational actor maximizes the minimal gain which can be expected from different strategies. In the game presented in Table 4.1 this will lead to a unique Nash equilibrium where both players use mixed strategies.

---

[27] Heap, Hollis et al (1992) discuss interactive choice in depth.

[28] Perrow (1984) discusses the relevance of absolute rationality for risk analysis.

[29] The infinite regress between *ego* and *alter* is stabilized at a fixed point where for each agent all steps in the sequence yield the same probabilities.

4.1.2 The Labor Market as a Game

Game theory can be used in many ways to analyze the workings of how economic and other institutions work. There is some danger, therefore, of getting lost in the intricacies of abstract model structures, losing sight of the historical reality which lends plausibility to the notion of a risk-society. A crucial element of this reality is the labor market, as it confronts individuals and families with major existential risks.

As an illustration, consider an economy in which money wages are fixed once a year in a central bargaining process between workers and management. Once wages are fixed, workers determine – in a complex informal process – how carefully and effectively they will perform their tasks and how far they will try to improve their personal skills. Assessing whether the wage deal is fair or not plays a major role in this process. As a result, high or low productivity may result, and wage costs per unit of product will vary accordingly. In turn, managers react to worker behavior. If wage costs rise throughout the economy, managers can reduce real wages by raising product prices. Moreover, if managers feel that wage costs are increasing too fast, they may generate unemployment by slowing down investment. Each year, the whole process repeats itself.

This situation involves two collective actors facing a wide range of risky prospects. Whatever each actor does, the outcome is always co-determin-ed by the actions of the other actor. RAP can be used for the study of this kind of situation by relying on the tools of game theory. In particular, this example can be analyzed as a repeated non-zero sum game with two actors.[30]

According to the "folk theorem" for repeated games, in such a situation any conceivable distribution of the total payoffs can be an equilibrium as long as workers are not driven systematically into starvation or managers into bankruptcy. (The folk theorem owes its name to the fact that it crept into the literature as an obvious result about repeated games without being accredited to a single author.) This remarkable result calls for a closer look at the concept of equilibrium.

It seems obvious that a rational actor will try to achieve a Nash equilibrium if one exists. But, as the folk theorem shows, such attempts need not lead to a unique outcome; there may even be an infinity of possible Nash equilibria.

This result is not due to the fact that only two players are present. If a competitive labor market is represented by curves of demand and supply, they need not show a unique point of intersection, but may instead share

[30] For a related model with an explicit dynamic structure see Edenhofer and Jaeger (1988).

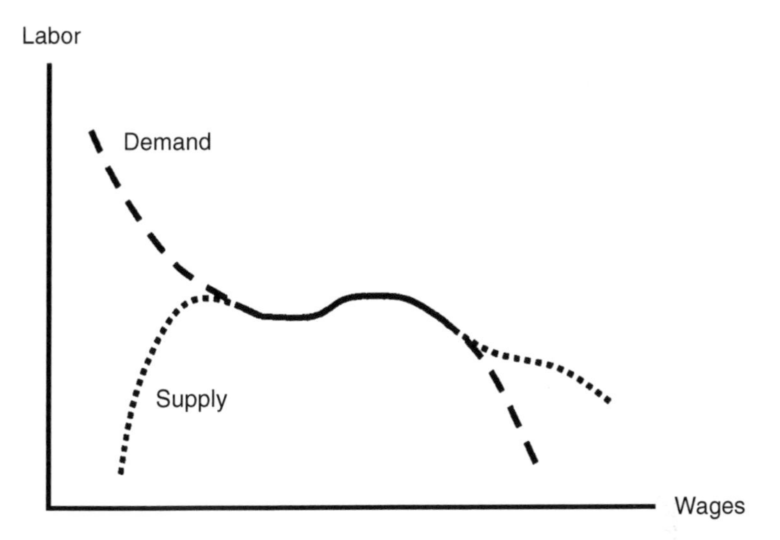

Figure 4.1: A labor market

a whole set of common points, as illustrated by Figure 4.1. This may be due to the fact that demand for labor need not fall with rising wages, because wage increases can create additional demand and because rising wage costs can lead to rising prices for various products, including labor-saving machinery. It may also be due to the fact that the supply of labor need not increase with rising wages, because wage increases can enable people to work less in order to enjoy more leisure time, to get a better education, to take care of children, and to enjoy life. Basically, labor markets display a kind of uncertainty which is a challenge for RAP, but which looks quite tractable in terms of social rationality.[31]

4.1.3 Conceptual Issues

Game theory provides powerful metaphors for the study of risky choices by interacting agents. An important example is the well known prisoners' dilemma (Rapoport and Chammah 1965). Two prisoners held in isolation can reduce their punishment if one of them testifies against the other one, but they will increase their punishment if they both testify. As rational actors, they are bound to achieve the worst possible outcome, because for each

[31] Elster (1989b) gives an instructive analysis of this kind for Scandinavian wage bargaining. See Perrow (1984) for the relevance of social rationality for risk analysis. We will return to the contrast between absolute and social rationality in Chapter 7.

the option of not testifying is less attractive than the option of betraying the other. If the actors are allowed to weaken their absolute rationality, some may find ways out of the dilemma. In this vein, repeated prisoners' dilemmas have been used for studying the evolution of cooperation (Axelrod 1984).

Another use of game theory is to clarify conceptual complications that impair our understanding of actual problems. This goal can be pursued with highly artificial games which have the advantage of drawing out specific aspects of our analyses of interactive choice. As an example, consider the "prisoners' coordination problem" discussed by Sugden (1993) and represented in Table 4.2.

In this problem, the famous prisoners of game theory get another chance to reduce their expected punishments, because the district attorney has changed her mind. She offers to reduce their sentences to one year in prison if they both confess. If only one confesses, however, she will charge both of them for the maximum penalty of ten years. If both stay silent, they will be released, because their crime cannot be proven.

Common sense suggests that both prisoners should remain silent, but as rational actors they have a problem. Within RAP, it is obvious that Prisoner A should remain silent if he can expect Prisoner B to remain silent, but Prisoner B should expect Prisoner A to remain silent in order to remain silent himself. The two rational actors are caught in an infinite regress that cannot be resolved by absolute rationality. The game has two Nash equilibria, and RAP provides no means to choose between multiple equilibria – even where one of them is plainly superior according to the

		Prisoner B	
		Stay silent	**Confess**
Prisoner A	**Stay silent**	0 / 0	-10 / -10
	Confess	-10 / -10	-1 / -1

Table 4.2: The prisoners' coordination problem

Pareto criterion. Of course, in a case as simple as this, common sense leads the way to the obvious solution. But in more complex cases of real life, RAP may blind us for analogous, but less obvious, solutions.

Interestingly, with this kind of pure coordination, bounded rationality often is superior to absolute rationality. As Schelling (1960/1980) argued in his seminal work about "the strategy of conflict," actors can coordinate themselves by relying on focal points. A focal point in a coordination problem is a possible combination of strategies, which can capture the mind of the players for reasons irrelevant in absolute rationality. In our case, the possibility of jointly avoiding punishment might well have such a quality. Bounded rationality can rely on focal points in the prisoners' coordination problem. By pointing to the best outcome for all players, bounded rationality may enable both actors to tacitly cooperate by remaining silent.

However, instead of looking at bounded rationality as an imperfect, but practically relevant, version of absolute rationality, it may be more sensible to look at it as an embryonic form of *social rationality*. The players involved in the prisoners' coordination problem may ask themselves what social rule is appropriate if they try to think as a team, and the rule to stay silent will then become their point of reference. Overcoming the difficulties of RAP with the uncertainties of interactive choice may lead to a view of human agency which treats social rules as necessary presuppositions of social action, and which studies how individual choices and social rules interact in the management of economic as well as other risks. An important step in this direction is a closer look at communication processes.

4.2 Risk Communication

Risk communication emerged from research in risk perception showing that public or lay concerns about hazards did not correspond with those of experts. Experts and governmental regulatory bodies were managing hazards based on supposedly "objective" assessments of potential risks. They defined risk as the probability of a hazard event occurring multiplied by the magnitude of the hazard. Risk professionals found it difficult to accept that the public would prioritize hazards differently from themselves. The field of risk communication developed as a means to research how expert assessments could best be communicated to the public so that the public would base their decisions on "true" estimates of risk, and in order to reduce the tension between public perceptions and expert judgment.

How is risk communication understood from a RAP perspective? The following statement by Covello, Slovic, and von Winterfeldt seems to best capture the spirit of risk communication:

> Risk communication is defined as any purposeful exchange of information about health or environmental risks between interested parties. More specifically, risk communication is the act of conveying or transmitting information between parties about (a) levels of health or environmental risks; (b) the significance or meaning of health or environmental risks; or (c) decisions, actions, or policies aimed at managing or controlling health or environmental risks. Interested parties include government agencies, corporations and industry groups, unions, the media, scientists, professional organizations, public interest groups, and individual citizens. (Covello, Slovic et al 1986, p 172)

Thus, risk communications fits into classic definitions of communications as a purposeful exchange of information between actors in society based on shared meanings (DeFleur and Ball-Rokeach 1982, p 133). Purpose is required to distinguish the sending of a message from noise in the communication channel. "Message" implies that the informer intends to expose the target audience to a system of meaningful signals, which in turn may change their perception of the issue or their image of the sender. Acoustic signals without any meaning constitute noise, not communication.

Risk communication obviously implies an intentional transfer of information. Therefore, one must specify what kind of intentions and goals are associated with most risk communication efforts. The literature offers different sets of objectives for risk communication, usually centered on a risk management agency as the communicator and groups of the public as target audiences (Covello et al 1986, p 172; Kasperson and Palmlund 1988; National Research Council/Committee on Risk Perception and Communication 1989; Sandman et al 1987; Zimmermann 1987, pp 131-132). A serious controversy exists over the general purpose of risk communication: Should it aim at changing behavioral responses, or should it be confined to the exchange of information about pending dangers and potential remedies? (Wilkins 1987, p 80).

Most authors in the field clearly favor the former proposition (Covello et al 1986, p 172; Keeney and Winterfeldt 1986; Lee 1986, p 151). Hence, the list of legitimate objectives, according to this consensus, should not only include the transmission of information, but also persuasive messages intended to trigger behavioral changes of individuals, as well as social group responses. Accepting this premise, risk communication can serve many purposes, ranging from awareness to behavioral changes. For our purposes, the variety of objectives can be collapsed into three general categories (National Research Council/Committee on Risk Perception and Communication 1989; Zimmermann 1987, p 131). Risk communication can attempt to:

- make sure that all receivers of the message are capable of understanding and decoding the meaning of the messages sent to them;
- persuade the receivers of the message to change their attitudes or their behavior with respect to a specific cause or class of risk;
- provide the conditions for rational discourse on risk issues so that all affected parties can take part in an effective and democratic conflict-resolution process.

4.2.1 Senders, Transmitters, and Receivers

The traditional approach to the study and analysis of risk communication is based on the communication model of information transfer among sources, transmitters, and final receivers. Although the model was originally developed in the late 1940s (Lasswell 1948; Shannon and Weaver 1949), it is still the most prevalent framework for communication studies to date, and is often recommended by risk managers (Thomas 1987). In a recent review of 31 communication textbooks, P.J. Schoemaker concluded that nearly half of the books used the Shannon and Weaver model (Schoemaker 1987, p 120). Another approach was the transactional view that emphasizes the creation of shared meaning among senders and receivers. The two approaches can obviously be combined.

According to the classic sender-receiver model, a message is composed by the communication source and then sent to a transmitter. The transmitter decodes the message and recodes it again for its target audience. The new message is then forwarded to the final receiver, who decodes the message and deciphers its meaning. The receiver may respond to the message by sending out his/her own message, either to the original sender or to other constituents. S/he may also feel compelled to take direct actions in response to the message(s) received. The original source may collect or process the receivers' responses. Feedback messages may pass through a transmitter station before they reach the original sender. The original messages, and even more so the feedback messages, are distorted with background noise when they are sent through several channels via transmitters and signal amplifiers (see Renn 1991b for a detailed discussion of the signal amplifying process).

The sender-receiver model has drawn fire for promoting a mechanistic understanding of communication and for emphasizing a one-way communication route (Kasperson and Stallen 1991; Otway and Wynne 1989). Yet, if the model is used only as a sequential illustration of the transfer of messages from one party to another, and if the roles of sources and receivers can be mutually exchanged, it can serve as a powerful tool in the analysis of communication processes. It is a heuristic tool to structure the communication process, not an empirical model of how communication is factually organized in a society.

Within the classical risk communication model, sources for risk-related information are basically scientists or scientific institutions, public agencies such as environmental protection agencies, interest groups such as industries or environmentalists, and, in the case of hazardous events (physical changes caused by hazardous activities), eyewitnesses. These primary sources code information in the form of reports, press releases, or personal interviews, and send them to transmitters, or occasionally directly to the final receivers (Renn 1988, pp 101ff).

The next step in communication is the coding and recoding procedure at the transmitting stations. The media, other public institutions, interest groups, and opinion leaders are potential transmitters for risk-related information. A press release from EPA may stimulate industry to hold a press conference or to write an open letter to the agency. Interaction among social groups, in particular among adversaries, often takes place through the media and not via direct communication. The goal is to mobilize public support and to initiate public pressure (Peters 1984, 1990a). The last step is the processing of the recoded messages by the receiver.

Again, it is helpful to distinguish between different types of receivers. The media usually serve the general public, but many journals are targeted to specific audiences within the general public. Specialized journals are either appealing to professional standards (science communities, business circles, risk assessors), avocational activities (culture, sports, traveling, etc.), or value groups (environmentalists, religious groups, political camps, etc.). The information will be framed for each audience in a different manner to assure their attention and to please their expectations.

The common thread running through most risk communication studies is that public understanding is hampered by the complexity of the risk concept (Short 1984). Furthermore, the multi-stage coding and recoding process during the transmission of messages accounts for numerous errors and misconceptions conveyed to the final receiver and to the transmitter when there is feedback. Transmitters and receivers reduce complexity by simplifying the message and focusing on those aspects they regard as relevant. This is part of the communication reality in modern societies, and provides the social framework in which messages are sent and received.

The communication process can be compared to a free market system in which goods are produced, transported, purchased, and consumed. One may expect that, over the long run, most good products will find their market niche, whereas most bad products will eventually fail to meet the market test. Similarly, messages that contain important information may be more likely to reach their destination, but many trials may be needed to assure success. In addition, packaging can help to sell the message faster and to overcome the obstacles on the way from the source, via the transmitter, to the final receiver. The package can help if the message is worth transmitting, but even the best package may fail in the longer term if the message is poor, dishonest, or simply irrelevant.

One of the most influential models of how risk communication may change people's attitudes is the "elaboration-likelihood model of persuasion," developed by Petty and Cacioppo in the late 1970s (overview in Petty and Cacioppo 1986). The major component of the model is the distinction between the *central or peripheral route of persuasion.* The central route refers to a communication process in which the receiver examines each argument carefully and balances the pros and cons in order to form a well-structured attitude. The peripheral route refers to a faster and less laborious strategy to form an attitude by using specific cues or simple heuristics. When is a receiver likely to take the central route and when the peripheral route? According to Petty and Cacioppo, route selection depends on two factors: *ability and motivation.* Ability refers to the physical possibility that the receiver can follow the message without distraction; motivation refers to the readiness and interest of the receiver to process the message.

Three conditions must be met to satisfy the criterion of ability: the information has to be accessible; the receiver must have the time to process the information; and other sources of distraction should be absent (Baird, Earle et al 1985). Several factors influence the *motivation* of a receiver to process the information actively. The information content has to be relevant (referring to personal interests, salient values, or self-esteem), and it should trigger personal involvement (with the issue, the content, or the source). Both motivational factors are reinforced if the receiver has some prior knowledge or interest in the subject, or is in need of new arguments to support his/her point of view. The model would also suggest that high ego involvement (i.e., the match between message content and the recipient's functional or schematic predispositions) increases the likelihood that the central route be taken.

The *peripheral route* is taken when the issue is less relevant for the receiver and/or the communication context is inadequate to get the message across. In this case, the receiver is less inclined to deal with each argument, but forms an opinion or even an attitude on the basis of simple cues and heuristics. These peripheral cues may be integrated into the source-receiver model and assigned to each step in this model (*source-related, message-related, and transmitter-related cues).* In addition, the context in which the communication occurs provides additional cues for the receiver to generate interest in the message (*context-related cues*).

Aspects of the message source, such as credibility, reputation, and social attractiveness, are important cues for adoption by receivers. It also helps to have the message sponsored by multiple sources (Midden 1988). Message factors include the length of a message, the number of arguments, packaging (color, paper, graphic appeal, and others), and the presence of symbolic signals and cues that trigger immediate emotional responses (Kasperson, Renn et al 1988).

The transmitter of a message may also serve as carrier for specific cues – perceived neutrality and fairness, personal satisfaction with the transmitter in the past (this magazine is always right), similarity with the political or ideological position of the transmitter, and the social credibility assigned to a transmitter, all major elements in the formation of opinions or attitudes. In addition, specific channel-related aspects, such as visual impressions from the TV screen, are readily accessible cues.

Social context variables that serve as peripheral cues are often neglected in the discussion of peripheral routes. The social climate for trust and credibility and the image of institutions in a society may evoke a specific predisposition to accept or reject the arguments of a source (Lipset and Schneider 1983). With respect to the risk arena, impressions of expert controversy and the presence of competing messages are cues that may initiate a skeptical, or cautious, reception mode (Slovic 1987). Other variables can be added to this category, as, for example, the plurality of transmitters or the social reputation of specific media.

Once attitudes are formed, they generate a propensity to take action. As demonstrated by many attitude studies, the willingness to take action, however, is only partly related to overt behavior (Allport 1935; Fishbein and Ajzen 1975; Rokeach 1969; Wicker 1979). A positive or negative attitude is a necessary, but not sufficient, step for corresponding behavior. A person's decision to take action depends on many other variables, such as behavioral norms, values, resources, and situational circumstances. Hence, the communication process will influence the receiver's behavior, but a multitude of sources, transmitters, and situational factors influencing personal behavior render measurement, let alone prediction of the effect of a single communication activity, utterly difficult.

The weak correlation between attitudes and behavior is one of the major problems in risk communication aiming to change behavior (for example, for emergency responses). Most analysts agree that it is difficult enough to change or modify attitudes through information, and even more difficult to modify behavior. Some success stories (Fessenden-Raden et al 1987; McCallum 1987; Pinsdorf 1987, pp 47ff; Salcedo et al 1974) in the area of health risks (for example, reducing cholesterol and pesticide use), as well as some failures (Mazur 1987; Sandman, Weinstein et al 1987) to promote actions (for example, protection against indoor radon), suggest that three factors are crucial for increasing the probability of behavioral changes:

- the *continuous transmission of the same information*, even after a positive attitude has been formed towards taking that action (need for constant reinforcement);
- the *unequivocal support of most relevant information sources* for the behavioral change advocated in the communication process (need for consistent and consensual information);
- *adoption* of the behavioral changes *by highly esteemed reference groups or role models* (social incentive for imitation).

Emergency responses may in addition require actual exercises or practices before the desired behavioral responses are internalized (Covello, Sandman et al 1989). Behavioral changes, particularly if they involve painful changes of habits, are rarely triggered by information alone. Rather, information may be effective only in conjunction with other social factors, such as changes of social norms, roles, and prestige.

4.2.2 Risk Debates

Several levels have been distinguished in risk debates (Rayner 1987; Renn and Levine 1991; Funtowicz and Ravetz 1992a). On the most basic level, the focus may be on factual evidence and scientific findings. This requires a technical style of presentation and the involvement of experts in the communication program. The second level is professional judgment and experience. Past record of reasonable decision making, personal experience, and social recognition of performance are major elements of debate. The third level involves personal identification with sets of values and lifestyles. Communication on this level relies on finding and establishing shared meanings of the risks and their management.

Communication on the first level, if successful, serves the purpose of convincing an audience that the factual knowledge compiled by independent scientists supports the case of the communicator. First level debates require factual evidence to induce attitude or behavioral changes, but attitudes are rarely based on factual information alone. Rather, they focus on vested interests, distribution of risks and benefits, and the adequacy of the proposed solution in terms of economic and social compatibility. Although scientists and many risk management agencies are most comfortable with technical debates, they should be aware that attitudinal or behavioral changes require not only technical information, but also communication efforts on the second and third levels.

The second level of debate (clinical expertise) relies mainly on personal and institutional judgments and experience, not only on technical expertise. A debate on this level requires input from interest groups and affected populations. Typically, the issue in conflict is not so much the magnitude of the risk, but the distribution of risk and its tolerability, *vis-à-vis* the potential benefits the risk source is able to provide. Trust in this situation cannot be accumulated by demonstrating technical skills and expertise, but by compiling evidence that the communicator has been cost-effective in the allocation of resources, and has been open to public demands and requests. *Competent management* and *openness towards social demands* are the two major factors in providing incentives for attitudinal or behavioral changes on this level.

If the participants in a risk debate focus on values and future directions of societal development (third level), neither technical expertise nor institutional competence and openness are sufficient conditions for inducing any attitudinal or behavioral changes. They can only result from a more fundamental consensus on the issues that underlie the risk debate. As long as value issues remain unresolved, even the best expertise and the most profound competence cannot overcome the distrust people will have in the task performance of the acting institution, and consequently they will hardly see strong incentives to change their attitudes or behavior.

Preoccupation of society with environmental problems, perceived ambiguity of technical change, erosion of trust in public institutions predestine the evolution of many risk debates to third-level controversies, and to become issues of lifestyle and worldview. Risk issues have evolved from technical to institutional debate and, furthermore, to struggles over worldviews and lifestyles. The resulting conflicts often produce conflicting evidence, further erosion of trust, and personal frustration. In this situation, risk communication is forced to meet an almost impossible task – developing a framework of mutual trust and confidence so that conflicts can be reconciled in a rational and democratic way. This leads to a serious challenge for all RAP-based approaches.

Early risk communication, as exemplified in the public relations model, was an example, *par excellence*, of the rational actor paradigm (RAP). Risk communication, in this model, was seen as a *sell* – experts, in their role as regulators protecting public health or as employees of industry or special interest groups, were informing citizens about the appropriate concerns about risks. Recipients of the risk message were assumed to be individuals who formed their own judgments and attitudes on the basis of this message. The fact that people are embedded in social networks was largely ignored. The fact that people did not think of risks in the same way as experts was, likewise, largely ignored. Risk communication exercises were seen as competing with partisan messages from media and interest groups, who sought to manipulate public opinion through conveyance of partial truths, or even falsehoods, to individuals.

Recognition of the relevance of individuals' attitudes toward risks was accompanied by reliance on another premise of the rational actor paradigm: that individuals were free to select action courses based upon their own subjective expectations of future consequences associated with alternative choices. Relevant action might, for example, include choices about exposure to risk (food additives, cigarette smoking, x-rays, sunlight) or whether to bring political pressure on risk sources (regulation, development).

Risk professionals often argue that risk communication in the public relations model is well intentioned, that they wish only to reduce the overall risk to society. But this model of risk communication does not admit that *ends* can be rationally discussed; that is, the goal of reduced human harm is simply taken for granted. The traditional model of risk communication reaches its limits when a more thorough debate is required.

Risk communication as a systematic tool emerged partly out of the need for government to justify to citizens its actions with regard to risk management (Covello, Slovic et al 1986). It was developed mainly by leading scientists in the field of risk analysis, under the auspices of the National Academy of Sciences in the United States in the mid-1980s. Since this view of risk communication was developed by the scientists who did the risk assessments, it is not surprising that they had a rather restricted picture of what risk communication was all about. This view has been challenged in the professional community, and today it is common to hear criticisms of the standard view. Be that as it may, the standard view retains a substantial following in industry and in state, local, and federal government.

The success of risk communication has been limited, however. This seems related to the following problems (Otway and Wynne 1989):

- RC evolved out of an attempt to understand the acceptability of risks, framed without consideration of context.
- Based mainly in individual psychology, RC does not pay sufficient attention to social contexts in which people live, especially relationships among individuals.
- Intertwined activities, such as research, consultancy, and practical efforts to shape policy or public reactions, often involve actors who play more than one – and possibly conflicting – role.
- We know that lay people often define risk differently from experts. It is of little help to assume that experts' definitions are more "correct" than those of laypersons.
- There are other valid ways of knowing than the standard ways of science and mathematics.

- The perception of RC comes from the point of view of the authorities (the senders).
- Risk messages can never be unambiguous or unbiased. Someone will always interpret them in an unexpected way.
- Senders may not always be honest, competent, reliable, or responsible.
- The standard view seems unmindful of the fact that the relationship to authority established in the RC model could be misused to trivialize risks and to unwisely urge public acceptance of certain risks.
- While the value of two-way communication is mentioned with regard to conflict resolution, this amounts to little more than lip service in a framework that does not support it. Furthermore, the concept is undeveloped, and usually only guidelines that are obvious are mentioned.

The basic difficulty of RC lies in the attempt to base an analysis of human communication on the RAP approach that starts with isolated individuals. As we have seen, this approach has been successful in providing a series of influential methods and techniques of risk analysis. However, attempts to study problems of risk and uncertainty in a RAP framework have run into a multitude of serious difficulties. It is always possible to try to overcome these difficulties with a new and more sophi-sticated elaboration of RAP. Many such attempts have been undertaken, but they have again and again run into new difficulties. It seems high time to explore the possibility of alternative approaches to risk analysis that avoid the atomistic view of human rationality proposed by RAP – and its offshoots, like RC – without losing the insights provided by this general theory. Such approaches will be discussed in Part II. Next, however, we will discuss how the practical interaction of major players in the risk field has been discussed by RAP-based risk analysis.

4.3 Risk and Social Movements

4.3.1 Resource Mobilization

In many instances, environmental and technological risks became prominent because of social movements that emerged around them. For this reason, the study of risk can hardly be divorced from social movements. In this field, too, we find RAP looming large. The rational actor paradigm came into full blossom in social movement studies with the rise of Resource Mobilization (RM) theory in the 1970s. Since its emergence, RM theory truly revolutionized social movement studies and now dominates the field (McAdam 1982; McCarthy and Zald 1987; Oberschall 1973; Tilly 1978). According to RM theory, risk is a factor that rational actors take into consideration when making judgments about possible actions. People who sense they are at risk, and that their risk could be reduced or mitigated through participation in a social movement, are likely candidates for participation.

Resource mobilization theory developed out of critiques of *collective behavior*, *mass society*, and *relative deprivation* theories, approaches that dominated the early period of the field.[32] These approaches tended to view social movements as an oddity, and to search their origin as a function of personality traits or specific attitudes of individuals. At the heart of these models was a preconception with the subject as a non-rational, irrational, or even pathological actor. Certain characteristics were thought to sensitize the individual to the negative impacts of social changes, resulting in grievances or a feeling of deprivation that pre-disposed the person to action. Costs – or risks – of participation were ignored.

RM theorists were troubled with explanations that painted the participants of social movements as deviants, unrepresentative of the public at large, or as listless beings subject to the determinations of external pressures. In opposition, RM theorists asserted that individuals rationally chose to participate (or not participate), based on considerations of expected personal costs and benefits. This stood in stark contrast to collective behavior approaches, which attributed emergent collective action to the "group mind", for example, the crowd effect (Le Bon 1960); to mass society approaches, which claimed people were alienated as a consequence of a new capitalist mode of social integration and would not participate (Kornhauser 1959); and to relative deprivation theory, which emphasized perceived tensions between rights and realities (Davies 1962).

[32] At the micro level, the question of movement emergence is often reduced to that of individual recruitment. What are the factors that predispose a person to participate in collective action? Early inquiries sought answers in the personality traits or attitudes of individuals. Psychological factors such as character traits (for example, the authoritarian personality: Adorno et al 1950) or stressful states of mind (alienation, Oedipal conflicts, relative deprivation) were investigated. The most dominant of these was the theory of *relative deprivation* (Davies 1962). This approach suggested that individuals conceptualize a gap between the trajectories of what they feel they are entitled to have and what they perceive themselves as having. This relative sense of deprivation was posited to develop frustration and fear and a mental state of anxiety in the individual's mind. Such people would be predisposed to participate in social movements, not necessarily because doing so would yield the rewards that could close the gap, but because the activity was therapeutic for the anxiety. Relative deprivation explanations were not used to explain risk oriented social movements.

Attitudinal explanations contended that people participated in a movement because they sympathized with its goals or values. This approach relies on an assumed behavioral link between attitudes and action, but this link is uncertain. Both the psychological and attitudinal explanations have been empirically tested and are thought to be important factors in determining individual participation, but with little predictive power (Klandermanns 1984).

RM theory used RAP to offer an alternative explanation to the traditional approaches, based on two assumptions. First, it rejected the conception of the actor as non-rational, and assigned the subject an instrumental rationality. Second, it adopted a conception of the actor as an atomized individual, who is free to act, who anticipates expected consequences of possible actions, at least in some rudimentary way, and who acts in a way that is instrumentally and strategically rational.

RM theory made two chief contributions to social movement studies. First, by applying a straightforward version of rational action theory, it explained individual decisions about whether or not to participate in a social movement by rational considerations of expected costs and benefits. Second, RM theory emphasized the organizational level of social movements, specifically, the key role that social movement organizations (SMOs) play in the development and fulfillment of a social movement. Zald and McCarthy have emphasized that social movements need to be understood at an organizational level in terms of how SMOs relate with other organizations and how the SMO organizational infrastructure enables or hinders the social movement (1987). In an argument clearly building on the rational actor paradigm, these authors likened social movements to economic markets, even discussing Social Movement Industries (SMI) and explaining SMO survival as a process of market specialization, a process guided and managed through Social Movement Professionals – people akin to corporate managers. This gave rise to the idea of the professional SMO.

Resource mobilization theory has several variants, but these share certain common features. (McCarthy and Zald 1987) emphasize the ways SMs mobilize and routinize the acquisition of resources (money, facilities, legitimacy, labor); the importance of the professional SMO (replacing the concept of the classical SMO, which relied on volunteer staff and resources of constituents); the role of SMOs in securing outside contributions of resources for use by the movement (Jenkins 1983, p 533); and the explanation of individual involvement using the concepts of costs and rewards.[33] The literature has addressed two phases of SMs – their emergence and development. It has also addressed the need to bridge the micro and macro levels of analysis and to focus on those contexts in which individuals interact under a frame of organization or where leadership emerges.

RM theory has been severely challenged, however. Once seen as liberally progressive (because it accorded the average person rational

[33] Tilly (1978) emphasizes the structural contexts in which opportunities for success are presented. Oberschall (1973) emphasizes the links between individuals and the importance of preexisting structures. McAdam (1982) focuses on the political process aspects for mobilization, on how opportunities are created by regime weakness and instability or support.

cognition), it is now criticized for its overly strong emphasis on instrumental action (focusing on means) and its hesitance to recognize the importance of social factors, such as solidarity in personal calculations of whether or not, and in what manner, to participate.

To document these criticisms and to explore how the rational actor paradigm has coped with them, we return to risk. Risk as such is not a core concept in social movement theory. Social movement theory seeks to explain how collective action is possible. As an undesired side effect of human actions, or as a characteristic of the natural world, risk is a factor influential in collective action. Hazards and risks influence behavior, whether constructed socially or whether physically present. Three questions may be asked:

- What role does risk play in influencing the decision of individuals to participate in a social movement, in shaping the emergence of a social movement, and in the development of a social movement?
- How have these SM activities been understood from a RAP perspective?
- What are the shortcomings of these RAP-based interpretations or explanations?

Risk-oriented social movements may target a specific hazard or potential hazard or they may target a more abstract level of risk-related action. Specific sources can be a preexisting hazard (as with the discovery of an existing risk burden, for example, superfund sites, food additives, or radon), an emerging hazard (as with AIDS, industrial accidents, or ozone depletion), a proposed hazard (as in disputes over the siting of noxious facilities, such as the Clam Shell Alliance of Boston's opposition to the construction of the Seabrook nuclear power station). Such hazard-stimulated movements tend to be spontaneous, have a single issue mandate, and a limited lifespan.

Other movements aim at risks in general or at the very practice of risk analysis. This includes the critical response to bias and ineffectiveness in hazard identification, to methodologies employed in risk assessments, and to the process of policy making about risk management (mitigation and compensation). These movements tend to coalesce into Social Movement Organizations (SMOs), to be more spatially diffuse, and to have longer lifespans. Examples of these groups include the Citizens' Clearinghouse for Hazardous Waste (founded by Lois Gibbs in the wake of the Love Canal toxic site), the National Toxics Use Reduction Campaign, consumer groups, and child protection groups. Other SMOs that are not explicitly risk-oriented may evolve to risk-oriented social movements if their leadership perceives it as strategically wise (a way to maintain the

SMO and keep their jobs as SMO professionals, or a way to pursue other, more central goals) to do so.

Explanations of social movements have tended to focus either on the question of emergence: How did the social movement come into being?, or on the question of maintenance and development: How did the movement sustain itself and try to achieve its goals? But the bulk of the work in the field has been done on the topic of emergence. Further differentiation is made according to the level of analysis. Micro-level approaches assume that the motivations and the incentives for individual action determine the emergence and maintenance of the social movement. Macro-level approaches place the emphasis on social structures and dynamics as the driving and sustaining forces of social movements (McAdam et al 1988).

4.3.2 RAP and Resource Mobilization Theory

To understand the relevance of the rational actor paradigm to studies of risk-oriented social movements, it is helpful to focus on three central concepts: (1) the model of individual decision making as instrumentally rational; (2) the model of the SMO as a rational actor; and (3) the relevance of micro-mobilization settings and macro factors in influencing or distorting the rational calculations of individuals and the organization.

The Individual as a Rational Actor. Mancur Olson (1965) first applied the concept of the rational actor as spelled out in rational choice theory to the phenomenon of collective behavior at the micro level. RM theory accepts that it must cope with this fundamental challenge (McCarthy and Zald 1987). Olson premised that people who rationally choose to participate in collective action will assess the expected costs and benefits of participation before deciding whether or not to participate (re-stated as risk-ratio by Oberschall 1973). Olson elaborated this approach to explicate the problem of the *free-rider.* He argued that an individual deciding whether or not to participate in a movement with the aim of obtaining a *public good* is predisposed to selecting the "rational" option of not participating. A *public good* is something that, if one person in a population attempts to consume it, cannot be feasibly withheld from all other members of the population (Olson 1965, p 14). Clean air, for example, is such a good. If a social movement produces or maintains a public good, a single individual is likely to make little difference in the ability of the movement to achieve its goal. Therefore, such an individual can take a free-ride and receive the benefits of the movement without expending any effort. The free-rider problem is hypothesized to become worse when the goal is more remote and requires more steps to its attainment.

Recognizing that people do, nonetheless, decide to participate in social movements and disregarding those who do so for altruistic reasons as "unrepresentative," Olson surmised that some factor must have altered the cost-benefit analysis of participants (Olson 1965, p 149). Selective incentives, according to Olson, are those rewards that are capable of tipping the cost-benefit analysis in favor of participating. As long as people are rational, they will only choose to participate if the movement can offer them some sort of incentive to participate (free food, a magazine, opportunity to win a raffle, etc.). According to Olson, negative sanctions for not participating (withholding rewards, negative stigma, etc.) may also have the same end effect on an individual's decision. In short, Olson predicts a high percentage of people will opt to become free-riders in any collective effort to acquire a public good or dispel a public bad. The main reason for opting to free-ride are individual calculations of personal costs and benefits associated with the two action alternatives.

Social Influences on Rational Calculations. In a naive version of the rational actor paradigm, actors are individuals who make calculations about possible actions in a social vacuum. Some RM theorists realized that this premise does not accurately reflect reality. One of the contributions of resource mobilization theory has been to recognize that people construct their rational calculations in settings that are rich with interaction. To explain such interaction, they developed the concept of the *micro-mobilization setting*. These settings – informal friendships, *ad hoc* committees, coalitions, pre-existing groups oriented around an entirely different subject such as a church – are settings where individuals come together and in which some form of organization can emerge and produce mobilization. They contain the established structures of solidarity incentives on which most social behavior depends. There are, of course, macro influences on these micro-mobilization contexts.

Micro-dynamics inside micro-mobilization settings influence individuals' calculations of expected utility from potential action courses mainly in three ways:

- *Frame alignment.* Ideological rationales are constructed to legitimize the movement. Frame alignment facilitates individuals developing a positive attitude toward the movement.
- *Value-expectancy.* This RAP-based model of individual decision making focuses on how people make choices about taking action, given their encounter with a legitimizing framing of the movement (Klandermanns 1984). According to the model, in assessing costs and benefits, individuals consider three kinds of expectations: 1. about the number of participants; 2. about one's own contribution to the probability of success; 3. about the probability of success if many people participate.

When all three are high, the person will most likely participate. Of course, this is susceptible to the criticism of all rational choice theory; the calculations are assigned entirely to the individual, ignoring the importance of the collective settings in which the utilities themselves are assembled (i.e., what a person cares about), and in which the computations are made (how the expectations come about – other people can influence your personal calculations).

- *Resource mobilization.* This emphasizes the organizational aspects – members, infrastructure, leaders – for mobilizing resources that are provided by the micro-mobilization setting.

Since Olson, other kinds of factors influencing the rational calculation have been identified. Gamson and Fireman (1979) proposed that group solidarity – a sense of belonging and cooperation – is more important than selective incentives, especially among those who are already members. Repeated calls to action are not accompanied by cost-benefit calculations once an individual has made an initial commitment. Purposive incentives, i.e., value commitments, may be created (Jenkins 1983). Moreover, if uncertainty due to complexity is too high, strategic reasoning falls apart; even the bounded rationality assumed by Olson is exceeded. Mitchell (1979) suggested that "public bads" (radiation, polluted water, and air), because they are unavoidable and repulsive, are themselves strong enough motivations (without stronger "public goods" or selective incentives) to induce participation. According to this view, many risks are simply "public bads," and people may have intrinsic motivations to get rid of them.

Oliver (1984) notes that optimism about others' actions is a key factor when the action has an accelerating impact on achievement of the public good. On the other hand, pessimism about others' actions plays a larger role in an individual's decision about whether or not to participate when there are diminishing marginal returns for participating (when the job is short-lived, or the majority of the work comes at the front end, when the payoff is small). Some individuals may acquire the perception that their participation does matter: "If I don't do it, no one will."

Walsh and Warland (1983) tested hypotheses about free-riders based on Olson's theory. They studied the SM that arose in the Three Mile Island (TMI) communities after the accident. SMOs were already in place before the accident happened, but afterwards they made renewed efforts to mobilize individuals. They found evidence to support Olson's hypothesis. Eighty-seven per cent of the people who opposed TMI opted not to join a social movement that sought to shut down the undamaged TMI-1 reactor and monitor the cleanup of TMI-2. Not all of the hypotheses ventured by Walsh and Warland were validated, however. They discovered, for example, that free-riders had higher levels of certain kinds of solidarity relationships than did the activists.

The results by Walsh and Warland (1983) and Oliver (1984) pointed to two types of calculations that were differentiated by the macro conditions surrounding the decision:

- In large populations with diffuse public goods or bads, individuals assume that other people will participate. They compute costs and benefits to themselves based on some estimations about the likelihood and extent of the movement's success or failure and their expected role and experience were they to participate.
- In smaller populations, people must pay more attention to the possible impact of their personal non-participation on the performance of the movement. This requires more strategic reasoning. Pessimism plays a large role.

It was evidence such as this, as well as other critiques of RM theory, that led to the development of the new concept, noted above, called the *micro-mobilization setting* (Gamson, Fireman et al 1982). After all, it is in these social contexts that individuals come together under some fabric of leadership and organization to initiate a social movement. This reconceptualization tempered the model of rational choice with the influences of social contexts. That is, the individual is not depicted as making rational calculation in a vacuum, but as a social being influenced by other people, structures and settings, and a personal biography.

Finally, it should be noted that one of the spin-offs of the RM critiques of the traditional approaches has been a revival and revision of the relative deprivation model, called the model of suddenly imposed grievances (Useem 1980; Walsh 1981). Much as structural strains produce macro-level conditions that disturb individuals, this model proposes that sudden dramatic changes can impact strongly on public awareness and attitudes, shock and appall them, and induce them to act. The conception of the subject as a non-rational actor is preserved from earlier relative deprivation theory. The innovation of the model is recognition of the need for a pre-established vehicle for the social movement to ride. To engage in sudden action requires a framework for coordinating activities already in place. Because of the sudden nature of some hazard events (nuclear power station failure, tornadoes, earthquakes, industrial accidents) this model is especially relevant to risk-oriented SMs.

4.4 Organizations and RAP

4.4.1 The Rationality of Aggregated Actors

SMOs are essential for risk-related social movements because organizations are a ubiquitous feature of modern life. No assessment of RAP-based risk studies can ignore the role of organizations in today's society. They dominate the landscape not only of social collectivities, but also the daily experience of a large majority of citizens in modern societies. Eight out of ten of us are directly dependent on organizations for our livelihood, making organizations critical for well-being. The acquisition of knowledge and skills, and the subsequent ability to command a position within an organization, are likewise critically dependent upon educational organizations. And many of our daily requirements, enjoyments, and travails take place within or because of organizations. Put most forcefully, in today's societies "all social processes either have their origin in formal organizations or are strongly mediated by them" (Perrow 1986b). For many observers, modern societies are "organization societies."

A distinctive feature of the long, evolving path of modernity was the emergence of the formal organization as a dominant means for combining and directing human actions. It was the natural concomitant, according to Weber, of the progressive rationalization of society. It was also a tangible manifestation of the "relentless progress" begun with the age of Enlightenment. Along with the emergence of science, market transactions, moneyed economies, double-entry accounting systems, and other rational institutions came formal organizations – replacing the variety of traditional forms of authority and coordination existing through Medieval times. Max Weber located the rationality of the organization in bureaucratic structure: structure built upon a foundation of rational-legal principles; rational because it operates according to formal rules, and legal because its identity – as *persona ficta* – exists not in physical reality, but as a legal entity, backed by legal sanctions.

Key features of the rational-legal form of bureaucracy include: universalism rather than particularism in the treatment of employees; developing stable, routine tasks and coordinating them; reliance on explicit standards of performance; and establishing and enforcing rules and regulations binding for all members of the organization. Instrumental human activities were bureaucratically organized for purposes of efficiency. By imposing rules, structure, and coordination on human tasks, uncertainty over outcomes was reduced and so was the chance that any of a countless collection of risky contingencies would obtain. As a consequence, things would run smoothly, efficiently. Indeed, for Weber the rational-legal form of bureaucracy was the most efficient form of organization known to man.

Its principal cost was a hierarchically organized division of labor where round pegs could be found in round holes and square pegs in square holes.[34] While doubtlessly efficient, there was, for Weber and others, the potential for an abuse of organizational power, threatening to incarcerate its members in an "iron cage."

Weber's focus on rationality was, apparently, confined to patterns and structures, not to actors or entities. Thus, for example, due to the influence of the Protestant ethic, the rational outcome of social actions – entrepreneurial success and capital accumulation – was the unintended consequence of actions framed in religious terms. The rational outcome was embedded within patterns of religious practices. Weber was skeptical about individual rationality itself, imputing to the individual actor his own version of "false consciousness."[35] While he incorporated rational elements and rational forms of action into his "ideal types", this was a methodological device to represent general patterns, not necessarily an assumption about the rationality of individual actors. He was likewise mute about whether organizations, though rationally structured, acted rationally. This was true for both personae of the organization, the *personae vivae*, or real actors within the organization, or the *persona ficta*, the organization as an individual entity.

It remained for the "decision-making" school of organizations, especially in the work of Simon and March, to address the question: Do organizational actors, real individual or legally constructed entities, truly engage in rational action? Does RAP shape their criteria of action, and does RAP account for their behavior? An answer to the question was prefigured within the Human Relations Movement of organizations with the work of Chester Barnard (1968). For Barnard, organizations were rational entities, while individuals were not. A similar assumption was clearly seeded in the organizational work of Weber, but left undeveloped. March and Simon (1958) addressed the rational actor question, not via Barnard's assumption,

[34] It was, of course, Frederick Winslow Taylor who married, with his "scientific management," two of the rationalizing trends that evolved with the growth of modernity: science and bureaucracy. The systematic analysis of the various organizational tasks and their subsequent division into finely graded activities would improve efficiency, resulting in better economic performance. Part of the performance improvement would be due to the reduced labor costs commanded by a deskilled labor supply.

[35] Weber argued in *Economy and Society* that "in the great majority of cases actual action goes on in a state of inarticulate half-consciousness or actual unconsciousness of its subjective meaning. The actor is more likely to 'be aware' of it in a vague sense than he is to 'know' what he is doing or be explicitly self-conscious about it...Only occasionally...is the subjective meaning of the action, whether rational or irrational, brought clearly into consciousness. The ideal type of meaningful action where the meaning is fully conscious and explicit is a marginal case" (1968, pp 21–22).

nor via Weberian obliqueness, but with a clearly developed analytic strategy. March and Simon intended to explain both the actions of members making up organizations and those of the organization itself. Understanding of the former (individual actions) would be used as a basis for understanding the latter (organizational behavior).

Organizations can neither exist nor make decisions without people; these are a defining feature of organizations. It seems plausible, therefore, to assume that an understanding of organizations should begin with the individual.[36] Furthermore, a common activity of both individuals and organizations is decision-making, and decision-making became the root concept for developing organizational theory. The strategy would begin with individuals as the foundational unit of analysis and then build upon that foundation until the organizational edifice itself appeared.

Presumably, the edifice would be the outcome of a logic that ordered the building blocks into an organizational structure of coherence. The presumed goal, once accomplished, would result in a grand celebration of RAP, for here the paradigm provided the orienting perspective leading to remarkably fruitful theory: fruitful because it was logically structured from first principles and because it could explain the actions of micro and macro actors within a consistent logic. Logical and empirical difficulties, however, conspired against the theory. As a result, the presumed goal of integrated coherence was never realized. Nevertheless, this theoretical tradition yielded a variety of insights – many of them now regarded as classic gems that have become contingencies in the application of RAP.

Most humans intend to act rationally most of the time. We have a wide variety of tastes, desires, and goals that require the evaluation of alternative means for satisfaction. With a clear vision of goals, and with thoughtful assessment of the best means for achieving those goals, we are seen to be capable of rational decisions. But, despite sincere intentions to act rationally, we often don't. Limitations in our powers to reason rationally – cognitive limitations – are one of the reasons. Another is that there are often unacceptable costs, such as time and money associated with the search for alternatives, requisite with the attainment of the one best rational outcome. A rational decision is often defined as the one resulting in an optimal outcome, where optimal refers to the most preferred of all available alternatives. Often, the criterion of "most preferred" is the alternative that maximizes some good, such as money. Frequently, however, we do not seek and choose the optimal solution, but choose one that we consider satisfactory; we do not optimize, rather we "satisfice". March and Simon's

---

[36] This is, of course, a common analytic assumption of theorists from across a variety of academic disciplines who are inspired to develop their theories within RAP. For a recent statement on this assumption, see Elster (1989a).

example is "the difference between searching a haystack to find the *sharpest* needle in it and searching the haystack to find a needle sharp enough to sew with it" (March and Simon 1958, p 141).

With RAP in hand and real-life individuals in sight, Simon and March set out to explain the rational action of individuals in organizational settings and to develop a theoretical explanation of organizations that used the application of RAP to individuals as its foundation. Along the way they discovered that humans systematically violated key assumptions of RAP theory. We try to act fully rationally, sometimes do, but most often engage in a sub-optimal form of rationality: bounded rationality.

How can the rational actions of organizations be explained within RAP? First, by ignoring the limitations of individuals via simplifying assumptions. Second, by removing the actor's volitions to engage in satisficing or other sub-optimal decisions – bounded rationality – by circumscribing the individual's decisions, or, as Perrow (1986b) puts it, by shaping the premises of their decisions. In the first instance, individual complexity was abstracted by shifting the unit of analysis to the organization. In the second, the decision focus was shifted from individuals to premises the organization imposes upon individuals intending to make rational decisions. These premises are based on the conjunction of the principal features of classic bureau-cracy: division of labor, routinization, universal standards of performance, an authority structure, communication channels, and indoctrination (pro-cedural and ideological).

The upshot of the analysis was that the organization, the *persona ficta*, was seen to be a rational entity, an actor seeking to perform rational decision-making. How did this view differ from the classic bureaucracy of Weber? There were three chief differences. First, the organization was viewed as an active rational entity, not simply a hierarchical authority structure resulting from rationalization of social processes. Second, the propensities, capabilities, and limitations of individuals in the organization were explicitly recognized, thereby deepening the analytic framework and signaling dangers in theory construction. Third, by explicitly recognizing individual limitations posed by bounded rationality, a chief function of the organizational form was highlighted: the imposition, top-down, of rational constraints on sub-optimizing individuals so that rational outcomes can be generated. The organization can achieve optimal solutions even if individuals do not search out optimum alternatives. And it can do this with the rational control of individual behavior. Perrow aptly put it this way: We can see "...in the organization a means of controlling the individuals in the interests of the goals of the leaders of the organization. The organization is more rational than the individuals because order is imposed on members by those who control the organization, and the order is in the interests of goals and purposes established and guarded by those in charge" (1986b).

Thus, an unaltered RAP had to be abandoned with regard to Simon and March's first objective: to demonstrate that individuals are rational actors who obey the provisions of RAP. For the second objective – the demonstration that organizations are also rational actors – RAP served the purpose well.[37] Organizations could be analyzed and understood as rational actors seeking optimal solutions.[38]

RAP has underpinned economic theories of organization, old and new. Classical and neoclassical economics devoted little attention to organizations, *per se*, by a singular focus on the individual. They abstracted away organizational complexities and variation by treating organizations as if they were entrepreneurs. As entrepreneurs, organizations were expected to engage in rational decision-making, seeking optimal alternatives. Thus, the theoretical tools for understanding entrepreneurial behavior would serve for organizations too.

Newer economic models of organizations have breathed new life into the individual. Consistent with the assumptions of neoclassical theory, the individual is both presumed to be a rational being seeking to maximize her utility and the unit of analysis. As such, individuals are driven by RAP.

Two recent models are agency theory and transaction-costs economics. *Agency theory* conceptualizes all of social life as a series of contracts. The contracts take place between the buyer of goods or services, the principal, and the provider of the goods or services, the agent. The fundamental assumption of agency theory is an expression of RAP *par excellence*, assuming that contracting individuals attempt to maximize their utility, defined as rewards minus efforts. This, of course, is precisely the same assumption going back to classical economics that focuses on a decontextualized *Homo oeconomicus*. What, then, is different about agency theory? Because it is formulated around dyads (the principal and the agent), it asks a further question: Under what conditions is one's utility maximized by egocentric, self-regarding behavior, and under what conditions is one's utility maximized by unselfish, other-regarding behavior? According to

[37] Since pointing to bounded rationality as a human cognitive limitation, a growing body of cognitive research has demonstrated a wide variety of other limitations, leading some to label humans as "cognitive cripples." A number of these limitations, such as framing effects and preference reversals, are a direct threat to the foundations of RAP. Do "rational" corporations, likewise, suffer from cognitive limitations? To date the question remains unaddressed, though work in progress (Jackson and Rosa 1994) is making progress toward doing so.

[38] Our focus of attention has been on singular actors, individuals and organizations, with the goal of identifying their status as rational actors in a variety of theoretical formulations. In a seminal paper, DiMaggio and Powell (1983) shift the level of attention to ask whether, in a given field, organizations come to act according to principles of collective rationality. According to the authors, organizations in a given field are subject to external factors, including coercion by the state via laws and regulations that subject them to "isomorphic pressures," such as imitation. As a result, over time organizations within a field come to resemble one another more and more: they evolve toward isomorphic identity (cf Powell and DiMaggio 1991).

agency theory there is no innate propensity to engage in one of these behaviors over the other. The propensity is shaped by organizational structure. Thus, agency theory seeks to understand how such structures shape utility maximizing behavior – behavior in service to the tenets of RAP.

Transaction-costs economics (TCE) is a recent revival of ideas originally expressed by John R. Commons early in this century (1924), and taken up and refined somewhat by Ronald Coase in the 1930s (1937). Its recent revival is at the hands of Oliver Williamson who, via training and academic appointment, was highly influenced by the "decision-making" school of organizations (Williamson 1979). The main goal on the intellectual agenda of TCE is to account for the shift, over the past century, from many organizations to a few large ones. Efficiency is seen to be the principal and most systematic cause for these monumental organizational changes, and the attended concentration of economic power. But, for TCE the efficiency derives, not just from the typical good management practices of the rational maximizers – buy cheap and sell dear – in theoretical open markets, but from strategic actions designed to reduce market uncertainty.

Like agency theory, TCE emphasizes contracts, the establishment of relationships via rational-legal principles – but with a twist. Part of the twist, revealing the influence of the decision-making tradition, is that bounded rationality prevails. Another key – a dismissal of a central premise of rational economics – is that free and open markets with many producers and suppliers often do not exist. This is due in large part to the fact that organizations seek to achieve efficiency by reducing uncertainty in their operating environments. One way of accomplishing this is to engage in long-term contracting; another is to integrate vertically and horizontally. All of these practices achieve the same outcome: a reduction in transaction costs resulting in greater organizational efficiency.

4.4.2 The Rap on the RAP Actor

Behavioral paradoxes for RAP: Wave I. It is easy to see that, when it comes to uncertainty, the logic of RAP is a logic of consequences. To act rationally is to choose the alternative for which the consequences have the highest expected value. Two key elements in this choice process are the knowability of consequences and the stability of preferences. Both have attracted theoretical scrutiny. In order to properly assess consequences, "rational choice involves two kinds of guesses: guesses about future consequences of current actions and guesses about future preferences for those consequences. ...Theories of rational choice are primarily theories of these two guesses and how we deal with their complications" (March 1978).

Theoretical scrutiny of the first guess enjoys considerable history and critical depth.[39] It was first taken up by Herbert Simon (1955; 1976). Simon began to challenge the "perfect knowledge" assumptions of the theory, initiating a line of investigation that continues to the present day. The challenge was focused especially on the informational and computational limits on human choice.[40] Simon's launching pad in his quest to understand problem-solving behavior was RAP – especially the formulations of neoclassical economics. In attempting to harness the orthodox (neoclassical) economics version of RAP to the behavior of actual problem-solving actors, Simon showed (later refined in collaboration with March (March and Simon 1958) that RAP bestowed agents with powers they cannot realistically possess.[41] Agents were endowed with computational powers far too sophisticated for all but the most specially trained or the most extraordinary of actors, for example, with informational storage and retrieval powers that approached those of computers with infinite memory. Thus, Simon discovered that RAP's abstract, calculating, strategizing, maximizing actor, once held up to the mirror of actual behavior, was, in fact, "too rational." It would not be far wrong to claim that, in view of these findings, RAP seemed best suited to explaining the behavior of a race of non-existent superhumans rather than the behavior of ordinary humans struggling to solve problems.

In order to better understand problem solving, the overly rational RAP actor needed to be replaced with a more realistic actor. Simon found such an individual: the boundedly rational actor. This "real life" actor tackles choice behavior and problem solving with a procedural rationality that imposes limitations on both expectations and costs. Information is acquired through an active search by the actor, an activity demanding of attention. But attention is a scarce resource; it is costly and time consuming. Recognizing this and recognizing one's own limits to the processing of that information, actors place a boundary on the amount of information they are willing to collect.

That boundary is, in part, determined by expectations that actors have of the problem solution. These expectations are their aspiration level, their

[39] As pointed out by March (1978; 1991b) an examination of the second guess was delayed and still has been less considered than the first.

[40] The long and distinguished career of Herbert Simon has ranged freely over a variety of topics from psychology to economics to organizational theory to artificial intelligence. His grand intellectual agenda, despite its staggering variation, is underpinned by a singular concern: how to understand the actions of problem-solving agents – be they individuals, groups, or large organizations. How, in fact, do intendedly rational actors go about the business of making decisions, and what shapes their decisions?

[41] Throughout their work, separately and collaboratively, Simon and March have adopted an unalterable methodological individualism, thereby obscuring the distinction between decision making by individuals or by organizations acting as though they were individuals.

problem solving reference point. Actors choose an optimal level of aspiration which then is used to gauge their problem solving efforts. According to Simon, actors know when they have solved their problem: it is when they become content that their optimal aspiration level has been reached. They search for alternatives only up to the point where their aspirational goals are satisfied, then they stop. In Simon's terminology, decision-makers "satisfice." Additional effort and information may improve an actor's chances of arriving at a solution that maximizes some outcome desirable to the actor, but by satisificing the actor chooses to forgo the additional costs of the added effort and attention needed for the incremental satisfaction, beyond aspiration level, that maximization may bring.

Thus, in contrast to the rigid presuppositions of RAP, especially the neoclassical economic formulation of utility theory, people develop acceptable rules of thumb to simply muddle through the process of problem solving. Against the formal deductive logic, the internally locked axioms, and the abstract and closed world of RAP and utility economics, human problem solving behavior is to a large extent irrational. It may be within the grasp of the actor to determine the future consequences of alternatives to be assessed, but seldom does the actor choose to do so. Bounded rationality is a way of describing and analyzing these human psychological limitations.

In the words of Amitai Etzioni (1991, p 3), "Neoclassical economists view man as a two-legged calculator, efficient, and cold blooded." This RAP view of humans abstracts away all human qualities, except instrumental rationality, thereby circumscribing human behavior in a set of overly rational expectations that are, in the face of actual behavior, unrealistic. Humans don't act this way. Human rational capacities are interfered with by – among other things – two other basic human qualities: feelings and emotions and an unremitting proclivity to take shortcuts that violate formal logic.[42] Emotions cloud the logic of calculation. Shortcuts do the same. The would-be rational actor – pressed with not only competing alternatives, and with not only baffling complexities, but also with emotive contexts and the ability to process only so much information – is deeply handicapped if compared with standards of absolute rationality. What remains is a problem solving actor who simplifies in the face of all of these complexities. The actor simplifies by inventing or borrowing rules of thumb that provide

---

[42] There may be evolutionarily sound reasons that humans act not only on the basis of rational direction but also in response to emotions. Survival may, in fact, sometimes depend upon it. For example, Etzioni (1991, p 6) points out that: "Sometimes emotions help rather than hinder decisions; for example, when we experience sudden fright, that fear enables us to escape a danger, where stopping and thinking could have wasted valuable time." Stopping and thinking in some contexts, too, would threaten our very lives.

an acceptable ("It's good enough") solution that is often sub-optimal from a utility maximizing standpoint.

Behavioral paradoxes for RAP: Wave II. The work of Simon, March and other researchers, building on the tradition of the psychology of decision making that they launched, raised serious questions about the first guess – the one about consequences – underlying rational choice theories. More than that, it also paved the way for a reorientation in psychology, away from Skinnerian behaviorism – the then overwhelmingly dominant perspec-tive in the field – and toward systematic investigation of human cognition.

From an evolutionary and comparative perspective humans are, of course, unique among living species by virtue of our complex reasoning powers and our powers for reflection; but even more defining of human uniqueness is the human capacity to reflect on our reflections. As a consequence of this double reflexivity we can reorganize our thinking to be sequentially ordered, internally coherent, and validly deduced. It is these capacities that define rationality in the sense of RAP. Properly exercising these capacities attracts the term "rational" to the actor doing so. Yet, except under the most restricted conditions, humans violate these condi-tions of rationality. Simon and March, and others who built upon their foundations, have found that actors may have every intention acting rationally, but typically fall short of that intention. The rationality they exercise in making decisions is persistently bounded or otherwise limited.

Thus, the capacities for rationality are one of the defining features of humanness; yet, seldom do humans fully exercise those capacities. What accounts for this paradox? With the reorientation in psychology toward the study of cognition, cognitive psychologists – Kahneman and Tverksy (1979), Slovic, Fischhoff, and Lichtenstein (1980), Nisbett and Ross (1980), and others – sought to find an answer. Through a cumulative series of ingenious experimental studies, they think they have found it: cognitive architecture. The reason actors are boundedly rational is contained in the architecture of human cognitive processes. The empirical evidence demonstrates a variety of systematic biases in how people think. What Simon and March had observed about human problem solving – that people take shortcuts and use rules of thumb[43] in order to satisfice – Kahneman, Tversky, Slovic, Fischhoff, Lichtenstein, and others found to be a fundamental part of our cognitive structuring. The human mind uses a variety of shortcutting techniques, heuristics, as a way of organizing

[43] Shortcutting is the cognitive complement to choices based upon emotion. Both merge the two guesses of calculated rationality into one guess. This short-circuits the time and space between sensation or perception and choices. It also short-circuits the amount of information apprehended, thereby resisting information overload. As with emotions, shortcutting may also have evolutionarily sound reasons such as shortening the reaction time in responding to danger.

and processing perceptions and information. Under suitable conditions, these heuristics produce "rational" perceptions, but faced with the kind of situations envisaged by RAP they often produce biased ones (Gigerenzer, in press).

The second guess – about future preferences – required of the rational actor has been far less considered than the first one, although it has been addressed, in part, in the work of Kahneman and Tversky. In standard versions of RAP-based special theories, risk preference is typically assumed to be a fixed trait of an individual or collective decision-maker.[44] In particular, decision-makers are usually described as either risk averse or risk seeking over a whole range of choices. Kahneman and Tversky discovered that decision-makers do not generally exhibit a fixed proclivity toward risks, but are alternatively risk averse or risk seeking depending on situational factors. One important situational factor is whether the risk is "framed" as a loss or a gain. Kahneman and Tversky also discovered that the risks we see depend on where we are standing – figuratively speaking. In particular, the expected value function comprising the range of available choices is generally concave for gains while convex for losses, meaning the displeasure of losing is greater than the pleasure derived from a gain of the same amount. Thus, when we "stand" to lose, that prospect heightens the decline in our expected value. On the other hand, when we "stand" to gain, that prospect dampens the increase in our expected value.

Classical theories derived from RAP, such as expected utility theory, combine a normative model of decision making (what people ought to choose to maximize their utility) with a behavioral model (people are rational and actually make such choices). Normatively used, RAP is a prescriptive model purporting to describe optimal behavior. When used to describe the actual behavior of individuals or social institutions, RAP is a descriptive model. The upshot of the two waves of research reviewed above, each examining different facets of human capabilities for rationality, is a lingering and troubling paradox for RAP. Humans too frequently and too systematically deviate from the axioms of rational behavior, a troublesome pattern for RAP insofar as it wishes to predict and explain actual choice behavior. The cumulation of these inconsistencies led Tversky and Kahneman to conclude that

> ...the logic of choice does not provide an adequate foundation for a descriptive theory of decision making. We argue that the deviations of actual behavior from the normative model

---

[44] "For the most part, theories of choice have assumed that future preferences are exogenous, stable, and known with adequate precision to make decisions unambiguous" (March 1978).

> are too widespread to be ignored, too systematic to be dismissed as random error, and too fundamental to be accommodated by relaxing the normative system.... We conclude from these findings that the normative and descriptive analyses cannot be reconciled. (1987, p 68).

Thus, the normative-behavioral paradox is bad news for RAP. It is good news for those who see in these results other human potentials. In violating the axioms of RAP, the human decision-maker may be revealing other forms of intelligence: intelligence that economizes decision effort, acts without reflection in the face of danger, is respectful of the moral order of things, is aware of the social embeddedness of decisions, and relies on habits that have served well in the past.

The foundations of RAP, as repeatedly noted, require that actors make two guesses, one about uncertain future consequences and another about their uncertain future preferences. Over three decades of cumulative research has shown that actors, though intendedly rational about both guesses, fail to demonstrate these conditions of rationality. They fall short on the first guess because their rationality is bounded by their willingness to satisfice rather than maximize. They fall short on the second guess because actors often do not know their preferences, or maintain sets of conflicting preferences, or find their preferences morally repugnant, or are constrained from acting on them due to tightly drawn social and organiza-tional contexts (March 1991a; 1991b).

Bounded rationality combined with preference flexibility and ambiguity challenges the central core of RAP thinking. It already also led to a revised vision of human rationality as "limited rationality": that is, to a neo-rationalist position. The neo-rationalist position[45] rests, not on the omnisciently rational actor, but on an intendedly rational actor of limited, sub-optimal choice performance (see also Section 3.6). External features of the environment, contingencies in the exercise of choice, and cognitive limitations inherent in thought processes conspire to reduce the rational actor (thereby challenging classical RAP) to a puzzling, emotive, slogging, wending, muddling, "sort of" rational actor.[46]

[45] Due largely to the work of Simon, March, their students, and to a subsequent generation of cognitive psychologists, such as Tversky, Kahneman, Slovic, Fischhoff, Lichtenstein, Nisbett, Ross, and others stimulated by similar concerns.

[46] Work in progress by Dietz and Stern addresses more fully the criticisms of RAP emerging from the two waves of criticism of the classical version described here. Their work is leading to a boundedly rational model of choice that, recognizing both the strengths and weaknesses of classical RAP, attempts to preserve its framework while simultaneously allowing for individual values and social influences: that is, for the influences of social context well known and demonstrated in social psychology and sociology. See, for example, Dietz and Stern (1995).

Rap to Risk. The modern field of risk analysis (Starr 1969; U.S. Atomic Energy Commission 1975) emerged after at least a decade and a half of cumulative research in cognitive psychology and the decision sciences. That research, as noted above, demonstrated systematic behavioral deviations from RAP models of human behavior. What the evidence made clear is that there are numerous limits to rationality in decision making. March (1991b), with characteristic brilliance, summarizes how actual decision making contradicts the expectations of RAP: "Because there are various costs to gathering information and interpreting it, decision-makers follow rules; because there are costs to developing one's own rules, decision-makers copy the rules of others; because there are costs to changing rules, rules are mostly stable – in short, because there are costs to acting rationally, it is rational not to do so."

From this, it was a short, direct bridge to connect the idea of heuristics and other cognitive limitations to questions of how humans perceive risks and make choices about them. Ironically, however, the field of risk analysis eschewed the apparent connection, all but ignoring the troublesome exceptions to RAP models, and the constraints bounding rational choice, choosing instead to examine risks through the sanitized lens of classical RAP. This remarkable irony was exacerbated when the same sanitized RAP models guided policy research, especially early ideas of how to conduct risk communication programs.

Thus, while the adoption of RAP to risk research may have been appropriate as a normative calculus in some circumstances, its adoption as a model of behavior, as the evidence shows, was problematic. It is reasonable to suppose that in some context RAP can properly act as a guide to intelligent risk choices. But it does not accurately portray how people actually make those choices, and it becomes misleading as a normative tool in situations that cannot be handled as optimization problems.

II. LOOKING FOR OTHER WORLDS

5. CHALLENGES AND ALTERNATIVES TO RAP IN THE STUDY OF RISK

Up until this point, we have focused on documenting features of the rational actor paradigm and explicating nature of its applications in the topical area of risk. Our purpose has been to convey an appreciation of how pervasive is this paradigm. The decision to draw upon the risk literature was not arbitrary. Managing risk represents a fundamental ontological challenge for modern societies, with immediate, often severe consequences for individuals and social systems alike. RAP, as the hallmark of over four hundred years of progressive Western thought in the social sciences, has been pervasively employed to reconcile risk problems produced by modern life. As demonstrated above, applying RAP to risk problems has required a significant amount of ingenuity and creativity. While some have attempted to apply RAP's concepts and procedures in a more or less straightforward manner – for example Starr and Whipple (1980), Häfele (1990), Morgan (1990), Nordhaus (1994), Okrent (1996), Wigley et al (1996) – many others have recognized the need to make revisions or other adjustments to the paradigm. Although these variants preserve fundamental principles of RAP, they also point to limitations.

While maintaining our attention on the theme of risk, we now step outside the domain of RAP. In this chapter our goal is to reconstruct key features of the sociological arguments against the rational actor paradigm and its special theories as they are applied to topics of risk. The text starts with a procedure for risk management that has considerable practical potential, but which is incompatible with RAP. We then look at patterns of risk management that are well rooted in organizational practice but which again, although in a different manner, are hard to reconcile with RAP.

This leads us to ask for alternative approaches to risk. To find such approaches, we look at two theoretical approaches specifically designed to understand risk. They are, respectively, the social amplification of risk framework and arena theory. From there, the chapter examines two approaches to risk grounded in broad views of society that explicitly differ from RAP: the cultural theory of risk, and applications of social systems theory to risk.

5.1 Cooperative Discourse about a Risk Problem

A major alternate to RAP for assessing and managing risks is cooperative discourse. This approach is best introduced with a practical example. In late July of 1992, the director of the building department of the Swiss canton[47] of Aargau, along with the department head for solid waste

[47] The Swiss miniaturized equivalent of a U.S. state.

management and a team of social science scholars from the Swiss Federal Institute of Technology (ETH) in Zurich met to discuss public participation in waste planning. The cantonal government recognized a need to get beyond the traditional "decide, announce, defend" (DAD) approach to siting waste disposal facilities, by seeking advice on how to design waste management policies in a way that achieves effective and environmentally safe disposal of solid wastes while adequately integrating public values and concerns.

Among the social scientists in that meeting were two of the authors of this book, O. Renn and Th. Webler. Fresh from the United States, they had come to Switzerland to continue their research into public participation in risk decision making, specifically to design and implement cooperative processes that met operating principles of fairness and competence. They proposed to the cantonal officials a procedure for public participation in the decision making process, which would come to be known as Cooperative Discourse. Renn and Webler had experimented with similar techniques in the United States and Germany (Renn and Webler 1992; Renn et al 1993). One of their main goals was to show that social scientists can do more than analyze why things go wrong. They can take an active role as a catalyst for social change by designing and implementing procedures for collective decision making in accordance with the preferences and visions of those who may be affected by the decisions.

The Cooperative Discourse model directs attention toward achieving procedural criteria. Because modern democratic societies with pluralistic value systems tend to emphasize procedural techniques for collective decision making over substantive interventions, focusing on procedures has become popular within the social sciences. Achieving substantive agreement is difficult and often impossible by either abstract reasoning or empirical fact-finding. This is especially true for collective decisions about risk, since people disagree about what combination of benefits and risks is acceptable (Cvetkovich et al 1989, pp 262ff; Fischhoff et al 1981; Shrader-Frechette 1991, pp 29ff).

If agreement cannot be reached regarding appropriate substantive principles to use as a universal yardstick for evaluating decision options, scholars and politicians tend strongly to apply an egalitarian principle of procedural equity (Luhmann 1968). That principle asserts that there should be equal opportunities for all interested and affected parties to influence the decision making process. Agreement that procedural fairness should be confined to equal representation of each affected party has gained widespread acceptance in democratic systems (Rosenbaum 1979).

Torn between skepticism, hope, and enthusiasm, the cantonal officials finally agreed to experiment with the Cooperative Discourse model. They selected a decision context that posed a serious challenge to the politicians as well as to the research team. The objective was to site a new landfill in the southeast region, a place already hard hit by water and soil pollution from leaking existing landfills.[48] To avoid the appearance that this process was merely a ruse for the canton to get what it wanted by pacifying the public, the member of the state cabinet who backed this process gave his personal guarantee, in public, that if the procedure led to a consensus the canton would definitely not site a landfill in any community where it was not wanted. Furthermore, as long as the procedure led to a consensus, any community could drop out of the process at any time and not be penalized with the landfill. On the other hand, the power of the government to impose a solution if no consensus emerged clearly was a heavy constraint on the voluntary process. Although this constraint facilitated consensus, it also meant that the participatory effort to select one or more sites had to meet and overcome the challenge of potential NIMBY ("Not in My Back Yard") responses.

The building department was charged with the responsibility of designing the canton's solid waste disposal plan, but it was the municipalities that were responsible for implementing the plan.[49] In fact, however, the communities had neither the resources nor the inclination to initiate or implement such planning. Instead, they officially asked the canton to assume the responsibility for design and siting. The building department proposed, and the cantonal government approved, the construction of a one million cubic meter landfill in the southeast region, occupying a site of ten to twenty hectares and to remain in operation for about forty years.

Prior to the implementation of the Cooperative Discourse citizen panels, the building department characterized the need for new disposal facilities and chose potential sites through a mapping-elimination process. Federal and cantonal laws restrict siting landfills in parks, wetlands, inhabited areas, and geologically unsound areas. These areas were removed from the map, leaving 32 potential sites. To further narrow the list of sites, the department developed a set of "preference criteria" and rated each site. Six categories of criteria were used: geology, hydrogeology, utility requirements, settlement-recreation, land and nature protection, and existing use value of the land. These steps were done without consultation with

[48] Aargau has an official policy to incinerate 100% of its solid municipal waste. This is accompanied by mandatory recycling of compostable waste, batteries, appliances, and metal, and voluntary recycling of newspaper (curbside pickup by volunteers), glass, cans, and PET plastic containers (at local self-service bins located near residential areas, usually quite accessible even without a car).

[49] In the past, Aargau was composed out of three political districts: Fricktal (Northern), Bernese Aargau (Western), and the Reusstal (Southeast). These divisions are no longer politically distinguished, but the distinction lives on in regional planning.

the communities. Renn and Webler entered the process just as the results of the mapping elimination process were made public. The product of Phase 1 of the selection process was the selection of 13 potential sites.

In Phase 2, the main task was to limit the choice to from three to five eligible sites. These sites were to be selected and prioritized as a result of the geological surveys and discourse recommendations. This was accomplished in 1993. The following phase included detailed geological investigations in conjunction with citizen panels at each site. One to three sites were finally chosen for development and a legally binding licensing procedure was initiated. This procedure includes a formal environmental impact statement, a public hearing, and a final vote by the cantonal parliament.

Between November 1992 and October 1993, the Cooperative Discourse model was implemented. Four citizen panels were formed, each composed of representatives from each potential site community, selected to represent different social, cultural, and economic interests and values. Their objectives were: first, to develop criteria for comparing the different sites; second, to evaluate geological data collected during that period; third, to eliminate unsuitable sites from further consideration; and fourth, to prioritize the remaining sites with respect to suitability to host a landfill. The citizen panels met these objectives during the allocated time.

In late October 1992, the town councils of the communities in which the potential sites were located received a letter informing them that the canton was undertaking a process to find a single site or several sites for a new landfill. The councils were invited to send one member each to serve on an oversight committee. The oversight committee consisted of one member of each town council and the director of the cantonal building department. Sites located near boundaries were represented by both communities. All but one community opted to send a member of the council to the oversight committee. That town also abstained from sending representatives to the citizen panels. The oversight committee had the legitimate right to make the final recommendation to the building department. In addition, they were asked to inform the public about the site selection process, to review and critique the participation process, and to select the representatives from each of their communities for the citizen panels. The oversight committee also had the legal power to remove sites from further consideration. They were obliged to remove a site if the geological surveys proved the site violated legal requirements (such as standards for natural barriers). During the first part of the process two sites were removed on geological grounds.

Communities selected panel members in different ways. Some town councils published a notice in the local newspaper and asked for volunteers. Others used a formula, balancing political parties that made up the town council or, in one case, balancing men with women. Others attempted to balance the delegation with strong interest groups, such as environmentalists, farmers, and land owners (who could be victims or benefactors depending on their point of view). Far from random, these selection techniques enabled each community to select its own representatives in a manner they deemed appropriate.

Ninety-six citizens participated in the Cooperative Discourse process. Each panel included twenty-four citizens, with each potential site community being represented in each panel. All the panels were given identical tasks:

- review the past mapping–elimination process;
- review and interpret technical feasibility analyses that were undertaken by engineering companies parallel to the deliberation period;
- consider social, political, ecological, and economic impacts and equity issues, including benefit sharing packages;
- develop criteria for evaluating sites;
- suggest three to five eligible sites; and
- develop a priority list of sites for further investigation.

Having four parallel but independent panels, each with an identical task, enhanced the legitimacy of the results. If all four groups arrived at the same outcome, it would suggest that those outcomes were independent of group composition and special group dynamics. If all groups independently came up with similar recommendations, it was unlikely that dominant personalities or other idiosyncratic effects were the main cause for agreement.

Each of the panels met for three hours on one weekday evening, every two or three weeks over a period of six months. All four panels unanimously adopted the suggested rules for discourse and asked the research team to moderate their meetings. Between January and June 1993 the panels met seven to nine times and attended a two-day workshop in which the final recommendations were articulated. Not a single person dropped out of the process. Both regular meetings and special events (such as site visits, which took two Saturdays in succession) and the two-day workshops were equally well attended. Despite the fact that meetings were held from 7:30 p.m. to 10:30 p.m. on weekday nights, the panels decided to schedule more meetings than originally planned. Some panel members had to travel considerable distances, requiring up to a 45 minute drive. Attendance may have been helped by the fact that the panel members scheduled the places and times for meetings. Invitations typically were mailed early, and meeting places were often rotated throughout the region.

During the first half of the citizen panel process, most emphasis was placed on informing citizens about the problem of waste disposal and educating them about the potential risks and problems a landfill can create. They received a brochure in question-and-answer format prepared by a member of the research team and validated by a team of technical experts. Several experts were invited to talk about technical or economic issues, and the results of the geological surveys were conveyed to the panelists by the principal investigators of the engineering companies that made the surveys.

After learning about the solid waste problem, the landfill design, and the geological data, the panelists constructed value trees, i.e. a hierarchical set of criteria and indicators to evaluate each potential site on each indicator and to compute a single quantitative assessment of each site. As noted in Section 3.3.1, this procedure is taken from MAU (Multi-Attribute Utility) analysis as described in Watson (1982) and von Winterfeldt (1986). Before making final assessments of each site on each criterion, the panelists had the opportunity to visit each site, talk to the geologists at the site, and ask questions of local representatives.

In June 1993 the research team conducted a group delphi (Webler et al 1991) with ten experts on landfills and asked the group to provide its best scientific estimates on all those indicators demanding physical measurements or highly professional judgments. The results of this group delphi were given to the panelists. During a two-day workshop at the end of July, the participants rated each site using their group-defined criteria on the basis of their personal impressions, the written and oral information, and the results of the group delphi. Each of the four panels accepted the priority list developed by the MAU procedure and issued a consensual recommendation. Even those representatives who came from the town at the top of the list voted in favor of the recommendation. This was true for all four panels. It was obviously possible to accomplish a consensus even when some members of the group had to face the prospect of receiving an unwanted facility.[50]

Although all four panels recommended the same first priority site, they differed in the order of the remaining priorities. To resolve this conflict, each panel appointed five representatives to a superpanel. The superpanel met in September 1993 and issued a consensual list of five sites. This list was later approved by the oversight committee and forwarded to the building department. In December 1993, the result of the participation process was made public and the canton government entered Phase 3 of the process – initiating further geological tests at the selected sites.

Results clearly suggested a dominant option was present, permitting deliberations to converge. All panels agreed that the site near the town of

[50] At this point compensation had not been discussed. The panels preferred to leave it to the host community to negotiate its own compensation package with the canton.

Schinznach was their first choice. This site, however, was not the most favored choice of the building department. The technical director of the siting process had been asked to conduct his own analysis and to make a priority list before the panels made their final decision. The department's first choice was another town which ranked at best third on one of the four panels' lists. The main reason for this rank order was the high importance of geological stability that the officials from the building department assigned to their choices.

Throughout this process, the citizens of the selected towns for hosting a landfill were cooperative in conducting the necessary surveys and progressing through the various planning phases. Contrary to almost all predictions, the participants were willing and ready to accept a clearly unwanted facility (only three per cent believed that the facility would bring more benefits than disadvantages to the host community) because they were convinced that such a facility was necessary and that their home town was the most suitable, based on the criteria they themselves had developed and accepted as relevant.

Sharing responsibility and providing a structure in which normative agreement could grow were two of the essential reasons why the participants were ready to serve the common good in violation of narrowly defined self-interest. The point of this exercise was not to elicit altruistic behavior; it was, rather, to promote the formation of subjective, but *collective*, judgments about various risks, and subjective, but *collective*, preferences for risk allocation. Explaining this kind of individual or collective behavior is particularly troublesome for the RAP approach to risk. RAP is faced by additional difficulties in the study of risk management by formal organizations, to which we now turn.

5.2 Risk and Organizations

A central task of organizations is the estimation and management of two types of risk: risks in the organization's environment that challenge its operations, and risks associated with the products or operations of the organization. The first type is, in part, an issue of organizational risk perception, while the second is one of safe internal operations. Some organizations are confronted by one or the other of these types, while others must deal with them simultaneously. Because of their numbers, their size, and the resources they command, organizations hold immense power in society. Due to their ubiquity and power, organizations have the prerogative to frame and define public issues, including risk issues. "Increasingly, organizations – often very large organizations such as national and international business firms – and institutions such as governments and regulatory bodies set the parameters and the terms of debate of risk decisions" (Short and Clarke 1992b).

Despite this obvious and pervasive reality of organizational life, and the obvious role of organizations in public risk, the topic of risk and organizations remains barely examined.[51] An initial step toward filling this serious gap was undertaken by Charles Perrow in his now classic analysis, *Normal Accidents: Living with High-Risk Technologies* (1984). But the field has developed slowly, if at all. An important catalyst to its future growth is the work of Short and Clarke (1992a), representing pivotal spade work toward fostering the further development of this important topic.

Left unaddressed in that work, however, is the question of why this gap developed in the first place. We suggest that it developed, in part, because of the general tendency – among lay persons and experts outside the sociological and macro-analytic traditions – to view social phenomena in individualized terms. This bias, a form of psychological reductionism, is manifest throughout the risk field (by lay persons and experts as well). As Perrow (1986a) points out, among high risk organizations approximately 60 to 80 per cent of all accidents are attributed to operators, when a realistic figure is likely to be about half that. The same bias emerged with the release and subsequent promulgation of the Kemeny Report (United States President's Commission on the Accident at Three Mile Island and Kemeny 1979) after the Three Mile Island nuclear accident in Pennsylvania in March 1979. The key conclusion of the much publicized report, as it was inter-preted by vested interests, the press, and other parties, was that the accident was due to operator error. Close examination of the report, how-ever, reveals that human factors were only one element in the accident: organizational and management practices were of key importance and specified as well, a conclusion replicated in the report of the Challenger Disaster seven years later (United States Presidential Commission on the Space Shuttle Challenger Accident 1986). Such gaps between causal explanation and empirical reality apparently reflect what might be called "psychological attribution error."

In the emerging research on risk and organizations, RAP is notable for its absence. RAP-views of organizations, discussed in Section 4.4, have little to say about the way risks are actually generated by organizations. The stage thus was set for analysis of the relations between organizations and risk that eschews RAP, an analysis launched by Perrow (1984) in his study of industrial accidents.

Weber's analysis of bureaucracies provides the starting point. Because bureaucracies are able to outperform other forms of organization, a comprehensive process of bureaucratization characterizes modern society.

[51] Clarke (1989) insists that organizations are so important in determining the incidence and acceptability of risks that they are the appropriate focus of attention, the appropriate unit of analysis, for understanding risk – not the undifferentiated publics of perception studies.

In Weber's view, this was a lamentable but ineluctable fact about social life. It meant that in the long run this "iron cage" would provide little room for personal autonomy. Perrow adds two important arguments to Weber's thesis. First, he emphasizes the importance of the technologies on which bureaucracies rely, some of which are so sophisticated that they cease to display the behavior of linear systems. In linear systems, the smaller a cause, the smaller the effect. Therefore, such systems are amenable to control by human action. In non-linear systems, an arbitrarily small cause can produce an arbitrarily large effect. Therefore, the behavior of non-linear systems is often unforeseeable and also uncontrollable. Under such circumstances, bureaucracies that rely on non-linear technologies are accident prone; they produce "normal accidents."

Non-linear systems are pervasive throughout much of the modern world. They make the world an exciting place. But that excitement poses problems for technological reliability. Perrow proposes that the appropriate way to deal with such systems is to take care that they are only loosely coupled by human activity. By so doing, the unavoidable risks associated with non-linear systems can be limited and their amplification through man-made chains of organizational systems avoided. Many bureaucracies, however, are based on tight coupling of constituent parts.

In a related argument, Perrow (1991) argues that the efficiency of bureaucratic organizations is due to their ability to externalize social as well as environmental costs. The problems so generated then give a rationale for the establishment of additional bureaucracies – of health care, the welfare state, environmental protection, etc. – which pursue the same process at a higher level.

This analysis fits well the nightmarish spirit of Weber's description of modern society as an iron cage. Perrow (1991), however, sees an opportunity for a way out. Recent experiences of small firm networks successfully challenging bureaucratic corporations indicate that Weber was too pessimistic when he regarded bureaucratization as the ineluctable fate of modernity.[52]

Subsequent research inspired by Perrow's path-breaking effort has been uninhibited in its rejection of RAP, often in conjunction with explicit and articulate criticism of RAP's canons (Clarke and Short 1993). While the seeds of rejection were planted by Perrow's "normal accidents," its cultivation came with a variety of in-depth case studies that followed (Clarke 1989; Vaughan 1996).

Context matters: This is the organizing principle of the range of studies included in Short and Clarke's (1992a) *Organizations, Uncertainties, and*

[52] See Piore and Sabel (1984) for the pioneering study on this subject.

Risk. Indeed, context matters so much that organizations, institutions, and the elites running them devote considerable resources to the shaping of contexts, in efforts designed to deal with risks. Focusing on context reveals that cost-benefit comparison, the allegedly rational tool of action, is often more myth than practice. Focusing on organizations unlocks the importance of power, a constitutive element of organizations virtually ignored by all theoretical applications of RAP. Indeed, as Clarke (1989) and others have convincingly demonstrated, formal risk analysis – an operational tool of RAP – is often not applied to establish quantitative estimates of risk, its intended purpose, but to rationalize management decisions already made. The formalities of risk analysis are often only afterthoughts to political and economic considerations. A remarkable instance of this was a cost-benefit analysis supporting the highly problematic location of the gas tank on the Ford Pinto, conducted after production decisions had already been made (Kinghorn 1984).

The more general point, illustrated repeatedly across a variety of contexts, is that the cognitive heuristics and biases that shape individual risk perceptions, amply demonstrated in cognitive psychology, are in themselves shaped by organizational and institutional contexts. The front stage is where "flawed" risk perceptions take place, apparently due to cognitive limitations, but the stage itself has been set with conditionalizing props by a variety of elite actors and decision processes. It is these actors and processes that beg of theoretical investigation.

A common institutional thread running through the varied contexts in the Short and Clarke collection of case studies is the law (see also Short 1989b). Tort law, legal codes, and regulations are omnipresent, powerful, and unavoidable. They impose structure and procedure on a wide variety of organizational forms and stimulate strategic actions by organizations. Strategic interactions of this kind often take place between managers and regulators, with definitions of risk itself hanging in the balance. Risk is, thus, one product of the "definition of situation" resulting from such strategic interactions. Laws and regulations shape, to no small degree, local contexts in other ways as well.

Despite the unmistakable strides this collection of studies makes in moving the framework of analysis up the ladder of aggregation, their approach is quite individualistic, both inter- and intra-organizationally: the former due to the failure to consider interactions among organizations, the latter because they fail to make explicit the implications of social stratification.

The task of analyzing the topic of risk and organization without sharing the limitations of RAP has barely begun (Tilly 1992). One of the issues deserving further study is the paradox of high reliability organizations. "Working in Practice but not Theory" – the provocative title of a seminal article by a group of University of California-Berkeley social scientists (LaPorte and Consolini 1991) – captures the essence of the results of a coordinated suite of studies completed by the group over a period of several years. The focus of these studies is on organizations whose operations are technologically complex and inherently risky, often with catastrophic potential. These organizations perform with such remarkable smoothness they have been named High Reliability Organizations (HROs). The findings are remarkable because they fly in the face of dominant beliefs about technological systems: that high risk technological systems are inherently dangerous and human actions cannot remove the danger entirely. This is the first component of the HRO paradox.

Examination of HRO safety records reveals near-flawless operations. They are shiny examples of organizations that know how to properly manage risky technological systems. But how? The answer seems to lie in bureaucratic structures that are hierarchical and rigid during routine operations (the majority of time), but flat and flexible during times of exigency or crisis. The results are impressive, but they have yet to find their place in an integrated theory. Instead, we are left with rational performance absent a specification of the presumably rational structure producing it. This is the second component of the paradox.

5.3 The Concept of Social Amplification of Risk[53]

The studies discussed in the previous section analyze risk as a social phenomenon in the context of organizations. We now turn to an approach which looks at risks in a larger context, involving many different organizations and also other kinds of institutions.

The notion of the *social amplification of risk* is based on the thesis that the social and economic impacts of an adverse event are determined by a combination of the direct physical consequences of the event and the interaction of psychological, social, institutional, and cultural processes. Proposed by Kasperson, Renn, and colleagues (1988), the goal was to develop a framework for explaining social responses to risk that were not adequately captured in competing concepts, such as the approaches of psychometrics (see Section 3.6) or cultural theory (see Section 5.5).

[53] Since one of the authors of this book (Renn) has been largely involved in developing the social amplification framework and another author has used the framework in several of his own papers (Rosa), the present section draws heavily from these earlier publications, in particular Renn et al (1992).

Social interactions can heighten or attenuate perceptions of risk. By thus shaping perceptions of risk, risk behaviors are also shaped. Behavioral patterns, in turn, generate secondary consequences that extend far beyond direct harm to humans or the environment. Liability, insurance costs, loss of trust in institutions, and alienation from community affairs are a few such examples. Secondary effects such as these are important because they can trigger demands for additional institutional responses and protective actions. They can also – in the case of risk attenuation – impede the installation of protective actions.

Mediated interactions such as these and the secondary outcomes they create typically lie beyond the reach of RAP approaches. This is because of the rich complexity of social dynamics, the high uncertainty of outcomes, and the presence of multiple actors acting independently but reacting to each other in a dynamic way. The social amplification of risk framework presents a way of coping with these uncertainties by examining the collective effect of multiple actors in a given scenario.

Whereas a RAP-based approach would revolve around the individual actors, the social amplification approach places foremost attention on signals (Renn 1991b). Signals are created as a response to a risk event. Several signals on one topic form a message. Signals and messages may be created by individual actors or they may emerge from a group process (such as when a newspaper publisher writes an editorial). In either case it is the content and ordering of the message that is of analytical interest. By comparing these properties of messages about a risk event in society, we learn how actors selectively interpret (through social interaction) facts and anticipated consequences. This provides a basis for testing hypotheses (albeit only retrospectively, since there is not enough evidence to suggest generalizations beyond the immediate case study) about how the privileging of information can influence the social construction of risk.

Social amplification is useful for selecting, ordering, and classifying phenomena and for suggesting causal relations that can be investigated empirically. It provides a heuristic for the analysis of risk experience. One can also think of it as a dynamic framework that allows for systematic interpretation of empirical data while attempting to integrate differing perspectives on risk. Several empirical applications have been reported, and the results used to refine the framework (Burns et al 1990; Freudenburg 1989; Kasperson et al 1989; Machlis and Rosa 1990; Renn 1991b; Renn et al 1992). Machlis and Rosa described social amplification as a "framework that, like a net, is useful for catching the accumulated empirical findings, and that, like a beacon, can point the way to disciplined inquiry" (Machlis and Rosa 1990, p.164).

5.3.1 The Process of Amplification

Amplification includes both intensifying and attenuating signals about risk. Thus, alleged overreactions of target audiences receive the same attention as alleged "down-playing". The idea of the social amplification of risk starts with either a hazard event (such as a chemical spill) or the recognition of an adverse effect (such as the discovery of the ozone hole). Amplification or attenuation occurs as individuals, groups, or institutions select characteristics of the event or observation and interpret them according to their perceptions, mental schemes, and political motives. These interpretations are organized into a message, which is passed on to other individuals and groups (Renn 1991b). As individuals or groups collect and respond to information about risks, they act as "amplification stations". Amplification differs among individuals in their roles as private citizens and in their roles as employees or members of social groups and public institutions.

In the amplification framework risk is conceptualized partly as a social construct and partly as an objective property of a hazard or event (Rosa 1998; Short 1989a, p 405). This avoids the problem of conceptualizing risk in terms of total relativism or total determinism. The experience of risk is not only an experience of physical harm, but also the result of a process by which individuals or groups learn to acquire or create interpretations of hazards. These interpretations provide rules for selecting, ordering, and explaining signals from the physical world. Both processes may have physical consequences. Hazards may directly impact health. Communication about risks may result in changes in technologies, methods of land cultivation, or the composition of water, soil, and air.

With respect to the *individual stations of amplification*, the perception and amplification process can be subdivided into several steps. This cognitive process has to be supplemented further by emotional and subconscious processes that filter incoming messages and co-determine their evaluation (Lee 1986; Renn 1984, pp 111–115)

Decoding and evaluation processes determine the receiver's selection of significant information. Components of the decoded message that are inconsistent with previous beliefs or contradict values to which the receiver feels attracted tend to be ignored or attenuated. Signals are intensified if the opposite is true. The process of receiving and processing risk-related information by individuals is well researched in the risk perception literature (Covello 1983; Renn 1990; Slovic 1987). But an individualistic perspective is not sufficient; individuals also act as members of larger social units that co-determine the dynamics and social processing of risk. For example, under conditions of ambiguity concerning risk, individuals often seek information and guidance from trusted others.

These larger social units are called the *social stations of amplification*. Individuals in their roles as members or employees of social groups or institutions do not follow their personal values and interpretative patterns alone, but perceive risk information according to the rules of organizations or groups with which they are associated. Rules may be derived from a variety of sources: professional standards (characteristic for scientific communities, interest groups, media editors, political institutions, etc.); institutional interests, functions, and foci; role expectations pertaining to the specific position of the receiver; and interpretation of those role expectations by the holder of the position.

Role-specific reception factors are internalized and reinforced through education and training, identification with the goals and functions of the respective institution, belief in the importance and justification of the produced output, and positive rewards (promotion, salary increase, symbolic honors) and negative punishments (downgrading, salary cuts, disgracing).

Behavioral and communicative responses are likely to evoke secondary effects that extend beyond the individuals directly affected by the original hazard event. Secondary impacts include:

- enduring mental perceptions, images, and attitudes (e.g., anti-technology attitudes, alienation from physical environment, social apathy, or distrust in risk management agency);
- impacts on the economy (e.g., drop in business sales, residential property values, and tourism; increased liability and insurance costs);
- political and social pressure (e.g., political demands, changes in political climate and culture; social disorder);
- changes in the physical nature of the hazard (e.g., feedback mechanisms that reduce or heighten the potential impact of the hazard);
- repercussions on other technologies and activities (e.g., higher or lower level of acceptance).

Secondary impacts are, in turn, perceived by social groups and individuals so that a further stage of amplification may occur, producing third-order impacts. Impacts may spread or "ripple" to other parties, distant locations, or other risk arenas. Each order of impact may not only disse-minate social and political effects, but may also trigger (in risk amplification) or hinder (in risk attenuation) positive changes for risk reduction.

The social amplification of risk framework conceptualizes signals as the basic unit of social communication (Renn 1991b). Signals in social interactions establish a relationship between the information source, potential transmitters, and receivers. In social communication, signals must be meaningful, otherwise they are regarded as noise. A cluster of meaningful signals pertaining to the same topic is called a message.

Any change in the order of signals or alteration of the signals may change the meaning and thus the message. Changes in messages usually occur during transmission and constitute an important part of the process of social amplification. As metaphor, several specific amplification mechanisms of the framework can be identified:

- volume effect (intensifying or attenuating messages)
- filtering effect (intensifying or diluting information)
- muting and adding effect (deleting or adding information)
- mixing effect (changing the order of presentation)
- equalizing effect (placing the message in different contexts)
- stereo effect (receiving the same or similar messages through different channels)

The *volume effect* describes the resonance a message receives after being channeled to transmitters and receivers. In an analysis of media coverage of Love Canal and Three Mile Island, Mazur argues that the massive quantity of media coverage not only reported the events, but also defined and shaped the issues (Mazur 1984; see also Peltu 1989). Even in the absence of any distortion of information, the pure volume effect has an influence on the perception of the seriousness of the message and may determine the political agenda of social groups and institutions. Conversely, the most important event may not even be recognized as anything noteworthy if it is transmitted in "low volume," e.g., on the back pages of newspapers or during unpopular broadcast periods.

A *filtering effect* is the degree to which partial information is highlighted or down-played. Again, the basic composition of the message may be the same (signals are neither deleted nor added), but the understanding of the meaning may be strongly influenced by partial amplification.

The third effect, *deleting or adding information,* does not necessarily change the meaning or the message. If transmitters act as translators for converting scientific or agency jargon into "ordinary language," they actually enhance the message and make it more intelligible to the final receiver. More frequent, however, is the case that messages are changed by adding or deleting signals. This may be done intentionally in order to create a different image than implied in the original message, or accidentally by attempting to shorten or lengthen the message or to make it more comprehensible to a lay audience.

Changing the order of information in messages *(mixing effect)* is almost routine in the news media. Conventions about what to report first (who, when, what, where, why) reflect the anticipated interest of final receivers by journalists. Most of these rules of journalistic editing seem to be intuitively plausible, but as a routine procedure they may well affect the meaning of a message. An article reporting on a conference about the greenhouse effect due to carbon dioxide concentration may be perceived as outdated or less alarming if the conference itself took place three months before the report appeared and if this information was placed in the first sentence of the article (in accordance with the "rules").

The context in which the message is embedded *(equalizing effect)* is also an important factor in signal amplification. Whether the message is part of the news section of a newspaper or part of a commentary conveys a different degree of objectivity with respect to the content of the message. Lee has pointed out that messages hidden in fictional writing or entertaining movies such as *The Day After* or *The China Syndrome* may be powerful agents in creating images of objects, in particular if those images are constantly reiterated in literature and film (Lee 1986, p 171). The creation of images through symbolic illustrations such as cooling towers, flashing lights in the control room of a nuclear reactor, dramatic background music, workers with gas masks, or mushroom clouds, is part of a sublime message transfer that connects frightening or joyful symbols with an object or person.

The *stereo effect* relates to the multitude of channels in transmitting messages. One message or topic may penetrate the information market and dominate newspapers, television, journals, and other publications. In addition to the mere volume effect, the multi-channel coverage of an issue assigns more importance and credibility to the message. In particular, if the stereo effect is well orchestrated and information sources manage to use different channels in a complementary manner, the message will more likely reach the target audience (Pinsdorf 1987, p 47). There is also an increased likelihood that the message content will be intensified.

5.3.2 Signal Amplifiers in Social Communication

Describing the effects of signal amplification demonstrates not only the merits, but also the limitations of using the metaphor of electronic signal theory. Although each of the social amplification effects can be expressed in terms of volume, filtering, equalizing, mixing, muting, and stereo, they only make sense if the denotations of each term are adjusted to the social context. In particular, transmitters in social communication have hardly any resemblance to electronic amplifiers. Instead, social transmitters are active human agents in the communication process, with their own independent agendas, rules, and goals.

The function of transmitters and the interrelatedness of message and messenger have been the predominant criticism against the metaphor of amplification and the use of information theory for analyzing risk communication. In his review of the original paper introducing the concept of social amplification in risk analysis, Rayner criticized the social amplification metaphor as a mechanistic understanding of social communication, which cannot be characterized as a system in which messages and messengers are separate entities. The changes wrought in the signal bearers constantly transform the instrument, i.e. society (Rayner 1988).

The proponents of the social amplification framework have responded to this criticism by stating explicitly that using signals as the basic unit of analysis does not imply that stations of signal processing are passive and mechanical transformation stations (Kasperson 1992; Renn 1991b). On the contrary: all actors participating in the communication process transform each message in accordance with their previous understanding of the issue, their application of values, worldviews, and personal or organizational norms, as well as their own strategic intentions and goals.

The social amplification metaphor has evolved as an umbrella framework that provides ample space for social and cultural theories. Theoretically it is based on the idea that social experiences of risk are only partially determined by the experience of harm or even expected harm. The amplification or attenuation process takes into account individual factors of perception (qualitative characteristics), social mobilization potential, effects of signal transformation, and cultural propensities to evaluate risk situations according to specific patterns. It is open to inclusion of other elements as well. Its contribution to the risk field is that it has demonstrated that under a common metaphor different, often competing, theories of risk in society can be reconciled and their specific contributions placed into a single framework. Yet it does not offer a solution when different theoretical approaches suggest different research procedures, different hypotheses, or different interpretations of results. The social amplification metaphor can therefore perform the role of an additive model rather than that of a truly integrative theory.

5.4 The Arena Metaphor[54]

Another approach to risk deserving of attention is the notion of social arenas. Social arenas focus on *influence* in decision making – a central theme among social scientists who study how society copes with natural and technological risks. The goal is to explain why different individuals or groups achieve different levels of influence in what is essentially a political process. According to the RAP approach of Resource Mobilization Theory (Section 4.3), there is in the conflict over risk decisions a competition for

[54] The concept of social arenas for risk debates has been articulated and developed by one of the authors of this book (Renn). Parts of this section have therefore been taken from Renn (1992c). They are not identified by additional reference citations.

social resources. Individuals and organizations can influence the policy process only if they have sufficient resources available to pursue their goals. Actors are presumed to be goal-oriented and to select the most effective means for mobilization of resources for goal attainment.

5.4.1 Actors in Policy Arenas

Social Arena Theory accepts this RAP presupposition, but goes beyond it by recognizing that the outcome of the struggle is determined not only by individual or group actions, but also by *structural arena rules* and *interaction effects* among the competitors. The political organization of an arena and the external effects of each group's action on another's constitute structural constraints that complicate explanations of how and why a particular outcome occurs.

A social arena is a metaphor to describe the symbolic location of political actions that influence collective decisions or policies (Kitschelt 1980) – symbolic because arenas are neither geographical entities nor organizational systems. Rather, they describe the political actions of all social actors involved in a specific issue. Issues can include political decisions such as siting of facilities or increased property taxes, social problems such as crime or education, or ideas such as civil liberties or evolution versus creationism. The arena concept attempts to explain the process of policy formulation and enforcement in a specific policy domain. Its focus is on the meso-level of society rather than on the individual (micro-level) or societal behavior as a whole (macro-level). It reflects the segmentation of society into different policy systems that interact with each other but still preserve their autonomy (Hilgartner and Bosk 1988).

The arena idea incorporates only those actions of individuals or social groups that intend to influence collective decisions or policies. An actor who believes in deep ecology and communicates the idea of equal rights of animals and humans to others is irrelevant for this analysis unless he or she attempts to change environmental policy, reform hunting laws, or restrict current practices in agriculture or animal laboratories. Intentional behavior to affect policies is certainly not the only way that policies are affected by public input (examples of other possibilities are given by public opinion polls or by media coverage), but these external effects are conceptualized as inputs into the arena rather than as elements of it.

That actors in an arena intend to influence policies is the only assumption made about intentions, goals, purposes, or motives. This assumption is important, however, because it provides the yardstick for evaluating social constructs that actors may use to define their causes and to pursue their goals. Under this assumption, success and failure can be measured (intersubjectively) by the amount of influence that the actor has been able to exert on the resulting decision or policy.

The center stage of an arena is occupied by the *principal actors*, i.e. those groups in society that seek to influence policies. Groups often focus on several issues at once and are hence involved in different arenas; others focus only on one issue in a single arena. Each arena is characterized by a set of rules: formal rules that are coded and monitored by a *rule enforcement agency* and informal rules that are learned and developed in the process of interactions among the actors. Among the formal rules are laws, acts, and mandated procedures; among the informal rules are regulatory styles, political climate of group interactions, and role expectations. In most cases the rules are external constraints for each single actor. Formal rule changes require institutional actions, while informal changes occur as a result of trial and error and may happen according to whether or not rule bending is penalized. Several actors may join forces to change the rules even if they disagree on the substance of the issue.

The *rule enforcement agency* ensures that the actors abide by the formal rules and often coordinates the process of interaction and negotiation. In many arenas the rule enforcement agency is also the ultimate decision maker. In this case, all actors try to make their claims known to this decision maker and to convince her by arguments or through public pressure to adopt their viewpoint. In an adversarial policy style (typical for the United States; O'Riordan and Wynne 1987; Renn 1989), rule enforcement agencies regard themselves more as brokers or mediators than as sovereign administrators who are advised by various social actors, (the typical European policy model; Coppock 1986).

Issue amplifiers, to take the arena metaphor a step further, are the professional "theater critics" who observe the actions on stage, communicate with the principal actors, interpret their findings, and report them to the *audience*. Through this communication process they influence the allocation of resources and the effectiveness of each resource to mobilize public support within the arena. The audience consists of other social groups who may be enticed to enter the arena and individuals who process the information and may feel motivated to show their support or displeasure with one or several actors or with the arena as a whole. Part of the political process is to mobilize social support by other social actors and to influence public opinion.

In contrast to traditional role theory or the theater stage metaphor (Goffman 1959; Palmlund 1992), the arena concept does not picture the actions on stage as a play with a script or actors as performing predetermined role assignments. Arenas are more like medieval courtyards in which knights fought for honor and royal recognition according to specified arena rules that determine the conditions for the fight, but leave it to the actors to choose their own strategies.

Accordingly, modern arenas provide actors with the opportunity to influence the policy process and to direct their claims to the decision makers. Their behavior is not necessarily defined by behavioral roles and routines. Actors may use innovative approaches to policy making, as well as traditional channels such as lobbying. However, arenas are regulated by norms and rules, which limit the range of potential options. Actors may decide to ignore some of the rules if they feel that public support will not suffer and if the rule enforcement agency is not powerful enough to impose sanctions on violators.

The outcome of the arena process is by no means preordained. On one hand, various actors may play out different strategies that interact with each other and produce synergistic effects (game theoretical indeterminacy). Strategic maneuvering can even result in an undesired outcome that does not reflect the stated goal of any actor and may indeed be sub-optimal for all participants. On the other hand, interactions in the arena change the arena rules (structural indeterminacy). Novel forms of political actions may evolve as actors experience the boundaries of tolerance for limited rule violations. Therefore, arenas often behave like non-linear systems; small changes in strategies or rules are capable of producing major changes in conflict outcomes. It is also difficult to predict who is going to benefit from potential rule changes induced by trial and error. Both characteristics of arenas limit the use of arena theory for predictions, but do not compromise its value for explanation and policy analysis.

5.4.2 Resources in Policy Arenas

The fundamental premise of arena theory is that, in order to be successful in a social arena, it is necessary to mobilize social resources. Resources can be used to gain attention and support of the general public, to influence the arena rules, and to gain advantage in competition with the other actors. In arena theory, resources provide the means that help actors to be more influential. (Arena theory has much in common with resource mobilization theory on this point.) Resources may be the ultimate goal of an actor but more often they are the means by which an actor can accomplish her specific intentions. Whether these intentions are egoistic or altruistic, overtly stated or hidden, is irrelevant for the success of resource mobilization unless the goals themselves are used in a debate to improve one's opportunities to gain resources.

The early functionalist school of sociology referred to social resources as "all persons or organizations, which can be of help to an individual or a social work agency in solving problems" (Fairchild 1955, p 291). As a means to mobilize resources, different functional segments of society use generalized media, i.e., instruments to mobilize support (Etzioni 1961; Parsons 1951; Parsons and Smelser 1956).[55] Parsons suggested four generalized media: money for the economic sector, power for the political sector, prestige for the social sector, and value-commitment for the cultural sector (Parsons 1963). Other authors suggest that these are actually the resources that groups want to mobilize, whereas the term "media" should be confined to the currency (exchange value) within each resource type (Luhmann 1982; Münch 1982). Based on this understanding of resources, the following five resources have been proposed by arena theorists (Renn 1992c):

- *Money* provides incentives (or compensation) in exchange for support or at least tolerance.
- *Power* is the legally attributed right to impose a decision on others; conformity is established by the threat of punishment.
- *Social influence* produces a social commitment to find support through trust and prestige.
- *Value commitment* induces support through persuasion, solidarity, and cultural meaning.
- *Evidence* can be used to convince persons about the likely consequences of social actions (Laird 1993).

Money is most frequently used in the economic sector, but, like all other resources, it is also instrumental in other sectors, such as the social system. Its medium of expression is the transfer of capital which in turn may provide incentives for other actors to show loyalty to the investor. Money usually buys compliance rather than convictions (except over long time periods).

The same is true for *power.* Power operates through coercion and requires compliance to rules and commands independent of the subjugated group's convictions or personal values. Authority and force are the two media through which power is expressed and the basis for power relationships established. The motivation to comply stems from the threat of punishment which may include physical force. The threat alone is often sufficient to produce conformity without formal sanctions being imposed. It should be noted that in arena theory, the term "power" is used in the classical (Weberian) understanding rather than in the modern

[55] For a discussion of generalized media, social structure, and cultural evolution see Jaeger (1994).

systems-oriented concept of power as "any type of influence to make others comply with one's intentions" (Coser 1956) or as "exclusive possession of infor-mation" (Münch 1982). Otherwise, power is difficult to distinguish from social influence or value commitment.

Social influence is a resource that operates through two media: reputation and social reward. Reputation generates trust in the specific actor even if the meanings of the actions are not understood by others. For example, asking Nobel Prize laureates to defend nuclear power or asking prominent actors to endorse toxics legislation does not mean that people are convinced by arguments that nuclear power is beneficial for them or that a new initiative restricting chemical use is desirable. People may believe these actors because they accept them as role models, identify them as experts on this issue, are convinced of their sincerity, etc. Social rewards, the second medium, constitutes symbolic reinforcements of behavior and generates social prestige. By analogy with money, social rewards can increase conformity and evoke support for one of the actors. Social influence is not based on shared values or meaning with respect to the issue in question, but on socially accepted incentives for assigning credibility to others and receiving social status through others.

Value commitment is a cultural resource for finding meaning and sense in the behavior of social actors and society as a whole. The two dominant media of expression are persuasion and meaning. If social actors are able to persuade other actors that their behavior is in accordance with their commonly shared values, interests, and world-views, they can count on the solidarity based on this communality. Shared meaning conveys a sense of purpose in life and creates a cultural unity that also extends into solidarity and a feeling of community. The resource value commitment has become one of the most powerful organizing principles in political debates as societal pluralism provides opportunities for individuals to be selective in choosing world-views and to change alliances if this is deemed appropriate.

Evidence can be the empirical basis for convincing people of real options. Evidence, however, is not identical with truth. Truth is an ideal that has validity for all people at all times, whereas evidence is the claim of truth that social groups or special subsystems of society make based on experience, collected data, or theoretical reason. Evidence is continuously being tested for validity using shared experience, proper methods for data collection, and accepted theoretical knowledge as yardsticks. Knowledge is revised when new compelling evidence is accepted into the culture. For example, if a group claims that a specific risk source is benign, repetitive occurrences of accidents with negative consequences may finally trigger a revision of this claim or evoke a substantial loss of social influence for this specific group. Evidence is not arbitrary in spite of the fact that it is relative and pluralistic.[56] Evidence in the arena concept serves as a power-ful social resource to convince people that the expected factual implications of one groups' claims are in their best interest, whereas the potential implications of the competing groups' claims are not.

The need to collect various types of resources creates an exchange market for resources via generalized media of expression. Social actors with a great deal of money may attempt to purchase social influence by paying highly reputable persons to join their cause. Likewise, groups that offer meaning and values use these resources for fund raising. High prestige groups may use rewards to honor charismatic leaders who in turn will provide value commitments. Other groups may use their power or money to hire experts in exchange for receiving evidence. Resources are partially convertible against each other and it depends on the context and the availability of other resources whether one resource can be exchanged against another. The exchange of resources is not a zero-sum game; a resource can be generated without taking it away from other groups. The generation and distribution of resources may result, however, in inflationary or deflationary developments (Parsons and Shils 1951). Too many medals, for example, diminish the social value of each medal.

Another limitation of resource exchange is the problem of legitimizing the use of resources outside their dominant application. The extensive use of one resource (such as money) outside its home sector is likely to come at a cost to the existing reservoir of other resources. For example, the use of money for compensation may lead to a decline in social influence and value commitment because the transaction is perceived as bribery. In a similar vein, the appeal to common values and convictions may be seen as a signal of weakness in the economic market, and potential investors may be more cautious to supply the respective actor with money. Gaining resources in an arena is a balancing act in which the need to exchange resources has to be weighed against the probability of losing both the resources one is willing to sacrifice and the resources one hopes to gain.

A further strategy to gain specific resources is to use one's established influence in other arenas to generate resources and to transfer them to a novel arena. Groups may enter an arena only for the sake of receiving resources that they can use for another political issue. Although arenas in modern societies tend to be structurally segmented and autonomous, the success of resource mobilization tends to depend on the perception of overall performance in several arenas. A company that wants to sell its products may enter an environmental arena in order to gain social reputation and value commitment even though the issue of this arena is

[56] See the model of graduated rationality in Renn (1981).

of low interest to its managers. Actors also like to use "hot" issues to piggyback their own claims to the targeted audience. This is one way to gain attention and social recognition. This strategic behavior of groups is one of the reasons that arena theory makes no assumptions about the substantive goals of the actors, but limits itself to the resources that actors try to mobilize as a means to influence the policy outcomes.

Arena theory attempts to explain responses of social groups to risk issues and to interpret institutional and political actions directed towards risk reduction and risk management. It makes no inference about the actors' intentions or motivations, but focuses on their ability to mobilize resources within the structural rules of the arena. If the resources require the overt statement of goals (which may often be the case, especially for value commitments), then this becomes part of the general strategy.

By side-stepping the question of motivation, arena theory borrows the concept of goal-oriented actors from RAP. That is, it assumes that social groups are organizations that intend to maximize their influence in specific arenas by choosing the most efficient means for reaching their goals. In contrast to many RAP-based sociological theories, however, arena theory does not assume that groups want to maximize their interest as such. At the same time, it does not predefine group profiles as done in the cultural theory of risk (see below). Whatever the goals of actors, they can only accomplish them by mobilizing resources. Availability of resources provides the bargaining power to influence the outcome of the policy process.

Arena theory avoids the relativism and solipsism of social constructivist theories as well as the structural determinism of Neo-Marxist theories and of many applications of critical theory. It provides an intersubjective anchor for determining success in a political arena. The fundamental axiom is that *resource availability determines the degree of influence for shaping policies.* As far as this axiom is appropriate, social arena theory provides an elegant and powerful instrument for the analysis of social issues in general and risks in particular.

Yet, the arena metaphor clearly has its limitations. It creates the impression of politics as a game in which players want to win and spectators want to be entertained. Although some political debates support this impression, others certainly do not. Many debates are characterized by a good faith effort of different actors to improve a situation or to resolve a conflict. The emphasis on social resources may obscure the fact that not all political actions are strategic and that people often mean what they say. In addition, the division between actors and spectators seems to support a concept of democracy in which elites fight for power and influence and the masses are used as instruments for these elites to gain relative advantages. Finally, the metaphor provides a conceptual framework which is compatible with many different, even highly contradictory assertions.

As a consequence, specific statements made within the framework of arena theory may remain somewhat arbitrary. This is certainly not the case with the approach of cultural theory, to which we now turn.

5.5 Cultural Theory of Risk

A general paradigm gives birth to specific theories that explain what we wish to understand. One of the tests of the robustness of theories is how well they explain events that seem counter-intuitive or are unexpected. If the explanations are not convincing on empirical or theoretical grounds (that is, they do not hold up to further scrutiny), then alternative theories may emerge. These theories may come from within the dominant paradigm, or they may represent features of a competing or emerging paradigm. The story of risk perception and the application of cultural theory to issues of risk are telling examples of this type of paradigm challenge.

Risk perception is the name given to a body of research tapping into people's recognition of and concerns about risk. It emerged as a topic of study after empirical studies found that people's behavior deviates considerably from the expectations of RAP, and that they have very different understandings and opinions about risk. It was pursued by cognitive psychologists using the techniques of psychometrics (see Section 3.6). One of the earliest, now classic studies was conducted by Fischhoff, Slovic, Lichtenstein and colleagues in the late 1970s (Fischhoff et al. 1978). These researchers were interested in perception of risk assessed in two ways: first, in comparison to actuarially-based probabilities, and second by comparing lay persons' assessments of risk with those of experts. Using a list of thirty hazards, a sample of lay people and experts were asked to estimate the risk of dying from each hazard. Findings indicated that experts used a much broader range of numbers to estimate risk than did lay people; lay people rated the risk to nuclear power relatively much higher than did the experts; and experts' risk perceptions correlated well with technical estimates of fatalities. Further studies confirmed these results and elaborated discrepancies among risk perceptions (Slovic et al 1980). Significant differences were found not only among experts and lay people, but also among the general population itself.

This raised two important questions to the community of scholars studying risk. What accounted for the differences in risk perception among people, and on what basis can judgments about the sufficiency of competing judgments be made? Although several approaches to these questions occurred within the rational actor paradigm, sometimes amended by cognitive psychology, cultural theory emerged as a competing perspective, leading, in turn, to the application of cultural theory to issues of risk.

Scholars working within the rational actor paradigm had a number of things to say about the problem of risk perception. One of the loudest messages came from the engineering and science community, reflecting one of the strongest presuppositions of positivist thought. It was argued that facts and values can and should be separated and handled differently. Facts are to be dealt with by science while values are the domain of politics (for a discussion of this approach in risk management, see Whipple 1992). Some of the outspoken scholars invoking RAP from this technical perspective, therefore, suggested that the most appropriate perception of risk is that which is most objective or value-neutral. Counting dead bodies or estimating failure rates based on reliability testing and probability mathematics, they argued, was the most objective way to estimate risks. This answer was tantamount to saying that the experts were right and the public was wrong. The straightforward political implication of this conclusion would be to disenfranchise the public from risk management.

Social scientists operating from RAP-influenced perspectives also pursued explanations that were less politically charged than those of their technical colleagues. In general, these scholars have tended to legitimize (as opposed to dismiss) the risk opinions of lay people. Because these approaches were discusses in Chapter 3, they are only briefly mentioned here. An early contribution was prospect theory, advanced by Kahneman and Tversky (1979). Prospect theory explained that many people do not perceive risk in a linear relation to magnitude of loss (or gain). Instead, they are less likely to accept either losses or gains, and probabilities associated with them, depending on several situational circumstances. Another path, taken by Slovic and colleagues, has been to relate characteristics of hazards with the judgements of respondents (Slovic 1992). Hazard characteristics such as familiarity or dread shape lay people's judgements. A third approach within RAP has been to look into political factors that might explain differences in risk perception. Here people perceive risks in the context of a political conflict of interests and one's interests shape how one perceives risks (as well as opportunities). Dorothy Nelkin and Michael Pollak elaborated on this approach in their study of nuclear power in Germany and France (Nelkin and Pollak 1979).

Cultural theory takes a quite different approach to problems of risk perception, beginning from the assumption that there exist within any society a number of competing rationalities. These rationalities account for people's differing perceptions of risk. "Rationality" here consists essentially of one's values and beliefs about how society ought to be. In the language of cultural theory these are called "cultural biases" or "ideologies".

Cultural theorists recognize three to five forms of cultural bias (or worldviews), each characterized by a set of beliefs or values about how social relations ought to be structured. An excellent example of applying cultural theory to the study of risk perception is the recent work of Slovic and Peters (1996). In a national telephone survey people were asked a number of questions designed to measure adherence to beliefs associated with different worldviews identified in cultural theory (see Dake 1991; 1992). They also asked questions about respondents' images and opinions about risks associated with nuclear power. Results showed a highly significant correlation between worldviews and perceived risks from nuclear power. Slovic and colleagues have also examined how perceptions of risk differ among cultural groups. They have found differences in risk perception among, for example, American and Hungarian students, and also among subgroups within countries, for example, French-speaking and English-speaking Canadians.

Cultural theory goes beyond RAP in several ways. First, it does not pose a duality among facts and values. Instead, it builds on the tradition of social construction of knowledge from the sociology of science to assert that people's understanding of the world – including factual knowledge – is mediated through value systems. Second, it perceives individual decision makers as acting within a social and organizational context. Steve Rayner has asserted that "institutional structure is the ultimate cause of risk perception" (Rayner 1992, p.86). Third, it argues that cultural diversity serves a protective function for society. Each cultural bias is blind to some things and sensitive to others. Together they form a shield that provides ontological security to society.

5.5.1 Acceptable Risk and the Origins of Cultural Analysis

Cultural theory leads to certain normative consequences which, though they must be judged in their own right, are quite different from the normative consequences of RAP. First, it creates some ambiguity about the role of the expert. Because experts ascribe to different ideologies and have different institutional affiliations, they may hold fundamentally different views of risk, as Dietz and Rycroft have shown (1987). Accordingly, there is no way to demonstrate that one expert's cultural bias is superior to another's. In practice, expert advice typically must be taken at face value. Second, cultural theory suggests that communication within society should be oriented toward the creation of shared meaning and trust as opposed merely to the articulation of quantitative expressions of risk benefits or tradeoffs (Rayner 1990). Third, cultural theory tells us that competing views may remain in conflict; indeed, there may be good reasons for conflict (Thompson and Warburton 1985). Differences often are not due to misunderstandings or to a lack of social consensus, but to irreconcilable differences between fundamentally opposed worldviews. Hence, decision making processes that place too much emphasis on reaching consensus may be a waste of time. Conflicts are endemic to society; and, indeed, serve valuable social purposes. Finally, and perhaps foremost, cultural theorists contend that decisions about risk management should not endanger cultural survival. This conclusion is based on the original work of Mary Douglas, who explained certain social policies as efforts to preserve cultural integrity (Douglas 1966).

One of the strongest normative messages to come out of cultural theory focuses on the way society makes risk management decisions and policies. Steve Rayner and Robin Cantor (1987), in a widely read paper, suggested that, given the moral necessity of accepting all cultural biases as legitimate, a further normative prescript for social decision making could be deduced. Their argument is presented in the context of a larger debate, and in such a way as to pose cultural theory as a contender to naive RAP-based technological approaches to risk management. The latter, they argued, fail to meet inter-subjective criteria for achieving a *fair distribution* of effects. Fairness, they suggested, is a more important moral consideration in American culture than is utility maximization. Justification for this assertion also comes from cultural theory, which posits four fundamentally different "cultural types" in society. Continuing:

> From the contrast between the four institutional cultures engaged in controversies over technological risk ..., it is clear that policymakers within each type of sociocultural framework will have great difficulty in understanding the fears and objections of others. (pp. 6ff)

But policy makers with different cultural biases *ought* to understand each other, because they need to collaborate in order to make effective risk decisions. It stands to reason that fair process must be a highly cherished cultural norm. This is why the authors proposed replacing the question "How safe is safe enough?" with "How fair is safe enough?"

The claim that RAP-based technological approaches to risk management often produce normative outcomes that are obviously unjust is a thesis with great power. If equity is a widely held social norm – and this seems to be almost taken for granted – then decisions must be judged to some extent in view of equity concerns. The so-called objective approaches to setting acceptable risk thresholds according to toxicological or epidemiological data and models often were not unbiased. They consistently disadvantaged vulnerable populations – often children and pregnant women, but also the elderly or hypersensitive individuals of all races and ages. These people routinely fall outside the normal distribution of population sensitivity to a hazard source. The mathematical models indicate the level at which 95% of the population will realize adequate safety, but unless the problem of fairness is explicitly taken into account, they are blind to systematic sorting of sensitive populations into the outlying 5%.

In sum, technological methods (e.g., actuarial or PRA) that embody fundamental principles of RAP in the most extreme form fail to acknowledge that the outcomes may be consistently biased. For this reason, they do not always meet socially established moral norms of fairness or neutrality.

RAP does not accommodate the importance of moral norms for making decisions about the acceptability of risks. Cultural theory focuses precisely on the link between what a culture treats as risky and what it treats as evil.[57] In cultural theory, risks are defined in relation to socio-cultural necessities. Cultures define as unacceptable risks that threaten basic assumptions upon which a culture is founded.

This fundamental claim can be well illustrated by one of the major sources of the cultural theory of risk. Douglas (1966) argued that the Israelites' Old Testament prohibition of eating pork was not due to the health risks presented by pork in hot climates. Although these risks can be assessed with the tools of contemporary science, they could hardly have been known by experience in the twelfth century B.C. Moreover, other cultures existing in similar conditions did not develop similar prohibitions. The true risk of eating pork, according to cultural theory, lay elsewhere.

Douglas's claim was that cultures enable human beings to find an orientation in the world by offering broad classificatory schemes. In the schemes of ancient Jewish culture, animals with horns and cloven hoofs, such as deer, sheep, and cows formed an important conceptual category, in contrast to animals with paws and without horns, such as dogs, foxes, and wolves. Pigs, however, blurred this distinction by having cloven hoofs, but no horns. Pigs warranted special treatment in this culture, because they challenged basic schemes of orientation.[58]

A similar argument can be made regarding the incest taboo. Human cultures organize patterns of human relations starting with elementary kinship relations. Binary relations like mother-child or brother-sister can

---

[57] This link inspired Short (1984) in his influential presidential address to the American Sociological Association, which played a key role in the emergence of contemporary sociology of risk.

[58] It is obvious that in the meantime the prohibition to eat pork has assumed other meanings as well. In particular, it establishes a social marker which helps to sustain the identity of a cultural group and to maintain patterns of social solidarity.

As an aside, we note that Douglas's analysis of classification schemes for animals may be relevant to contemporary debates about the risks of genetic engineering. Quite independently of any technical risks associated with the prospects of genetic engineering, these prospects pose a cultural problem which may well arouse strong feelings of distrust and anxiety.

be linked in chains: the mother of one's mother is one's grandmother; the brother of one's mother is one's uncle. Concepts of social relations break down if incest blurs the distinction between, say, father and brother. Quite independently of the long-term risks of impairing the genetic endowment of people by in-breeding, the threat posed by incest to basic schemes of cultural operation may explain why human cultures in general treat the risks of incest as unfit.

Actions threatening the cultural stability of one's community may be treated as sins by that community. The cultural theory of risk highlights such links between risks and sins. Costs and benefits are inadequate for analyzing these links. The commission of a sin may entail costs that people happily incur based simply on the expected benefits of such behavior. Sins are behaviors that people are expected to avoid, regardless of cost-benefit considerations.

On these grounds, the cultural theory of risk proposes an interesting analysis of the distinction between acceptable and unacceptable risks. Acceptable risks are those that do not pose a cultural threat. In such cases, comparison of costs and benefits for individual actors is still an option for guiding individual decisions. Unacceptable risks, in contrast, threaten the culture by undermining vital cultural presuppositions – unacceptable for the socio-cultural fabric as a whole. Their assessment does not allow individual actors to advance arguments based on assessments of costs and benefits, because they undermine the sense of ontological security provided by community culture.

One of the problems of putting cultural theory into practice may be that it is not easy to appreciate the meaning of conceptual upheaval. Are risks considered unacceptable if they jeopardize existing power relations? Are they unacceptable if they interfere with the natural cycle of social change? Moreover, it is unclear how precautionary action might be taken. Application of Douglas's observations of the Old Testament Israelites or twentieth century tribal cultures to the management of risks in modern technological cultures would require that fundamental cultural presuppositions of these cultures be unambiguously understood. It is by no means clear how one might go about doing this. Nor is it obvious that cultures lack the resilience to weather changes to their conceptual foundations.

5.5.2 Grid / Group Analysis

As we have noted repeatedly, one of the great limitations of RAP is its inability to explain how moral choices are made. As an attempt to provide an answer to this question, cultural theory is especially interesting, because it has often been applied to the field of risk.

At the root of cultural theory lies a typology of social structure that Douglas calls *grid/group analysis. Grid* is a variable that describes the nature of social interaction. High grid refers to interactions in which roles are greatly constrained and discriminated. Low grid refers to a condition where access to social roles is open to all. *Group* is a variable that characterizes how tightly individuals are incorporated into social units. High group would indicate a condition in which solidarity is strong and people have rich interdependencies. Low group characterizes a condition marked by individual autonomy and inter-individual competitiveness. Cultural theory starts with the assertion that these two variables are orthogonal, resulting in a two-by-two matrix that represents the complete realm of possible social relations. Each cell in this matrix is the home of a particular type of cultural bias that describes a social relationship. Cultural bias is not an organismic variable in the sense that age, gender, or race are. An individual may be of one bias in one situation, another bias in another situation. It is important to keep in mind that cultural biases describe features of individuals in social contexts.

According to cultural theory, the cell characterized by low group and low grid is the home of entrepreneurs, whose form of social interaction is typified as the market. Entrepreneurs have a low sense of solidarity (low group) and a low sense of social role discrimination (low grid). They are primarily oriented toward a competitive environment.

Directly opposed to entrepreneurs are bureaucrats (or hierarchists), whose preferred context of social interaction is in large, highly stratified organizations, such as the political administrative apparatus. Bureaucrats are highly sensitive to the discriminatory allocation of roles and social functions, and they share a high sense of solidarity with other members of their organizations. The archetypical example of the bureaucrat is the civil servant.

The distinction between entrepreneurs and bureaucrats corresponds to the distinction between the political and economic spheres, to which we return in Section 5.6, where social systems theory is discussed. It should be clear, however, that the cultural theory of risk does not treat culture as one of several social systems such as the economic or the political. Culture is a fundamental dimension of social reality, which is why it makes sense to talk about corporate culture, political culture, etc.[59]

These orientations lead to two very different risk cultures. For bureaucrats, what is really threatening are actions that undermine the orderly

[59] This leads to a different understanding of socio-cultural reality than the one used in many discussions of generalized media of exchange (see Section 5.4). While one may still see money and power as generalized media of the economic and political system, other domains require a different analysis (see Jaeger 1994).

rule of law or the successful operation of political administration. Health risks for individuals and commercial risks for firms are acceptable to the extent that they can be managed within systematic, formal procedures. Entrepreneurs, on the other hand, are threatened by disturbances to competitive action. Many risks are welcome, as they offer opportunities for such action. If a technology, such as electric power distribution lines, is found to present a serious health risk, entrepreneurs will see opportunity in restructuring the national electricity generation and distribution system. Entrepreneurs are likely to worry more about their financial investments than the health risks of new products. Bureaucrats develop regulations to limit risks of complex technologies and harmful activities. Entrepreneurs tend to see regulations as stifling competition and as making problems worse. They are more inclined to try to reduce risks of industrial and environmental accidents by technical and organizational innovations. The really unacceptable risks for entrepreneurs are institutional dynamics that threaten the regime of free enterprise.

Two additional orientations are posited by cultural theory. In the low grid, high group cell lies the home of the egalitarians whose social organization is typified by collectives. As with the bureaucrats, but unlike the entrepreneurs, there is a strong sense of group ownership or solidarity among egalitarians. But unlike the bureaucrats and like the entrepreneurs, egalitarians do not believe in highly discriminating social roles. The archetypical example of this bias used by cultural theorists in the risk field are environmentalists. In other areas of cultural theory the archetypical example is the religious cult or sect. Egalitarians treat risks as unacceptable when they are not evenly distributed, either within a generation or between generations.

This latter point is of obvious relevance to environmental risks and has fueled the debate about sustainable development. At least with regard to environmental and technological risks, the egalitarian risk culture seems to be much more risk averse than that of entrepreneurs or bureaucrats.

There is one last cultural bias to discuss in the grid/group taxonomy, but we pause to note a clarification. In their 1982 book, *Risk and Culture*, Mary Douglas and Aaron Wildavsky discussed mainly the three cultural types introduced above, but in a very simplified manner (Douglas and Wildavsky 1982). They grouped entrepreneurs and bureaucrats together and called them the "center," while egalitarians were called the "periphery". In their analysis, the dichotomy between politics and markets was depicted as the central feature of modern societies, one that they justified as legitimate and proper. They suggested that the natural antagonism between these two groups was healthy to modern culture. Environmentalists, in contrast, were not only on the periphery, but deserved to stay there.

The cultural bias labeled as egalitarian makes a strong distinction between ingroup and outgroup while objecting to hierarchies. Douglas and Wildavsky likened egalitarians to the social phenomenon of sects. In the context of many risk debates, the egalitarian cultural bias was exemplified by environmental groups, and so the environmental movement was described as sectarian. As with the dietary prescriptions of the Old Testament, the fear of environmental risks was explained by cultural traits of the fearful. Accordingly, the harm to human health and to the integrity of ecosystems which the environmental movement fears was said to be an irrational exaggeration. Their conflation of cults and environmental groups led Douglas and Wildavsky to make extremely prejudicial judgments about the irrationality of environmentalist claims, judgments that have harmed cultural theory many times over.

Other cultural theorists have lamented this exaggeration, which they felt revealed far more about the biases of Douglas and Wildavsky, than about the normative bias of cultural theory (Rayner 1992). The critical description of sectarian traits of the environmental movement is interesting in its own right, although it clearly describes only some aspects of the cultural orientation within this social movement. It is fair to ask whether similar critiques might be addressed at problematic aspects of risk cultures sustained by entrepreneurs and bureaucrats.

After this digression, we return to the grid-group scheme. The final cultural bias, with low group and high grid, are the fatalists – atomized individuals who have failed to thrive in either the market or large organizations and have not found an egalitarian community to their liking. They have not withdrawn from society, however. (Withdrawal constitutes a fifth bias, suggested by Michael Thompson (1980), "the hermits.") Occupying social positions that are highly constrained, but lacking a strong sense of solidarity with others, these people are described as apathetic, atomized individuals who play only a minor role in the cultural theory of risk.

5.5.3 Cultural Biases Clashing: Interference of Risk Cultures

One of the most promising contributions of the cultural theory of risk is that it acknowledges, accepts, and offers explanations for the clash of cultural orientations within modern society. While RAP's claim is that it offers the only correct rationality for understanding the world, cultural theory embraces diversity. It recognizes that the four cultural biases embrace different forms of rationality, and that making social choices about risks means finding ways to build common ground.

Brian Wynne (1992) and Ray Kemp (1985), two scholars coming from research traditions other than cultural theory, have recognized the potential for cultural theory to explain how understandings of the world are mediated through value and belief systems and forms of social organization. Wynne has argued that the validity of knowledge is conditional upon the model of rationality (1982a, 1982b). Thus, knowledge claims advanced by one cultural bias (or rationality) – say, fatalists – cannot be validated in the context of another bias – entrepreneurs, for example. This is highly relevant to the making of societal choices about risk, for it challenges RAP's claim that a single optimization routine can produce the favored choice under all circumstances. Optimization is a property of a very specific kind of rationality, not a universalistic principle endorsed by all.

Wynne's observation has an even more significant consequence for the practice of risk communication and conflict resolution. If knowledge creation is handled differently by each cultural bias, then the only way to create shared understandings about risk and agreement for action is to produce meanings that lie outside of the territory of individual cultural biases. Effective social policy depends on successful creation of shared meaning among – not within – cultural groups (Wynne 1992). Wynne makes two important points. First, he argues that shared social values as well as shared meanings are created through this kind of effective social interaction. Social values do not exist *a priori*, nor are they merely the intersection of individual values. Instead, they are created through social interaction. Second, he asserts that the only effective way to achieve this kind of interaction is through open dialogue. In this analysis, Wynne is recognizing, as have others before him (Plough and Krimsky 1987), that members of the public who are not experts in risk are valuable sources of information and social consensus. Local lay knowledge needs to inform risk management decisions, just as the results of technical analyses must. This is also the basic message of the newest comprehensive report on risk from the National Research Council (Stern and Fineberg 1996) and of recent research on participatory integrated assessment (Jaeger et al in press).

These normative conclusions stemming from cultural theory overlap significantly with those of critical theory (discussed in Section 6.5 below). They challenge RAP on the grounds that it ignores the making of moral choices in society. Cultural theory offers an alternative approach, one that originated in anthropological field research. While empirical studies have not corroborated all of the central assertions of cultural theory, cultural theory has proven to be an interesting scheme for explaining how people and groups make choices about risks.

An empirical example of such explanation is provided by Rayner's (1986) field study of American medical staff working in hospital settings involving low levels of radiological hazards. The degree to which these hazards were viewed as innocuous, on the one hand, or serious, on the other hand, was hypothesized to be a function of one's position in the typology of cultural biases.[60] That is, one's placement in the grid-group matrix determines the amplification or attenuation of risk seriousness. Findings generally supported the hypothesis, with high-status medical professionals and bureaucratic personnel de-emphasizing the dangers, and low-status workers and alternative medical practitioners emphasizing the dangers.

These findings hold key implications for RAP and for the design of institutional safety policies. RAP removes social actors from context, treating them as detached calculators wishing to achieve their goals via rational choice procedures. Implicit in RAP is the idea that social or institutional position is irrelevant to the preferences and expectations that drive these procedures. Rayner's results indicate otherwise. Safety policies are often implemented on an organization-wide basis, implying the expectation of uniform compliance and uniform impacts on organizational members. Again, the results suggest otherwise. The impact and response to safety procedures is contingent on a member's duties and position in the organization.

5.6 The Social Systems View of Risks

5.6.1 Risks and Double Contingency

Probably the most elaborate social theory offering an alternative to RAP is social systems theory. And in an important way, problems of risk and uncertainty lie at the very roots of this theory. Luhmann (1984) shows this nicely in his discussion of Parsons's concept of double contingency. Two actors, ego and alter, are purposive with goals as presupposed by RAP. But they are engaged in an interactive situation. The consequences of ego's actions depend on alter's action, and vice versa. As long as ego has no prior knowledge of alter's action, she faces a fundamental uncertainty that makes it hard to behave as a rational actor. Parsons argued that the problem of coordinating mutual expectations is solved in practice by the existence of social norms – that is, by informal rules that permit smooth interaction. Note that this setting is a general version of the

[60] Dietz and Rycroft (1987) studied Washington, DC risk experts, the "danger establishment", to determine the methodologies and types of risk analyses undertaken by the various institutions making up the establishment: industry, government, consulting firms, public interest organizations, and academia. They found that the organizations adopted styles of analysis consistent with the predominance of expertise within the organization, promoted that style of analysis as the most legitimate for developing risk policy, and argued for the delegitimation of opposing styles.

coordination problem discussed in Section 4.1 on game theory. As was stressed there, recent research on multiple equilibria in microeconomics has given new relevance to the sociological problem of double contingency.

Luhmann essentially follows Parsons's functional approach, but without sharing his belief in a platonic set of basic values governing the functioning of social systems. Instead, he is inspired by recent work on self-organization in non-linear systems. Accordingly, he treats social norms as contingent and transitory results of dynamic processes. In Luhmann's view, social systems constitute themselves by establishing norms that define boundaries between themselves and their environments. Neither these environments nor social systems are static, however. Social systems always face the danger of dissolution into their environments. The uncertainty associated with this danger can only be temporarily suspended, never eliminated from social reality.

Danger, in this view, is a possibility that implies a threat for some actors. In his work on risk, Luhmann then introduced an interesting distinction by defining risk as a danger that is chosen by an actor. This definition is important because in addition to the aspects of possibility and valuation, it also stresses the dimension of responsibility. Greater emphasis on responsibilities for risks may well be a hallmark of sociological approaches to risk and uncertainty (see Heimer and Staffen, 1998).

Risk in social systems theoretical language is not the opposite of safety (which is basically a social judgment of acceptable risk), but of danger (threat from the environment). Risk internalizes the possibility of experiencing an outcome other than the one hoped for. In contrast to RAP, in which risk results from the objective possibility that an undesired event occurs as a result of a chosen action, risk in social systems theory describes an attribution process (Luhmann 1993). One of the main tasks of social systems is to offer meaning to individuals even if the outcome of one's chosen action is different from the expected or the desired.

Risk offers a framework that distinguishes the unexpected from the undesired. The combination of unexpected and undesired outcomes is a potential threat to any system. Surprises are necessary for innovations, but they should be limited in time. If they appear too frequently, efforts are made to be more efficient in preventing surprises; but this may conflict with other goals such as achieving more successes or reducing threats (dangers) from the environment. An alternative strategy is to anticipate the undesired by expecting it. This does not make the event less dreadful or undesirable, but apparently creates meaning by showing that the undesirable effect has been taken into account as a potential cost in the interest of gaining some benefit. The knowledge may not lead to any physical change in terms of damage or dysfunction, but it justifies including options that may result in undesirable events without challenging the rationale of the choice itself. Taking a risky option can then be interpreted as a sign of rationality and prudence because it proves that a decision has been made on the basis of including uncertainties and the best available knowledge.

The essence of the argument is this: framing a decision as risky creates new contingencies. It provides a new legitimate set of decision options, but may exclude others. Functionally it offers post-fact justifications for decisions that were unsuccessful. Individual actions can be motivated by assigning meaning to a potential threat through reinterpretation as a potential opportunity. If outcomes are different from those desired, the rules of the game need not change. It is still more reasonable to repeat the same action and make the same choice when a similar decision situation occurs. Any system that manages to install risk as a meaningful concept in its members both creates and destroys contingencies by offering more decision options. At the same time, choice behavior is constrained by imposing decision rules even in the presence of undesired events. Risk offers immunization for systems that deal with high uncertainties in their decision structures and provides incentives for consistency in actions even when outcomes appear to be unfavorable.

This remarkable double function of risk – as a means to enrich the set of legitimate options and to reinforce predictability of option selection independent of success – comes with a high price. Undesired consequences are not restricted to the members of originating systems. For other systems these consequences may be surprises (dangers), because they are undesired and unexpected. Since surprises must be carefully managed and because they constitute pervasive threats, systems are extremely worried if neighboring systems show too strong a tendency towards high-risk decisions. In many cases, the same knowledge about potential risks and chances is also available in the neighboring systems; however, this knowledge is not always interpreted in the same way. What seems to be a minor probability for an undesirable event in one system may be evaluated as a major threat by another. What one system interprets as a fair chance is regarded as cynical play with human lives by another. Furthermore, risks and benefits are not equally distributed among systems (or within systems). Even if a neighboring system buys into the rationale of risk-opportunity balancing, it may share the risks but not the potential benefits.

The more harm is anticipated in an effort to provide meaning and contingencies in one system, the more other systems will be disturbed in their effort to provide meaning to their members. Internalizing dangers as risks and perceiving chances and risks as twins is not a fact of nature, but a social attribution. For any system that does not share this view, the inclusion of high-risk options in the legitimate pool of decision options seems insane and contrary to their central beliefs. At the same time, however, it needs to explain and justify – like any other system – the fact that intended outcomes and real outcomes do not always match. One way to do this is to blame high-risk systems for all the undesired outcomes within one's own system.

As a result, risk produces four pathways for conflicts between systems: first, it may enlarge the potential for undesired surprises for neighboring systems, which they cannot afford; second, neighboring systems may be affected by negative outcomes, but may not share the belief that risks are worth taking for accomplishing one's goals; third, neighboring systems may buy into the rationale of risk, but be more affected by the risks than by potential benefits; and fourth, neighboring systems may blame their own failures (occurrence of undesired consequences) on the high-risk policies of neighboring systems.

The dilemma of the risk debate is that the potential for conflict is high, but the chances for coordination are very small. In particular, when the fourth conflict (blaming other systems for self-inflicted failures) occurs, communication may not be pursued by either system. The high-risk system sees little benefit in communicating with another system that appears to link unrelated events to the high risk system's behavior. The low risk system has little interest in communicating, since the external threat provides them with a comfortable explanation for the observed discrepancies between promise and reality. It should be noted that it does not matter whether the representatives of the low risk system are aware of this mechanism, whether they use it deliberately as a strategy of self-immunization, or even whether they believe strongly in the alleged connection.

Modern sociological systems theory challenges RAP in three major ways. First, it asserts that complexity and social indeterminism render it impossible for any rational actor to predict the consequences of his or her choices. Second, social systems within a society construct their own reference system for explaining and changing reality without much understanding of competing systems with which they interact and share common resources. This self-referential bias leads to a selective perception of outcomes as a function of one's own organizational or cultural filters. Third, means-ends relationships are dependent on the respective reference systems. Thus, rational actors may come to different conclusions – even when sharing the same goals and the same knowledge. Reconciliation of conflicting courses of action among a plurality of rationalities requires a meta-rationality that cannot come from rational actor theories for at least two reasons. First, RAP posits the isolated, calculating individual, divorced from any social system. Second, from a systems perspective RAP is only an element of a larger and more comprehensive framework. In order to assess these claims, it is necessary to review briefly the fundamental elements of social systems theory.

5.6.2 Social Facts as Irreducible Reality

Contemporary social systems theory has its roots in the Durkheimian tradition of sociology. As the most important of the founding fathers of sociology in shaping the discipline as a theoretical alternative to RAP, Frenchman Emile Durkheim's writings (1893/1984; 1895/1982; 1915/1965) in the late 19th and early 20th centuries were crucially instrumental in establishing sociology as a separate discipline. They remain in a privileged position at the core of sociology.[61] Durkheim's work is, therefore, exemplary of the entire range of distinguishing features of the discipline, and of the crucial role of those features in establishing boundaries between sociology and the other social sciences. Because of his pronounced critical stance toward RAP, his work is, likewise, useful for highlighting the fundamental differences between the classic roots of sociology and RAP.[62]

The founding of sociology is usually traced to three principal figures, sometimes referred to as the "secular trinity": Karl Marx, Max Weber, and Emile Durkheim.[63] These writers, each within his own compass and from separate orientations, incorporated a fundamental critique of RAP in the process of developing their theoretical positions. The collective critique of the three has, until modern times, shaped sociology's uneasy and critical stance toward RAP to a degree that cannot be overemphasized. For much of its history, sociology criticized RAP for having put the wrong theoretical template to the right question.

All theories of rational action presuppose humans to be rational creatures. Few statements could be more obvious and, perhaps as a consequence, appear to be more trivial. Yet disguised beneath the obvious and the trivial is a troubling question: when is a creature human? Is a creature human because it is rational, or is it rational because it is human? The question becomes especially troubling in the context of evolution.

---

[61] Indeed, Collins claims that "Durkheim is the archetypal sociologist because institutionally he had to be the most conscious of what would make sociology a distinctive science in its own right" (1994, p 46) and begins his discussion of Durkheim's theories with the forceful words: "We now come to *the core* of the discipline. I am labeling it after the name of Emile Durkheim, its most famous representative. It is sociology's most original and unusual set of ideas" (1994, p 181, emphasis added).

[62] The details of Durkheim's theories, discussed here, and of Parsons's theories, discussed below, are beyond our compass and, therefore, less important than the essential orientation of both.

[63] Because the neologism "sociology" was coined by Frenchman Auguste Comte in 1839, it is often Comte who is credited with the founding of the discipline. Furthermore, along with other thinkers, such as another Frenchman, Claude Henri Saint-Simon, Comte was instrumental in separating social thought from social philosophy, thereby laying the groundwork for a scientific approach to society. Despite these early beginnings it was Marx, Weber and Durkheim who developed the foundations of sociology – such as its fundamental orientations – and who institutionalized the new discipline.

If humanness is defined as the point at which the ancestors of mankind acquired rationality, what is the defining feature of rationality that permits that evolutionary point of demarcation? Is it upright posture? Is it cranial capacity? Is it tool use? Is it something else? If, instead, humanness preceded rationality, how can humanness be defined independent of the cognitive architecture of rationality?

In either instance – rationality as antecedent or as consequence – the argument leaves unaddressed the following issue: what was the basis of social interaction, of social organization, of – in short – society before the conjoining of humanness and rationality? If proto-human species were human, but not yet rational, how, in the absence of rationality, was social life coordinated? Virtually no one would argue that organized social life was not evident for these species. And few would argue that interaction and coordination among these "approaching-humanness" species were due to instinct alone, or to some other hard-wired biological feature.

If rationality preceded humanness, then the proper scope of theories of rational action is not with humans alone but with all creatures capable of rational decision making.[64] But, on what principle would the criterion of rationality be applied to demarcate rational from non-rational creatures? If one reasons backward from "rational outcomes" – to be defined as rational are creatures who, for example, maximize their access to resources – tautology is almost unavoidable. Even, on the other hand, if such a principle could be stated nontautologically, it would still be burdened with an infinite regress: since species evolve, what would be the basis of social organi-zation for the proto-species just short of the demarcation point of rationality? For example, if the principle makes *Homo erectus* rational, what about *Homo habilis*? Or, if *Australopithecus* is rational, by our definition, what about *Australopithecus afarensis*?[65] The essential point is that regardless of where along the continuum of species one places that demarcation point of rationality, the question defies resolution.

These considerations, unwittingly or not, are foundational to thinking at the core of sociology in Durkheim's work.[66] Confronted with the need to develop simultaneously a theoretical account of aggregate social

[64] Modern rational action theorists (Herrnstein and Mazur 1987) do, indeed, argue that models of rational action can be applied equally to not only humans, but also to a wide variety of other species of animals. However, these theorists *do not* posit that these other animals are rational, but rather that since the model predicts animal behavior, we can think of that behavior "as if" it were operating on principles of rational action.

[65] *Homo erectus*, *Homo habilis*, *Australopithecus*, etc. are various categories of proto-human species in the line of evolution leading to humans (*Homo sapiens sapiens*).

[66] Although Durkheim never specifically elaborated the point as developed here, his undoubted familiarity with the writings of Darwin and his commitment to an evolutionary theory of macro social structure suggest his understanding of the same basic logic.

structure and action, on the one hand, and to demarcate the meta-theoretical foundation of the new discipline of sociology, on the other hand, Durkheim began by rejecting the two dominant intellectual currents of his day: utilitarianism and Freudian psychology. Utilitarianism was the forerunner of modern day versions of rational action theories, emphasizing the capacity for conscious, thoughtful, strategic decision making. Freudianism looked below the surface of superficial human rationality and claimed that the substructure of motivation leading to choice behavior was irrational in origin and content.

Durkheim – like Marx and Weber – objected to the utilitarian view of the world because it over-emphasized the rationality of human motivation and action, excluding emotive and social forces, i.e., consciousness (see Section 2.4). Humans, for Durkheim, were more than simply calculating machines. Growing evidence from traditional societies, evidence highly influential to Durkheim, clearly demonstrated that those societies functioned smoothly without overt rational calculation on the part of their members and without rational institutions, such as markets and legal contracts, for the coordination of individual actions. Despite sympathy for its emphasis on irrationality – the unconscious, the unfathomable, confining the irrational to individuals, not to the larger social milieu – he also rejected the Freudian view. Freudianism failed to infuse the social order with non-rational principles. Under such circumstances, any attempt to explain that order would ineluctably be forced into fanciful speculations – such as that of Freud himself with his argument in *Totem and Taboo* (1917/1946) and in subsequent writings that the origins of society could be traced to primal patricide.

While on one dimension (the rational-irrational) the rejected perspectives (utilitarianism and Freudianism) were presuppositional opposites, on another dimension (theoretical method) they were cut of the same cloth. Each was "individualistic", sharing a common perspective with a micro orientation: the individual was primary, society emergent – still another source of Durkheim's objections (1895/1982). For the utilitarians the individual was prior to society. In Durkheim's view, this had the whole matter backwards. A theory of society built upon individual actions was unavoidably and ultimately doomed to failure because it could not capture the essential verity that society is a reality *sui generis* – uniquely of its own type – irreducible to individual thoughts and actions.[67]

[67] Durkheim was keenly aware of and devoted to the task of distinguishing the sociological from the psychological level of analysis. This drove him toward a deep criticism of Comtian positivism. He was especially critical of Herbert Spencer – one of the leading sociologists of the day with a wide following in Great Britain and the United States – and his American follower, William Graham Sumner, because of their utilitarian thinking with its acceptance of the *laissez-faire* doctrine that modern society emerged from the competition of individuals in the market; that is, from economic exchange.

An individualistic approach basically denies the existence of "social facts", a category distinctly macro in character and content. And for Durkheim "the first and most fundamental rule is: *Consider social facts as things*" (1895/1982, p 14). We could understand social structure and social order, Durkheim argued, by looking at macro-patterns of "social facts," not by trying to aggregate the actions and behaviors of individuals.

Thus, we would need to think big. And we would need to recognize that the proper causal order starts from institutions that exist independently of individual psychological reality. Understanding of society, structure, social order was to be found, not in the actions of individuals, but in the institutional continuity of society as a whole. The seemingly rational actions of individuals were merely derivative from permissible patterns emanating from social structure. To properly understand the social world required studying it with the wide angle lens of a macro approach.[68]

Armed with this macrosociological lens, Durkheim trained it with theoretical rigor and empirical analysis of the question: "How is society possible?" Or, in another version: "What holds society together?" [69] An answer to these essential questions was already available, in the long line of thinking in the development of RAP from Hobbes to Locke, Hume, Rousseau and Kant: society existed because rational humans had entered into an original social contract.[70] Humans premeditatively agreed to exchange goods and services, to cooperate, and to coordinate the myriad of individual actions. Durkheim objected to this immaculate conception view of rationality. If humans were driven solely by self-interest, as utilitarian and classical economic thinking proposed, how could it be assured that any agreement reached among parties to a transaction would be honored? Why wouldn't self-interested motivation lead to calculated breaches of the agreement, to strategic cheating, to a variety of strategies for advantaging

---

[68] The affinity of Durkheim's metatheoretical and substantive foundations to German historicism and romanticism, especially Hegel's, can be found in *Rules of the Sociological Method* (1895/1982) in such passages as: The first duty of the state "is to persevere in calling the individual to a moral way of life" and "At the present day, the State is the highest form of organized society that exists" (1895/1982, p 69).

[69] Especially in *The Division of Labor in Society* (1893/1984) and later in *The Elementary Forms of the Religious Life* (1915/1965).

[70] In *The Elementary Forms of the Religious Life* (1915/1965) Durkheim reordered Kant's view that all perception and sensation and, therefore, knowledge is filtered through pre-existing structures in the mind. According to Kant the mind contains an *a priori* structure that categorizes and organizes the experience which enters it. Durkheim observed that these categories, which are *a priori* from an individual point of view, varied by society and varied over time. So, from a social point of view they could not be *a priori*, but must be *a posteriori* – emanating from the structure of society.

one's position? How, given the initial premises of rational action, could it be expected that any contract between parties would be upheld?[71]

The solution to these puzzles, Durkheim argued, could be found by looking below the surface of manifest human actions, by looking past the overt transacting actions of "rational" actors. It could be found in an underlying reality: a non-rational and subconscious reality. Traditional societies provided an especially apt setting for probing this underlying reality, and anthropological evidence was rapidly accumulating at the time of Durkheim's writings. Revealed there, across a range of these societies, was a precontractual solidarity: transactions between individuals were as evident as in Western societies, but the idea of formal or legal contract was absent from the repertoire of cultural knowledge. Few attempted to cheat, to deny obligation, or to jeopardize the cultural fabric for the sake of short-term individual advantage.

What enabled these exchanges? What assured the fulfillment of duty? The answer for Durkheim was: nonrational, emotive forces, such as trust, social bonding, moral obligation, and moral solidarity. These emotive forces were expressed through social rituals in face-to-face contexts, producing a tangible social solidarity ostensible to the participants who were committed to maintaining this solidarity. People felt obliged and duty-bound to fulfill contracts and trusted their comrades to do the same. Holding these simple, homogeneous societies together, said Durkheim, was "mechanical solidarity".

Reasoning in this way, Durkheim not only identified the importance of consciousness and emotive forces in social life, but also revealed and cemented into place a more fundamental and much larger point: instrumental rationality itself can only exist on the basis of a deeper foundation. For such rationality to emerge and to function smoothly in *any* society, not just traditional societies, there must first be precontractual solidarity – a moral template that subconsciously conditions the rights and duties associated with human interactions. Society is prior to the individual, held together by strong moral sentiments embedded in pre-rational attachments to religion, to the family, or to society itself.

[71] To argue that contracts would be upheld out of fear of sanctions by the state – via, for example, a military or a police – both begs the question and further illustrates Durkheim's point. It may be quite true that the various sanctioning apparatuses of the state are used to enforce contracts. But how did the state and its apparatus emerge in the first place? Utilitarians and other rational action theorists might argue that the state and its enforcement arms came about through social contract. But, now begging the question, what assurance was there that people would not cheat on this contract, too? The ineluctable implication for Durkheim is that rational action, such as contractual economic transactions, works only because it is preceded by non-rational institutions that make it work.

Accompanying the population growth and concomitant increase in the division of labor in modern societies was a fundamental change in the constitution of society, one characterized by a move toward greater heterogeneity and greater interdependence among social actors. This shift from homogeneous to heterogeneous social relations caused a change, not in the presence or absence of solidarity (since for Durkheim this was an unavoidable "given" of social life), but in its form: from the mechanical solidarity of simpler societies to what Durkheim called "organic solidarity" in more complex ones. With this argument Durkheim made a conscious, frontal assault on Enlightenment thinking – especially those versions of Enlightenment thought that criticized social institutions because they inhibited or constrained individual freedoms. From Durkheim's sociolistic vantage point it was the very presence of institutions that preserved the social order and guaranteed the rights of purposive action.

In modern societies moral solidarity may become submerged ever deeper below the surface of seemingly rational actions and, therefore, more difficult to discern. It is, nonetheless, present – and just as essential. "No matter how scientific, rational, businesslike and calculating we become, something like the mechanisms which produce moral solidarity in small tribal communities must nevertheless continue to exist. Beneath the surface, modern society still shows many traits of traditional societies" (Collins 1988, p 19).

Thus, in the final analysis, to the question: How is society possible?, the answer lies in the reality that society is based on a community of ideas, consciousness, and representations.[72] Commonly accepted and shared ideas are the glue of social order. Taking the reasoning a step further leads to the fundamental temperament of classic sociology: the idea of consciousness, summarized in the proposition that "social existence determines social consciousness" (Durkheim 1915/1965); that is, the most basic categories of thought, such as time, space, class, number, cause, force, etc. are derivative of the social conditions of human existence. The cognitive architecture of human thought, the individual's mental structure, is therefore determined by the structure of society. Indeed, even personality is derivative of that structure. As a consequence, the individual, for Durk-heim, was not a detached, autonomous, calculating creature, as posited by RAP, but little more than an abstraction submerged within group processes and in turn subordinated to society.[73]

---

[72] Durkheim's contemporary, the Russian geographer Prince Petr Kropotkin, extended the idea of cooperative solidarity beyond humans to include animals, as well, by pointing to the importance of mutual aid (1902/1914), not shared consciousness, as a universal glue holding social collectivities together.

[73] From the point of view of the school of symbolic interactionism, Durkheim's formulation is totally focused on "me's" to the full exclusion of the "I".

To whatever extent a rationally acting individual actor remained standing in sociology after the staggering blows dealt by Durkheim, that actor hit the canvas of sociological debate with the knockout punch delivered by Talcott Parsons. In his monumental work, *The Structure of Social Action* (1937) Parsons built heavily on Pareto,[74] and canonized Weber and especially Durkheim as the classic tradition in the field, explicitly displacing Marxism as a viable macrosociological and historical approach to social structure and organization.

Parsons constructed a grand theoretical system, a system whose foundation and structure were Durkheimian but whose contents were Weberian, especially with regard to Weber's idea of individualism. The result was a theoretical framework that provided for voluntaristic action by social actors, not the unfettered action of RAP, but action shaped by the reality of the physical and social structural environment.

Simultaneously, Parsons presented a sustained criticism of the foundations of neoclassical economics – the so-called marginalist school – especially the work of Alfred Marshall, and rejected not only utilitarian sociologists such as Comte, Spencer, and Sumner in the process, but also the foundations of RAP and the utilitarian individualism that guided their inquiry.[75] Parsons, in essence, had adopted from Durkheim the same essential foundational proposition that societies could not be held together on rational foundations entirely.

Parsons took Durkheim's notion of social order – the glue holding society together – as the foundation for a much broader and more deeply dug theoretical complex. Among the many developments was a more precise specification of the composition of that glue: social norms. In effect, Parsons translated Durkheim's often totally abstract concepts of solidarity and his baffling reification of society into conceptually more tractable

[74] Parsons was greatly influenced by Pareto, giving him extensive treatment in *The Structure of Social Action*. Indeed, Volume I of the work is subtitled: *Marshall, Pareto, Durkheim*. In part Parsons adopted the dualism of Pareto (1916/1935) between "logical" and "nonlogical" action and its parallel manifestation in the gulf between economics in the first instance and sociology in the second.

[75] Parsons's explicit concern for expunging the utilitarianism of Spencer from sociology is evident in the very first sentence of *The Structure of Social Action* (1937) where he begins with the question, quoted from one of America's most distinguished intellectual historians, Crane Brinton: "Who now reads Spencer? It is difficult for us to realize how great a stir he made in the world. [...] He was the intimate confidant of a strange and rather unsatisfactory God, whom he called the principle of Evolution. His God betrayed him. We have evolved beyond Spencer." Parsons builds upon Brinton's quote, saying: "We must agree with the verdict. Spencer is dead. But who killed him and how? This is the problem. [...] Spencer was, in the general outline of his views, a typical representative of the later stages of development of a system of thought about man and society which has played a very great part in the intellectual history of the English-speaking people, the positivistic-utilitarian tradition. What has happened to it? Why has it died? The thesis of this study will be that it is the victim of the vengeance of the jealous god, Evolution, in this case the evolution of scientific theory" (1937, p.3).

concepts, such as norms and roles, paving the way for operational measures of key elements of the Durkheimian scheme. The glue holding society together was not simply an abstract "moral solidarity," but, instead, a structure of normative prescriptions that guided social actors along a path of socially acceptable behavior.[76]

Merton (1968) and other proponents of structural-functional research in sociology showed that there was no need to get lost in the intricacies of Parsons' conceptual schemes; the general approach of social systems theory could be used in substantial social research about core problems of modern society. Risk was not yet perceived as one of these problems, but social systems theory laid the basis for an explicit alternative to RAP in the study of risk as well.

5.6.3 Uncertainty and Social Indeterminacy

Whereas RAP claims that the rationality of each individual actor creates the rationality of the larger social aggregates, modern social systems theory assumes the reverse. The rationality within larger systems produces the incentives and disincentives which establish and maintain individual rationality. This can include the deliberate "illusion" that one's individual action is driven by balancing expected positive and negative outcomes of a given set of decision choices. According to social systems theory, such an illusion serves the purpose of motivation.

In reality, or so this theory claims, the complexity of social actions and the multitude of potential interventions with its infinite number of possible outcomes prevent any reliable prediction of the connections between individual choices and observed outcomes. This problem of indeterminism prevails over the promises of game theory, the response of the rational actor proponents to the question of the effects of social interactions on goal-oriented deliberations. The world of ego and alter as proposed by game theory loses its predictive power if multiple actors are present and if each actor has different access to information, different expectations about what constitutes a successful strategy for obtaining his or her own goals, and different reference systems for evaluating outcomes and means. In such a social environment, prediction of outcomes is mere luck for the individual and constitutes a major societal effort involving vast resources and the creation of a specialized subsystem of sciences on a collective

[76] The central thrust of both Durkheim's and Parsons's writings was generalized by Janowitz (1978, p.57) as follows: "the central idea held by sociology's founders...that any frame of reference – be it economic or psychological – which stressed individualistic self-interest would not supply an adequate basis for analyzing social order and processes of social control."

basis – most often with elusive results. Furthermore, the impression that rational behavior is indeed possible or can even be called "rational" *vis-a-vis* a largely unpredictable world is a phenomenon that requires a sociological explanation rather than being taken for granted as a presupposition for sociological inquiry.

The phenomenon of social indeterminism for any given individual actor has been a subject of and a challenge for social theory from the beginning of sociology as a discipline. The social systems view tackled the problem by emphasizing the possibility of stable macro-systems composed by unstable micro-processes. Although individual actions were often executed in vain because actions by others interfered and prevented the original actor from obtaining his goal, the fact that actors continued to pursue strategies of rational action made the aggregate systems of social behavior more predictable than did the outcomes of individual actions.

Moreover, the sum of non-successful individual actions could still be interpreted as successful on an aggregate level. Through evolution, social institutions had developed in such a way that the pursuit of rational strategies by individuals was linked to supra-individual social goals (or functions). The "trick" of cultural evolution consisted in creating structures that forged a link between the *pursuit* of individual rationality and successful (or desired) outcomes of systemic rationality. This link would hold even when individual rational action did not succeed in its own terms.

Social systems theorists saw an inconsistency in RAP, because the observed futility of individual reasoning again and again coincided with the ongoing preservation of social functions. Systems theory argued that the socialization effect of motivating individuals to behave rationally (regardless of whether this was more successful than irrational behavior) resulted in functional rationality of the social system as a whole.

The structural-functional school extended the realm of individual actions to a whole portfolio of rational, pseudo-rational, impulsive, and habitual elements. Whether these elements made sense in terms of success (defined as obtaining a predefined goal) did not matter, as long as these elements provided sufficient meaning for the individuals within their framework of knowledge, norms, and values. As a result of this proposition, social systems need to convey meaning to individual actions.

This is true even for those cases where individual actions may lead to undesirable results as measured against the criteria of RAP. The institutions that provide meaning act themselves in the interest of larger societal functions. Key functions include allocation of scarce resources, the provision for social integration, individual and system reproduction, etc.

Social systems theory then allows each actor to act in obscurity and ignorance of the purpose of his or her action (particularly in organizational settings). How can social order obtain under such circumstances? It is because over time social systems evolved in which the sum of the (rational or non-rational) individual actions contributed to commonly accepted social goals. This explained habitual or non-goal oriented actions by individuals and provided a background for explaining the importance of socialization to social behavior. It was the process of socialization by institutions that linked individual action to socially conveyed meaning that, in turn, served higher functions of social systems.

5.6.4 The Challenge of Complexity

Modern systems theory – as applied to risk by Luhmann (1993) – is still based on this perspective. First, system rationality is not an automatic product of individual rationality (not even its pursuit). Second, providing meaning to actions is a broader framework for reducing choices for individual actions than simply goal-orientation (the meaning provided by RAP). In opposition to the structural-functional school, however, modern systems theory rejects the idea that a higher systems rationality provides functional meaning to individual action. Whether or not systems serve any social function is not clear. Systems provide structures of meaning to their members, meaning that resides in knowledge, norms, values, and worldviews. Social systems are the main engine that reduce complexity (by making individual behavior more predictable). At the same time, social systems create or preserve complexity (by offering complex organizational and functional differentiation through socialized roles, consistency in rule-oriented behavior and internalized expectations).

Orientation through meaning requires reinforcement with success. Otherwise, goals are permanently missed, causing ideology to fail over time. The two key elements of social systems are *meaning* and *contingency*. Social systems provide meaning (a sensible explanation for selecting one option over another) largely independent of expected outcomes. At a more subtle level they provide interpretations that can make nearly any outcome a success, by creating selection rules for present actions, which in turn create contingencies for future actions, which in turn create contingencies of the second order, and so on.

Structures of meaning created by social systems result in contingencies that support the declared functions of the respective systems over time. Because contingencies imply uncertainties but do not prescribe the actions of individuals, they make some types of actions more probable than others. Surprises are still possible; in accordance with all evolutionary theory, surprises are a necessary prerequisite for social change. Therefore, the main challenge for social systems is twofold. On the one hand, contingencies of social interaction must be reduced to a set of potential options for individual actions that meet the desired functions of the system. On the other hand, social systems must provide enough freedom of action for individuals that a sufficient number of innovative impulses are pursued so that systems remain adaptive to their changing environments. Adaptation to the requirements of the environment is dependent not only on generating innovative impulses within the system, but also on providing structures that can recognize changes in the environment. This implies first differ-entiation between system and environment (us and them) and the organization of communication between system and environment.

Environments for system vigilance can be other systems as well as the natural environment. Recent developments in systems theory claim that the growing complexity within systems and the constant creation of contingency chains (which differ from system to system) produce insurmountable barriers of communication between a system and its environment. Systems can no longer monitor all of their contingent environments. Furthermore, the self-organization of social systems is reinforced by the necessity for modern systems to provide a broad set of meanings without claim of universal application. (Religions or sects with universal claims are regarded as traditional relics or as counter-movements to the relativism of modern society.)

The fragmentation of formerly common cultural elements (such as language, values, basic knowledge) disables each individual member of each system from meaningful communication with members of another system beyond the exchange of services or products. At the same time, however, communication is essential because the consequences of the actions by the members of one system have repercussions on the availability of options for members of other systems. Even in those cases where systems have managed to provide a set of contingencies that meet all the requirements needed to fulfill the system's main functions, the actions of other systems interfere with the expected consequences and contingencies of the original system. Although contingencies provide potentialities for choices, thus offering some flexibility in coping with inferences from another system's actions, the growing inter-relatedness of modern systems in terms of influencing each other's outcome implies the need for greater inter-systems coordination.

Therefore, modern society is faced with a far-reaching crisis due to the convergence of two opposing trends. On one hand, modern systems tend to become more autonomous and self-referential in establishing their contingencies for actions. This necessarily interferes with the ability of any given system to transfer meaning from its own system to another. On the other hand, the actions of systems tend to become more interconnected and interdependent, necessitating increased effort for coordination and communication.

Systems theory abstracts from the level of knowledge and strategic reasoning of each individual actor, putting itself in the position of observing the observers. The system provides the structure that limits individual options but provides rules of consistency. Both the array of options and decision rules need to be justified, at least in part, independently from actual outcomes, since – from a system's viewpoint – anticipated success would not provide sufficient incentive for consistency and occasional deviance. The broader framework for coherence and integration is based on meaning and contingencies. Framing decisions as rational balancing of expected outcomes is one way of attributing meaning to individual decision makers. But it is not the only one, nor even the most effective one.

In this perspective, RAP provides one social strategy among others to explain two fundamental paradoxes of human existence. First, humans are intrinsically goal oriented, yet may have insufficient knowledge and foresight to predict the outcomes of their own actions. Second, individual rationality and the social good are not identical, yet need to be reconciled for making both operational.

Obviously, the approach of social systems theory is closely related to the role of risk and uncertainty in modern societies. The theory, however, provides little guidance for the management of specific risks, certainly less than the rich toolbox for risk management offered by RAP. But the idea that, with the dynamics of modern society, risk and uncertainty have come to play a new role in human life deserves careful consideration in its own right. Perhaps this is not just a matter of social theory, but of the worldview which has emerged through the historical process of modernization. This is the claim of scholars who relate problems of risk management to the dynamics of modernization – the focus of the next chapter.

6. MODERNITY AND BEYOND

6.1 Risk and Modernization

According to Beck's (1986/1992) influential work, modernization undermines ontological security in two very different ways. On the one hand, the use of scientific technologies by government and industry has led to large-scale technological and environmental risks. Although society is informed about them, mainly via the mass media, these risks usually cannot be understood without relying on scientific expertise. On the other hand, the manifold processes of individualization in modern society have eroded the reliability of the social networks in which human life is embedded. This is especially true of family relations, but also for neighborhoods and occupational networks – the social cement of ontological security.

In this view, modern society can be characterized as *risk society*, because living in modern society entails pervasive exposure to risk both in private life catastrophes and in the high drama of public disasters. In Beck's analysis, this new social fact has disrupted traditional social arrangements. All through industrialization and early modernization the main social conflict was between rich and poor. Once industrial production was well established, however, the distribution of wealth ceased to be the primary problem of society. Although doubtless still a very important issue, it no longer defines social conflict in general, because the distribution of risks is now orthogonal with class lines. Nuclear war, to take the strongest case, knows no social boundaries, threatening rich and poor alike.

Two main features of risk society are concerned with fairness, both involving displacement of risks: the first, displacement in time, the second, displacement in space. First, risks pose dangers that may be displaced in time to future generations. This means that people who initiate actions that produce time-delayed risks do not experience those risks, and they are not held responsible for their actions in this regard. Instead, unknown and uninvolved future generations must bear the consequences for the behavior of their predecessors. Aside from fairness, this clearly raises problems of uncertainty. How can society hold individuals or groups responsible for their actions when the future consequences of those actions are not foreseeable? Choices are not clearly delineated when at least the range of possible outcomes cannot be predicted and associated with some measure of likelihood. Second, risks increasingly transcend national and other political boundaries, making it very difficult to hold actors responsible for the outcomes of their action on distanced people.

Although modernization enabled a world of greater individual freedom of choice and the emergence of liberal democracy, it also imposed cultural constraints on individuals, based on scientism. In its dual roles, science is responsible, via new products and technologies, for the creation of risks, but also for societal dependency on science for knowledge and measurement of risks. At the societal level, therefore, it is often difficult to separate the undeniable achievements of rational behavior from increasingly irrational ways of dealing with uncertainty and risk.

In Beck's view, answers to the problems of risk society are to be found neither in traditionalism restored nor in a post-modern void. Rather, he advocates a reflexive modernization that would establish checks and balances within the process of modernization itself (Beck 1994). The ability to critically evaluate and creatively reshape the institutions of modern society is required if society is to continue to develop and evolve. This takes the form, not of some hypothetical speech situation as for Habermas (see Section 6.5), but of political and popular opposition to scientism, specifically by environmental movements and by popular opposition to noxious facilities and other risk sources.

The concept of reflexive modernization is especially important because of science's dual role as both a main cause of environmental risks and an indispensable tool for detecting and possibly overcoming these risks. One suggested mechanism then would be to implement institutions of adversarial research.[77]

Beck's view of the link between risk and modernity has been elaborated by Giddens (Beck, Giddens et al 1996; see also Giddens 1984). In this view social networks are not only pathways of interaction. They are also essential to maintaining a sense of ontological security without which human agency is impossible (see also Section 1.1). In order to assess the far-reaching implications of this assertion, it is necessary to consider a remarkable feature of the human brain. The brain includes parts that are similar to cerebral structures of animals in our biological lineage – parts that enable us to engage in all kinds of metabolic, behavioral and emotional processes. However, in the human organism these processes must be integrated with the help of the cerebral cortex. This integration is impossible for a human brain in isolation. We learn to walk, to talk, and to reason only if our brains become parts of networks with other human beings, with a shared culture.

Social networks provide patterns of routinized action, and, together with cultural rules, they enable human beings to handle the many uncertainties which are an essential feature of human existence. Ontological security is not an alternative to the awareness of uncertainty; quite the opposite. Because they are able to rely on many *certainties*, human agents can incorporate fairly smoothly the *uncertainties* of their existence.

[77] With regard to biotechnology, this idea has been advanced by Harlow (1986) with the proposal that separate research teams should be paid to improve our knowledge on the opportunities and the risks of specific forms of technological progress.

Uncertainty is, therefore, acceptable as a fundamental feature of the world in which we live. Although social networks can change, and even be transformed, there are limits beyond which the disruption of social networks may threaten a person's sense of ontological security.

It is instructive to compare Giddens' understanding of ontological security with Wittgenstein's (1979) reflections on certainty. Wittgenstein stresses that in a given situation there always are elements that cannot be put in doubt. If I talk with my colleague in the office, she may tell me that she is not sure whether some quotation is from Wittgenstein or from Dewey, and this may be a perfectly reasonable doubt. If she tells me that she is not sure whether we are on planet earth or on the moon, I may wonder whether she is making a joke. However, it would require very peculiar circumstances or a very peculiar argument to take this doubt seriously.

Wittgenstein used rivers as a metaphor to discuss the relation between certainty and uncertainty. Just as a river flows in its bed, so sentences that can be put into doubt are framed by sentences that cannot. Were it otherwise, we would always end up in an infinite regress. It may be uncertain whether a given surface is red, but there is no way to doubt that a red surface cannot be yellow at the same time. This relation yields a distinction between grammatical and empirical sentences (Fischer 1987). As a result, there is a kind of certainty which is marked by the impossibility of actually entertaining a doubt.

At a given moment in time, the river flows in a fixed bed. From time to time, however, the bed may be displaced by inundations, earthquakes, and other events. Nevertheless, the distinction between the river and its bed is maintained so long as the river exists. In a similar manner, human life is characterized by distinctions as to what is certain and what is not. An example of such a transition in human terms is the experience of religious conversion. Similar changes are involved in what, since Thomas Kuhn's (1962/1970) classic study, is described as a paradigm shift.

Kuhn's analysis of scientific revolutions describes how in the history of science grammatical sentences can become empirical, as well as the other way round. Wittgenstein, as noted above, hinted at this possibility by saying that the distinction between the river and its bed will be maintained even in situations where the riverbed shifts its location. The experience of modernity includes awareness of such transitions, not only in scientific discourse, but also in the discourses of everyday life. Repeated experiences of transforming grammatical sentences into empirical sentences, however, can undermine the ability to distinguish between them. This is what Giddens describes as a loss of ontological security. It is important to realize that grammatical sentences are not simply elaborations of empirical sentences. Grammatical sentences are, in Kantian terminology, conditions of the possibility of empirical description; they are presupposed by any use of empirical sentences.

What requires explanation, then, is not the existence of ontological security, but its loss. This has direct implications for the problem of trust, which is so relevant to contemporary risk debates. Human beings cannot survive the first years of existence without being embedded to a considerable degree in social relations that can be fully trusted. Trust is, so to speak, the default value with which human beings enter social life. How, then, does mistrust arise and how can trust can be restored once it has been lost?

As we have repeatedly emphasized in discussing the relevance of RAP to studies of economic behavior, it is instructive to ask whether Giddens's analysis of trust and uncertainty can be adapted to the economic realm. In this respect, financial markets are a highly interesting subject (Morishima 1992). While they are clearly amenable to RAP-based studies, the role of expectations in such markets also begs of sociological inquiry. In a study of German financial institutions, for example, Becker (1991) argues that the role of such institutions consists essentially in managing the risks which economic agents incur in order to perform their operations. He shows how these risks are defined not simply by analytical procedures, but by interactions in social networks knit together by trust relations. While Becker uses Luhmann's version of modernization theory as a conceptual framework, Ingham (1984) has used Giddens's approach to modernity in his remarkable study of the role of London as a financial center. Ingham shows how the dynamic reproduction of specific social networks has enabled the city of London to deal successfully with huge financial risks over several centuries. By now, this tradition has given the city major competitive advantages over other financial centers in Europe.[78]

Financial markets demonstrate that a sociologically based analysis of risk can be applied to economic phenomena. However, analyses like those of Becker and Ingham propose no alternative to RAP-based notions of supply and demand concerning crucial financial magnitudes like exchange rates and rates of interest. The use of sophisticated variants of modernization theory is an interesting complement to RAP studies of risk, but it is still a far cry from a true successor to such studies.

[78] It will be interesting to observe how these trust relations will evolve in interaction with European monetary integration.

6.2 Risk and Scientific Rationality

RAP is more than an innocent foundation for a conceptual approach that helps to make sense of risky choices. RAP makes some very serious epistemological claims, especially about how to deal with uncertainty. While these claims can be troubling in the context of the isolated individual actor, they often become unworkable when collective actors (such as decision making bodies) are involved.

RAP assumes that the future consequences of each action alternative can be assessed. On this basis, it states that the rational actor selects the alternative that leads to the greatest utility. In practice, both of these operations involve dealing with uncertainties. For individuals these uncertainties can be troublesome, but the ultimate decision rests with one's personal knowledge and subjective experience.

In the context of collective decision making, uncertainties present much deeper obstacles. Not only are the complexities greater, but the single authoritative actor who is capable of adjudicating between competing claims is now lacking. These conditions create fundamental issues of knowing and of societal operability. One thing a social system must do is to provide collective knowledge about potential consequences. By "collective knowledge" we mean knowledge that is sufficiently validated, according to the established customs in the society. In cases where uncertainties are high and such widespread agreement is absent, the social system must provide commonly accepted methods for coping with pressing situations. In Western, functionally differentiated, society these tasks are often performed by the subsystem of science.

Society requires an effective system for creating collective knowledge, in part to satisfy needs for ontological security and to enable closure of public debates. In modern society, this happens in two ways. First, there must be a fair and viable basis for expressing knowledge claims so that no interest position is systematically excluded from the policy process. Second, there must be a widely accepted protocol for resolving competing knowledge claims so that arguments can be tested and options narrowed. Such protocols are provided by scientific institutions.

6.2.1 Social Functions of Scientific Reasoning

The role of science in public debates about risk has been explored by sociologists of science. Helga Nowotny (1976) argues that, because the economic and political system demands reduced uncertainty, severe demands are placed on the scientific community to make aspects of economic and political actions that are uncertain less threatening. The scientific community obliges by assigning numbers (numerical probabilities) to potential adverse effects. In contrast to Beck and Perrow, Nowotny claims that the impetus for the quest for quantitative assessments of risks is based on the need for accountability and decision rationality, rather than a desire for low numbers (that is, numbers that will legitimize pre-selected actions). Uncertainty and scientific dissent leave decision makers vulnerable to criticism and attack. Numbers supply a rationale for decisions.

According to Nowotny, scientists respond to pressure to produce quantitative estimates of uncertainty for two reasons: First, quantification is consistent with the predisposition and training many scientists have toward reductionist thinking, problem solving, and quantitative mathematics and reasoning. Scientists are comfortable working in this domain, and in doing so, they believe they are fulfilling the expected role of scientists. This both enhances personal self-esteem and improves one's intra-community career options. Second, scientists legitimize their social role and their claims for more research funds by providing a valuable service to society.

Nowotny points out that the problem of attempting to wrench hard numbers from highly uncertain scenarios is complicated by the fact that the scientific community (or subsystem) itself is characterized by a high degree of specialization and abstraction. Both specialization and abstraction may lead to overconfidence in scientific assessments. Specialization may lead to the exclusion of relevant factors from analyses. Abstraction may lead to systematic underrating of idiosyncratic events. To strengthen her point, Nowotny emphasizes that most accidents in the past were caused by a unique combination of systematic faults and singular events. Perrow (1984), Freudenburg (1988), Clarke (1989), and others have made similar claims based on case study data. Singular events remain outside the realm of typical quantitative risk assessments because of their rarity. Failing to account for them leads to underestimation of risk.

In addition to these built-in biases, Nowotny raises concerns about the use of results of scientific risk assessments in the political arena. Here an additional bias enters the picture. All scientific results are valid only within a specified framework of assumptions and system limitations. These assumptions and limitations are rarely recognized and even more rarely communicated; and when recognized they are often not accepted by the receiver of the information. Many of the singular events mentioned above remain outside of the working assumptions or the system's capacity to consider and act. For example, aspects such as human error, natural disaster, or sabotage are frequently omitted from quantitative predictions of risk. The consequence is that risk estimates that are based narrowly on technical failure rates may understate the real risks of a new technology. Risk underestimation may then help to legitimate the new technology.

As noted above, the political system and the economic system reinforce risk assessment studies because they need a legitimate basis for making decisions among various options.

Risk studies fulfill two major functions: they transfer responsibility from the political system to science, and they can be used to appease potential opposition. In addition, the economic system relies on some quantitative estimate of the potential negative outcomes in order to assess whether an investment makes good financial sense. Reliance on such studies may be counterproductive, however, especially if actual hazard events occur with a frequency or magnitudes that exceed predictions. This is how the political system's use of science to legitimize technological choices can backfire and result in decreased credibility and trustworthiness of both scientific expertise and political leadership.

Nowotny's analysis has several implications for the theoretical foundations of RAP. Her major theme is not simply that risks may be over- or underestimated through scientific analysis, although she cites many examples in which this has happened. Her major point is that the scientific endeavor to quantify uncertainty leads to *the illusion of perfect knowledge* of the future. This creates a false sense of security and actually decreases the ability of societies and institutions to cope with surprises. Furthermore, it nurtures a way of thinking about the future that may turn out to be disastrous given increasing complexity and interconnectedness of human actions, technology, and the environment.

One might conclude from this analysis that the success of RAP is due to a considerable extent to its success as a self-fulfilling prophecy. The desire to have the future predetermined leads to corresponding actions by the economic and political systems which, in turn, shape the future in that predetermined direction. Weingart (1983) has followed up on this thought by linking scientific products and risk assessments to the needs of society to constrain the realm of potential options as a means to manage complexity.

Weingart's analysis focuses on the integrative role of scientific assessments in meeting latent social and political goals. His interests lie not in the epistemological question of whether scientific results reflect the truth about an objective world or constitute social constructions, but on the social functioning of scientific expertise.

Weingart argues that, by defining classes of behavior that are acceptable and unacceptable, science restricts potential options and provides orientations for actors in an extremely complex world. Acceptance occurs when claims can be justified with some minimal level of scientific argument. As science loses its privilege to speak the truth, however, it can better serve its latent *integrative* function. Because all interests in society are better able to use science in advocacy, science becomes a convenient tool for all to legitimize social claims. A one-truth one-science approach would necessarily be less adaptive to social needs and would almost certainly (as in the past) be rejected by those who feel threatened by what they perceive to be scientific bias. This pluralist approach to truth has the advantage that scientific argument becomes a basis for justifying and adjudicating claims. Ideally, all groups can use science without suffering systematic bias.

The use of scientific methods and arguments is not arbitrary, however. Although the internal rules of science may be too weak to establish or refute the validity of a plurality of scientific claims, Weingart argues that they are strong enough to *exclude* options and actions that contradict basic scientific principles. Broad acceptance of science means that it can serve a legitimizing purpose while remaining independent of social interests or opinions. This leads to the latent integrative role of science in the sense that options and classes of behavior are unacceptable if they cannot be justified by at least marginally acceptable scientific arguments.

Weingart's analysis is not centered upon the risk issue, but in his preface to Nowotny's (1979) book on nuclear energy (a risky technology) he observes that the anti-nuclear activists in the early 1970s were opposed to risk analysis because they felt it was an instrument of the nuclear establishment to legitimize its claim of relative safety. However, when new risk analyses were released – in particular the more critical studies by the Kemeny Commission (Lewis 1978; see also The Union of Concerned Scientists 1977; United States President's Commission on the Accident at Three Mile Island 1979) – these activists shifted their argument. Instead of categorically opposing the use of risk analyses, they criticized the assumptions and methods used for such analyses. Weingart reasons that this shift signified agreement with the scientific foundations and assumptions of risk analysis, even if not with the results of earlier studies. He argues further that acceptance of the scientific reasoning and argumentation was one of the main reasons for the success of the anti-nuclear movement, because they went on to beat their adversaries on their own ground – that is, the ground of technical risk analysis. As a result, the rationale for risk analysis was accepted by all camps.

Combining the analyses by Weingart and Nowotny we can conclude that the need for risk assessment is not a simple quest for objective knowledge, but a consequence of the need to reduce complexity and manage uncertainty. In an age of decreasing relevance of metaphysical explanations of the future, predictions based on personal experience or systematic knowledge gain in importance, as does the ability to eliminate unfounded expectations of future events.

Both science and the political system are the beneficiaries of this development. While earlier explanations of the social function of science emphasized its epistemological activities (discriminating false from true statements, thereby reducing uncertainty) and its products, newer analyses stress that it is the collective subscription to procedural rationality that enhances social and cultural integration. Because it is perceived to be fair to all stakeholder positions while effectively excluding evidence and action proposals that do not meet minimal criteria for plausibility, science is an effective problem solving tool in a democratic society.

This line of reasoning suggests that the dominance of RAP is due to its latent function of empowering a plurality of interests and values while effectively excluding some options and actions from further consideration. For this reason the discussion of the merits or pitfalls of the rational actor paradigm (or any other) cannot be confined to epistemological impacts and implications; it must address the paradigm's social and cultural functions as well.

Nowotny goes a step further, suggesting that through the mechanism of self-fulfilling prophecy the social and epistemological functions of RAP may actually coincide. Because people believe in the predictions, they make them become true. This feedback strongly reinforces RAP, since reality checks seem to confirm what all parties are more or less claiming.

6.2.2 The Use of Science in Policy Making

It is quite remarkable that nowadays all groups in the risk arena try to use risk analyses and that they rely to a large degree on scientific or quasi-scientific arguments. Sheilah Jasanoff's (1982) study of the role of scientific advisors in the risk arena is consistent with the main conclusions we have drawn from the work of Nowotny and Weingart. These conclusions also imply that decision making bodies have no veridical justification for selecting one interpretation over another. It is not sufficient that decision makers merely have the realm of potential options reduced based on scientific arguments. Reducing the number of options does not eliminate the need to select among them. Decision makers must select what needs to be implemented.

Social groups expect decision makers to base their judgment on scientific evidence – preferably evidence that has been obtained by independent investigators. But when evidence is contradictory or inconclusive, decision makers often lack a justifiable rule to select the best alternative. In this dilemma, exclusion rules are required. Exclusion rules are not for the selection of expertise, but for the *selection of experts*; selection by science courts, for example, or by meta-science rules concerning selection from among qualified science advisors. Neither the logic of legal systems nor the philosophy of knowledge provide generally acceptable rules for distinguishing good from bad science, however. Therefore, policy makers seek out scientists who are in agreement with their basic political views in order to legitimize their choices for formal advisory positions on the basis of criteria such as research experience or status within the community. Note, however, that such status has been acquired through the same networks that provide positive rewards for those who produce what the policy makers demand. This "incest" mentality provides constant gratification for those scientists who agree with the basic interests of policy makers and large stakeholder groups. It also provides legitimization for the policy makers to justify their choice of experts on the basis of status and other symbols. Jasanoff also observes a direct link between government intrusion into the guaranteed freedom of research in private (American) universities and the urgent need of governments to establish a network of prestigious institutions and individuals who are willing to serve as experts in decision making.

Dietz's and Rycroft's (1987) study on the role and function of "risk professionals" produced evidence that validates Jasanoff's claim of a devoted group of pro-governmental scientists. Furthermore, they showed that scientists have a self-image that is fundamentally opposed to Weingart's vision of plural science. Scientists in Dietz and Rycroft's study believed that science can provide definitive answers to questions of risk and environmental damage. Furthermore, scientists who disagreed with the "risk professionals" were treated as either ignorant or blinded by ideologies. This view is, of course, very attractive to policy makers because it provides them with a convenient argument that all other competing claims to truth are ill founded or ill motivated.

Adopting such a strategy, however, is certain to invite attacks from those who disagree with official policy makers. Some such outsiders may well have the means to force the government to listen to their claims, thus presenting a new dilemma for policy makers. On one hand they wish to rely solely on expertise to legitimate their policies. On the other hand, unacceptable political costs may occur when the primacy of a single scientific interpretation is challenged. Strategies for dismissing opposing scientists on the basis of their formal credentials and level of experience are not often effective. The public is less impressed by formal credentials and status than by the ability of experts to effectively express their point of view and is, therefore, likely to demand that all expertise be taken into account. Furthermore, as Dietz and Rycroft found, opposing experts may develop their own professional networks, and thus become more capable of initiating counter-offenses and of producing more public pressure on political elites.

Funtovicz and Ravetz (1992a, 1992b) provide a fruitful approach to the expert controversy question by developing a model of the production of scientific knowledge.[79] They emphasize two variables, systems uncertainty and decision stakes. Whereas systems uncertainty contains the elements of inexactness, uncertainty, and ignorance encountered in technical studies, decision stakes involve the costs or benefits of various policy options to all interested parties. They use this scheme to distinguish three kinds of science, each with its own style of risk assessment.

Low systems uncertainty and decision stakes describe applied science situations in which data bases are large and reliable, the opportunities and dangers associated with various decisions are limited, and the technical community largely agrees on appropriate methods of investigation. When both systems uncertainty and decision stakes are considerable, a different style of activity is expected – the mode of professional consultancy. This kind of activity involves the use of quantitative tools of risk analysis, supplemented explicitly by experienced qualitative judgment. Professional consultancy is highly relevant in the management of technical and environmental risks, as well as in the management of financial risks.

When decision stakes and systems uncertainty are high, Funtowicz and Ravetz advocate a scientific style they term *post-normal science*.[80] This kind of activity cannot be separated from qualitative judgments and value commitments. Inquiry, even into technical questions, takes the form largely of a dialogue, which may be in an advocacy or even an adversary mode. Although risk assessments that fall into this category are a limited proportion of the whole, they often are those of greatest political significance.

The role of science and the disorientation among traditional decision makers in the face of controversy has been the main field of investigation for Dorothy Nelkin (as in 1984). She orients her investigations from the observation that science has become a supermarket for rationalizing political decisions. The existence of a plurality of scientific expertise, however, jeopardizes the status of scientists as well as the credibility of politicians. Either of these developments can paralyze the political system and lead to a situation of indecisiveness and to the postponement of difficult decisions.

Nelkin argues that under these circumstances the political system turns to increased participation. It does so, not because it believes in the wisdom of the public, but because it realizes that it must gain public legitimacy to make any choice at all. And it must make its choice of actions transparent to the public. The sanctioned demand for accountability motivates the

[79] The following discussion of post-normal science is taken from Jaeger et al (1998).

[80] Earlier on, they used the label "total environmental assessment". This is clearly related to the emerging practice of Integrated Assessment (Jaeger 1998).

political system to involve interested and affected parties in decision making. But this also presents an opportunity for the decision making body to make its case. Part of doing this involves invalidating opposing evidence and expertise. However, since the parties advancing contradictory evidence and expertise are also participants in the process, political decision makers must proceed carefully. If they offend the participants, they risk attack and reprisals. Therefore, the political system adopts a strategy of trying to separate individuals and their interest positions from the discussion of the validity of evidence. Such a strategy, however, is rarely successful, says Nelkin. Participating groups are not so easily led astray. They enforce their own agenda and refuse to distance themselves from their own expertise. As a consequence, increased public participation need not resolve fundamental differences in the interpretation of evidence.

Nelkin's work, useful in highlighting the political implications of the pluralization of science, is important for other reasons, too. It points to the repercussions of the interaction between science and politics in debates about social theory. Few authors have asked how the role of the science system with respect to the risk issue affects social theory. Luhmann, Beck, and Giddens, covered in other sections of this and the preceding chapter, are the leading theorists to do so.

In view of the arguments in this section, several conclusions can be drawn: RAP provides a crucial service to society as it constitutes a framework that enables decision makers to reduce uncertainty (or, at least claim to) and to rationalize choices.

This service may not reflect a better understanding of the future than would intuition or prophecy, but it serves a need in modern society; the need to have a commonly accepted rationale for winnowing options or courses of actions while allowing for a plurality of outcomes, which can serve as arguments for justifying competing claims.

Through the feedback loop of self-fulfilling prophecy, social need and reinforcement through factual observation are linked. In practice this often makes the rational actor approach seem invulnerable. However, spectacular mispredictions or major accidents, even if they are theoretically covered by RAP-based approaches, such as probabilistic risk assessments, undermine public confidence in risk analyses and lead to growing alienation between public common sense and expertise used by decision makers.

The need for decision makers to select among courses of action often conflicts with claims of interest groups for a wider array of potential actions based on scientific reasoning. Choosing among equally legitimate courses of action becomes an almost insurmountable task since no meta-arguments exist for deciding the "best" choice. Either the choice must rely on political grounds, which may lead to legitimization problems, or on social selection criteria for expertise – or, more often, for particular experts. None of these options seems really satisfactory.

Scientists with a clear RAP orientation are found frequently among those who serve on governmental panels and advisory committees. By their algorithms for optimal choices among risky options, these scientists produce a modicum of rationality, a most valued commodity for most decision makers. In addition, RAP does not question goals or ends, tempting policy makers to incorporate their own ends into expert advice and selling the entire enterprise as an "objective" scientific result. Social studies of science provide interesting evidence on the role of science in general and of RAP in particular in contemporary risk management. They show that a more reflective ap-proach to scientific enterprise is required in order to understand how modern society copes with risk. At the same time, by treating RAP as a black box, they don't show how to develop a better understanding of decision making in modern societies than the one provided by RAP. But maybe the attempt to develop such an understanding is futile anyway because it is impossible to provide a general theory of human affairs.

6.3 Post-Modernism

Scientific rationality is a core component of modernity, and RAP attempts to provide rationality that is broader than, but consistent with, scientific rationality. In the preceding sections we have seen how the process of modernization leads to risks that undermine the naive belief in progress based on scientific rationality. This experience has led some authors to argue that the experience of modernity itself has come to an end – that we have entered the post-modern condition.

Lemert (1990, p 230) claims that post-modernism was "the single most popular intellectual fashion of the eighties" and deplores the fact that it was largely ignored in sociology. If post-modernism was nothing but a fashion such neglect would be quite justified. The post-modernist fashion, however, points to important processes of social change as well as to philosophical issues of considerable interest for the discussion of risk, and of RAP's role in understanding risk. Post-modernism does not challenge RAP on its own ground. Nor does it offer an alternative theory of human affairs that seeks to supersede RAP. Rather, it denies that such a thing as a general theory of human action is possible or necessary. This claim again elevates the anti-positivist idea of the German historical school – that there are no general principles governing human behavior – to prominence. Post-modernism belongs to no social science discipline (sociology, economics, etc.); it is, rather, a form of philosophical discourse operating at the level of worldviews.

While emanating principally, although not exclusively, from philosophical discourse, post-modernism holds many serious implications for the social sciences. Post-modernist writings challenge the hierarchical logic of definitions, where the meaning of one term is fixed by its relationship to other terms whose meaning is already well established. Indeed, post-modernist writings tend towards the opposite effect – seeking to undermine the familiar use of well established terms by relating them to more ambiguous ones. It would, therefore, be unduly naive to start a discussion of post-modernism with a formal definition of what is to be considered as post-modern, as would also be true of post-structuralism, deconstructionism, and other terms referring to this body of literature.[81]

It seems more appropriate to start with the writings of a series of French authors – Lyotard,[82] Baudrillard, Derrida, Deleuze, Foucault, Kristeva and others – who are credited with developing the discourse known as post-modernism. Most of them have credentials as philosophers, but many also dwell in literary criticism, psychoanalysis, history, and sociology. Meanwhile, similar writings have been produced in other countries. These authors observe that the belief in progress is the hallmark of modernity. The inclination towards a grand emancipatory narrative, a meta-narrative of progress constitutes the experience, or rather the illusion, of modernity. The loss of that illusion constitutes the condition of post-modernity.

Post-modernism does not replace the meta-narrative of progress with another meta-narrative, but with a plurality of partial narratives, an unending struggle of possible interpretations built on former interpretations. Progress consists only in dropping the illusion of progress. Lyotard described the very heart of post-modernism as "incredulity toward metanarratives" (Lyotard 1984, p xxiii), and urged, "Let us wage war on totality". Post-modern theories, be they philosophical, sociological, psychological, or aesthetic, can be characterized as rejecting monotheoretic approaches to totality and replacing them with explanations that build from a plurality of approaches. The rally cry of post-modernism seems to be: "Let a thousand meta-theoretical flowers bloom".

The condition of post-modernity was not meant solely to characterize the peculiar situation of a few French intellectuals. Rather, these intellectuals claimed that recognizing progress to be an illusion had

---

[81] Post-modernism is sometimes called post-structuralism because it was, in part, a reaction to the structuralism of Swiss linguist Ferdinand de Saussure and, later, to the structuralism of Claude Levi-Strauss. The term "post-modernism" is also sometimes treated in philosophy as a synonym for "deconstruction", "a term first proposed by Heidegger, but most often associated with Derrida, who was successful in gaining its acceptance in the philosophical lexicon" (Urmson and Ree 1991, p.256).

[82] Lyotard is credited with having coined the term (Urmson and Ree 1991, p.256).

become a wide-spread experience in contemporary society. They claimed to have identified a major process of social change, which they set out to accelerate. Kellner (1990) notes, "There is nothing like a unified, post-modern social theory". There is only a collection of various theories that share the belief that a transition from modernity is underway. In very rough terms, one might describe modernism as rational and rigid, and post-modernism as irrational and flexible.

In order to understand the post-modern critique of RAP, it is important to situate it within its specific historical context: "Post-structuralism is very much a product of the political and social events leading to and ensuing from May 1968 in Paris" (Lemert 1990, p.239). The protests, sometimes violent, by a coalition of students and workers were part of an enormous social effort to change capitalist society in a radical way. The fact that mass protest of students had merged with a militant workers' movement seemed to fulfill the revolutionary hopes entertained by Marxism, but it also defied the Marxist orthodoxy of the French communist party. This was a New Left, not orthodox French Marxism.

At the same time, these events were part of an international movement of protest against existing social and political conditions. There was sustained opposition to the Vietnam War in the U.S., a student movement in Germany, a workers' revolt in Italy, a fierce attempt to get rid of Soviet domination in Czechoslovakia. This broad movement was clear in its antipathy to capitalist society – including the state capitalism of planned economies – but had no coherent vision of an alternative to capitalism, though it certainly was searching for such an alternative. In intellectual terms, it was a revival of the Marxist idea that capitalism can and needs to be superseded by a less alienating kind of system while rejecting the Marxist practice of an economy planned by government authorities.

In France, as elsewhere, the movement portended far-reaching social changes. But it lost its momentum before any plausible alternative to contemporary capitalism appeared. In French intellectual circles, where Marxism had been very influential during the whole post-war period, a process of intensive intellectual self-reflection began. This effort might have grown into an extension of the Enlightenment, where the idea of overcoming absolutist regimes was gradually elaborated to the point of producing the conditions for far-reaching processes of social change. This would have meant criticizing the grave errors in the Marxist design of a better society in order to develop more appropriate ideas about how to overcome the weaknesses and absurdities of contemporary capitalism without destroying its historical achievements of individual freedom and welfare.

Many French intellectuals, however, reacted very differently. Their collective self-reflection led them to criticize and abandon precisely the kind of hope engendered by the Enlightenment. If Marxism had failed, then despair was to reign supreme. If the Marxist idea of progress had been an illusion, just as the capitalist one had been, then every idea of progress was an illusion.

In an even broader historical context, the terrible experience of the two World Wars within the span of decades, the development of the atomic bomb, and more recently the growing awareness of global environmental change are hard to reconcile with the optimistic outlook that characterizes Enlightenment thinking. Under these circumstances, the idea that the rational pursuit of self-interest by atomistic individuals is sufficient to ensure unending progress has lost much of its credibility.

Mechanistic thinking, which science contributed to modernity,[83] had already been undermined via philosophical discourse. Husserl (1936/1970) declared that Western thought had entered a deep crisis because of its inability to reconcile scientific knowledge with common sense. This tension had become hard to bear as basic tenets of scientific knowledge had been revised with the development of non-Euclidean geometry, relativity theory, and quantum mechanics.[84] Advances in physics and mathematics forced philosophers to reassess the strengths and weaknesses of mechanistic thinking. Even people without much understanding of classical physics learned that what had once been declared as basic features of the universe – the laws of Newtonian mechanics – were now downgraded to special cases of physical processes. This experience fostered doubts about mechanistic thinking which then resonated with historically grounded doubts about the rationalistic legacy of the Enlightenment.

Post-modernism has become a highly influential cultural trend because it simultaneously addresses historical experiences which undermine the belief in progress and the doubts and bewilderment brought about by the crisis in mechanistic thinking. This is especially evident in the case of architecture and urbanism. Modern architecture is one of the most fascinating products of mechanistic thinking. The attempt to build "a home like a car" (Le Corbusier), to design a kitchen like a factory, and a city like a machine, embodied the hope of progress associated with the rise of mechanistic thinking.

The experience of urban environments that were developed along these lines often became entangled in a nightmarish combination of social and ecological problems. Unintended consequences of human action brought about all kinds of perverse effects, giving considerable plausibility to post-

[83] See Section 2.1.

[84] Primas (1992) offers a bold attempt to discuss the implications of quantum mechanics for mechanistic thinking.

modern discourse. And there can be little doubt that what has become known as post-modern architecture has already produced some remarkable new buildings. The fact that these successes are confined to the aesthetics of single buildings, rather than leading to convincing solutions for whole urban settlements, however, suggests that the post-modern glorification of arbitrariness and irrationality may be too simplistic an answer to the problems of mechanistic thinking.

Mechanistic thinking clearly suffers from its negation of the possibility that there is any reasonable way to resolve moral disputes, and from its tendency to leave the description of chaotic interactions between people to arbitrary passions. This gloomy picture, which draws on basic ideas present already in Hobbes (1651/1968) and Hume (1740/1978), has been developed with considerable force by post-modern authors. They stretch basic ideas of mechanistic thinking – like the impossibility of reasonable agreement on moral matters – to the point where this worldview loses not only its optimistic outlook, but also its meta-theoretical foundation. A generalized relativism takes over general theory, and leads to the idea that RAP is not so much false as superfluous.

It is quite obvious that the rationalism of RAP is hard to reconcile with the post-modernist toolbox for the deconstruction of rational arguments. An illustration of how RAP may be challenged by post-modernist theories is the book *Beyond the Hype,* which proposes an alternative to RAP-based business management (Eccles and Nohria 1992). According to the RAP-based models, managers should make their decisions so as to rationally maximize profits over a range of well-analyzed alternatives. In order to implement these decisions, they should motivate employees through selective incentives. Although the authors do not paint their work as a challenge to the shortcomings of RAP, they propose a pluralistic approach that is in the spirit of post-modernism.

Managers often operate under conditions of great uncertainty. They often need to make decisions that are not well-grounded in any rational analysis of well-defined alternatives, but rather on intuitive use of unreliable knowledge. The skillful manager is like a chess master. She does not analyze the game many steps forward, because that would lead her to waste precious time in a hopeless effort. Rather, she looks at the next few moves and tries to find a path of action which leaves her ability to act flexibly intact (Leifer 1983). Eccles et al call this the principle of robust action. There can be little doubt that in many cases this is a more appropriate way of dealing with uncertainty than the rationalistic approach of RAP. Robust action, however, can hardly be called irrational. It seems more sensible to say that it replaces rational choice by reasonable decision.

When implementing decisions, managers need to recognize that treating employees as rational actors in the conventional way is dangerously misleading. Utility functions depend, in part, on the identity a person attributes to herself. Often, however, people are engaged in processes of identity formation and transformation. The art of management consists to a considerable degree in the ability to influence such processes while being involved in them oneself. Rather than taking utility functions as given, managers need to intervene selectively in the identity formation of employees, as well as their customers. Only when identities are taken into account is RAP an appropriate tool to deal with people. This complex element of effective management is not mediated primarily by incentives, but by discourse, rhetoric, and language games.

RAP, argue these authors, has not succeeded in advancing a good theory of management. Indeed, in a post-modern view, no theory can fulfill this need. Rather, the best understandings of management will have to come by combining different theories in specific domains. It may well be that these theories have differences that make them incompatible, and so we have gaps in our understanding of reality that cannot be explained by this or any assemblage.

According to Eccles et al, management is not concerned simply with maximizing given utility functions, but to a large extent with processes of identity formation in which such functions are defined and modified. The parallel to post-modernism comes in their description of these processes as never-ending attempts by individuals to present themselves as unique, as differing from anybody else. This description has wider relevance for criticism of RAP's assumption that individuals are the ultimate ontological reality, the atomistic building blocks of social reality. Post-modernist writings tend to refuse any view of an ultimate reality, and devote much critical effort to deconstructing the notion of an individual identity, which would in any case develop prior to social discourse.

This point has been elaborated by feminist writers like Sandra Harding (Harding 1992), who criticize any attempt to use gender identity as ontologically prior to social discourse, not only in order to justify gender discrimination, but even in order to define feminist emancipation. This careful attention to the implicit social presuppositions of what may appear as self-evidently given may be one of the most helpful lessons to be learned from post-modernism.

Post-modernists rightly insist that personal identity is always constituted in social relations. Identity formation happens in a dynamic process that is ongoing until death. At the same time, they direct the same deconstructionist critique towards any attempt to introduce an ultimate reality of collective identities. Communities are as socially constructed as individuals. The rejection of an ultimate reality of identities at any level of aggregation, and the recognition of the unending processes of multilayered constructions and deconstructions of the world, bring post-modernism into direct confrontation with RAP. At this point, a moral issue also arises. Is the deconstruction of unwarranted identities meant as a step in the process of constructing new, more sensible personal and collective identities, leading to new forms of social solidarity, or is it a pure display of social nihilism?

Left unchecked by some sense of social responsibility, this nihilist tendency in post-modernism leads to an endless game of intellectually deconstructing both individual and collective identities. In a society of organizations (Perrow 1991), this implies the absence of opportunities to build expanding networks of solidarity. The differences that people emphasize as they develop unique identities would make them helpless in the face of corporations, government agencies, global financial flows, and similar institutional settings. Large organizations follow their own dynamics, unimpressed by discourses that underline the socially constructed nature of their reality. In particular, post-modernism deals with the technological and environmental risks produced by organizations at the level of discourse only. Post-modern risk communication, then, becomes stalled in the deconstruction process. The consequence is the deferral of real risk reduction, and the most likely outcome a rising level of diffuse fear (Luhmann 1986).

Harding (1993) follows another route, using the deconstruction of gender identities to develop new forms of social solidarity, especially between women facing gender discrimination. A somewhat similar, if more old-fashioned, approach is taken by Marxist writers who use post-modernist arguments (Harvey 1996). These writers treat post-modernism as the ideology of late capitalism, and try to keep alive the hoary Marxist question of how capitalist society could be superseded by a different kind of social organization.

Such attempts to use deconstructionist arguments to foster new forms of solidarity, however, leave open the question of how RAP could be substituted as the main framework for the self-monitoring of contemporary society. Any conceptual framework claiming to be a superior alternative to RAP must meet a crucial test: Can it provide an explanation of price formation in interdependent markets which integrates the insights of RAP while offering additional insights? Perhaps post-modernist writers are correct in claiming that the far-reaching social change they perceive is currently under way. Even a capitalist "true believer" such as management guru Peter Drucker argues that we have already entered the epoch of post-capitalist society (1993). Drucker, however, explicitly argues that in this new epoch the theory of economic value provided by RAP needs to be superseded by a new understanding of how innovative markets really work.

This is very different from the post-modern claim that such understanding is impossible and that RAP should be replaced with a multitude of partial narratives that deconstruct personal and social identities. To the degree that such a displacement actually does take place in some intellectual communities, these communities exclude themselves from the ongoing scientific processes in which contemporary economic institutions are monitored, reproduced, and transformed.

6.4 Fundamentalist Revivals

Post-modernism reacts to the crisis of mechanistic thinking by stretching its core premises to yield extreme conclusions. This can be tied to Hume's (1997) idea that the choice of ends is necessarily irrational, a blind play of passions. If these passions happen to be benevolent, as Hume had it, then we are lucky; otherwise, we are left with the unwanted consequences, and there is nothing we can do about it.

An opposite answer to the same problem is the attempt to restore a guideline of fundamental moral values. In this view, the crisis of mechanistic thinking is due fundamentally to the demise of unquestionable moral values, and can only be resolved with the reestablishment of such values. Within highly industrialized countries an important form of this new fundamentalism consists of an emphasis on so-called ecological values. This form is linked to the phenomenon of egalitarian risk cultures, which we have addressed in Section 5.5. Labeling these cultures as egalitarian in the sense of cultural theory may be debatable, but the possibility that an entirely new worldview, crystallized around environmental issues, may be emerging in contemporary society, must be taken seriously.

Dunlap and colleagues (1992; 1993) have collected remarkable empirical evidence about an emerging ecological worldview. According to their studies, the new ecological worldview considers the idea of rational actors operating in isolation as a dangerous illusion. Instead, it stresses the interdependence of the flow of human actions and a multitude of life-sustaining natural processes which human actors can destroy, but not control entirely. A specific human action would then appear less as the result of individual rationality based upon an evaluation of alternative options and more as the outcome of interactions among people and between people and the environment. Clearly, such a worldview would fit neither with RAP nor with post-modernist discourse. Does it lead to new forms of ecological fundamentalism?

Examples of ecologists appealing to fundamental values are easy to find. In this orientation, biodiversity protection is not justified on scientific or consequential grounds, but treated as an intrinsically justified fundamental value which does not require discursive justification. Similar arguments are invoked regarding specific species, say the blue whales.

The existence of blue whales is a value *as such*. Engaging in debate about how this value is to be justified already implies that one is an enemy of the whales. Conversely, nuclear waste is treated by many ecologists as a fundamental bad in much the same way. The reference to fundamental ecological values obviously renders reasoned debate about many policy choices very difficult, especially in the field of risk management. To ecological fundamentalists, many risks are simply unacceptable, and no open debate about them is warranted. The bitterness of many exchanges dealing with the risks of genetic engineering is a case in point (see the discussion in Jaeger and Rust 1994).

Although fundamental ecological values often characterize contemporary conflicts about environmental policy and risk management, theoretical reflections about their existence and meaning are rare. A major difficulty for any such reflection lies in the fact that ecological activists typically have highly critical attitudes towards the social status quo, while fundamentalist arguments involve profoundly uncritical attitudes towards the moral status quo. The latter position is treated as naturally given, thereby raising questions about its philosophical status. It treats "ought" statements on a par with "is" statements – as if Hume's guillotine or G. E. Moore's naturalistic fallacy had never existed.

An interesting attempt to deal with this tension is offered by Tännsjö (1990), who argues that contemporary radical criticism of society represents a conservative style of thinking. Although his argument is explicitly Marxist, and it is easy to see that capitalism destroys many traditions that conservatives value, it is not so easy to see how ecological fundamentalism could flourish in today's pluralistic society.

The difficulty of a fundamentalist revival as the moral background for environmental policy becomes visible when forms of Christian fundamentalist revival are taken into consideration. A core component of much Christian fundamentalist thinking is stern rejection of abortion on the grounds that human life is a fundamental moral value. This attitude is rarely found in environmental circles, and it is hard to imagine an open debate about abortion between, say, a fundamentalist believer in eco-feminism and a representative of the Christian right. Both would invoke a moral sensibility to fundamental values, but from there they would draw deeply incompatible conclusions. The two lines of argument are similar regarding issues of genetic engineering and reproductive technologies, especially for human beings. In this arena, both parties are quite likely to invoke the fundamental moral value of human life to question man's tendency to misuse science and technology by tampering with creation and actually trying to play God.

But the style of critical inquiry by ecologists is quite alien to Christian fundamentalism. From time to time fundamentalist thinkers with differing orientations come to share points of view that enable them to argue particular cases in a careful and stimulating manner. After years of thinking in a Marxist framework, for example, Alasdair MacIntyre (1984) received great praise for a book on ethics in which he contrasted the lack of ethical orientation in modern society with the ancient conception of virtues. His provocative argument was that the authority of the Catholic church was not only the appropriate orientation for moral argument, but also that the emancipation of science from this authority was a main reason for the moral decay of modern society (MacIntyre 1988).

The power of Christian fundamentalism, real as it is, however, seems rather limited in comparison with the fundamentalist revivals taking place within the religious tradition of Islam. In many respects, Muslim fundamentalism is an explicit reaction to experiences of modernization. Barber (1995) goes so far as to claim that the two main forces shaping global society after the demise of communism are unfettered capitalism and Muslim fundamentalism.

In intellectual circles it is quite common to treat all these forms of fundamentalist revival with distain. Intellectuals do this at their own peril, however. Even if the pluralistic character of contemporary Western society is an irreversible achievement of the Enlightenment, and even if it should spread to the whole world, the available evidence strongly suggests that fundamentalist revivals will be a major component of pluralistic societies for a long time. And we should not underestimate the intellectual legacy on which such revivals can draw. While it is easy and sometimes necessary to inveigh against the dangers of fundamentalist revivals, it is more difficult, but perhaps more necessary, to try to understand the rationales behind them.

These rationales are provoked to a large extent by the crisis of mechanistic thinking, both intellectually and historically. But it is also important to note that fundamentalist revivals share with post-modernism the tendency to stretch certain premises until their consequences are extreme. In this sense, the new fundamentalisms draw on the tradition of conservative social thought.

An essential component of this tradition is Burke's (1790/1987) criticism of the French revolution, which to this day is a core element of political conservatism. Burke took the French Enlightenment to task for deluding itself into thinking it could construct a better society from scratch. This attitude had no respect whatsoever for the fundamental achievement of the status quo – namely, to be viable and to persist over time. The received values on which an established social order are based may seem questionable to the arrogant rationality of those unwilling to accept anything that cannot be explicitly justified. In fact, however, received values are indispensable components of the complex fabric that develops over the long course of history and that embodies the wisdom of many generations.

The conservative bent of this argument is obvious. Meanwhile, we may wonder whether Burke's argument has not been obliterated by the environmental and health risks engendered by the systematic use of science-based technologies.

Burke's argument can be linked to an even more authoritative strand of Western thought. One of the most profound answers ever offered to the question of the origin of values is found in the philosophy of Plato. Lengthy discussion of Plato's philosophy (and the footnotes to it provided by a venerable list of thinkers) is not necessary. For our present purposes, it suffices to acknowledge that the fundamentalist tendencies present in today's risk debates have a lineage which includes the very origins of Western philosophy. Plato criticized existing monarchies not in the name of democracy, but in the name of an ideal monarchy of philosopher-kings. The philosopher-king, he argued, would be able to contemplate ideal moral values as a basis of governance, rather than the real world of everyday experience. Reasoned argument would permit the philosopher-king to discover his faculty of contemplating these values, and he would then see that they do not follow as consequences from such argument: to the contrary, they provide its foundation.

The main upshot of the last two sections is an awareness of the fact that mechanistic thinking has come under serious attack from two opposite sides. Both are directly relevant to conflicts about how to deal with the many risks of contemporary society. By undermining mechanistic thinking, both post-modern authors and believers in fundamentalist revivals may foster a broader view of human society than RAP. At the same time, we should ask what these orientations have to offer in analyzing the functioning and malfunctioning of market economies. The answer to this question is straightforward – neither approach is of much help in understanding, say, the functioning of financial markets and the risks they engender. This suggests that RAP should easily survive these two attacks, unless they are complemented by more specific arguments about the operation of market forces.

6.5 The Relevance of Critical Theory

Given the limitations of post-modernism and of fundamentalist revivals, it is instructive to look at an attempt to develop a reflexive worldview that tries to combine some of the key issues of both. Critical theory in its current form draws both on the wisdom of the Platonic tradition and on the acceptance of the undeniable pluralism of contemporary society.

6.5.1 The Dialectics of Rationalization

What is today known as the tradition of critical theory began with the founding of the Institute for Social Research at Frankfurt University in 1923. Under the directorship of Max Horkheimer, the prominent theoreticians we recognize today as representatives of critical theory – Theodor W. Adorno, Walter Benjamin, Erich Fromm, Herbert Marcuse, and others – made many contributions to a new strand of social theory. Horkheimer and Adorno remained the most powerful and influential figures in the school until their deaths, but the stage was already set for transition. In the late 1950s Jürgen Habermas joined the Institute as an assistant to Adorno. Although he left Frankfurt in 1964 and the Institute for Social Research later disbanded, Habermas continued the work that was started, adding his own creative, ingenious reformulations. Habermas is now best known for his theory of communicative action – an attempt to formulate a theory of society and societal evolution of a caliber comparable to Parsons, Marx, and Mead – and his theory of discourse ethics, which is an attempt to formulate a theory of morality. Since much of Habermas's work challenges that of Horkheimer, Adorno, Marcuse, and others, it is common to speak about two different eras in critical theory – the Frankfurt School, or early critical theory, and the critical theory of Jürgen Habermas.[85]

The contributions of the Frankfurt School to a critique of RAP can be best summarized by Horkheimer's two theses: rationalization as a loss of meaning, and rationalization as a loss of freedom (Habermas 1984, p.346; Horkheimer 1974). To understand how these theses are related to RAP, it is necessary to understand the meaning and central role played by the concept of rationalization in German philosophy and sociology.

According to the Kantian view, the Enlightenment of Western society is a positive outcome of the increased development of scientific and moral ways of conceptualizing the world (rationalities). When critical theorists view the movement of human society in terms of changes in social rationalization, they are building on a history of German philosophical thought that began with Hegel. Max Weber was the first German sociologist to seriously take up the theme of rationalization (see 1983), and it is Weber's work that Horkheimer and Adorno (and Habermas) use as a stepping-off point. Weber portrayed the development of western society to the modern period through changes in the dominant form of rationality. His main achievement in this regard was to associate the emergence of *Zweckrationalität* (purposive rationality) with "The Protestant Ethic" (Weber 1904/

[85] We will not discuss more recent developments in critical theory, not because they are not valuable, but because they seem less directly relevant to the issue of risk and uncertainty.

05/1993). This life choice prescribed by Protestant Christianity was characterized chiefly as a methodical way of life. Weber argued that the emergence of world religions marked the break between a pre-modern world dominated by magic and myths and the emergence of modernity. He labeled the process of rationalization from a myth- and magic-centered way of knowing the world to a form of rationalization that was "purposive" as "disenchantment."

The Frankfurt School's critique of this rationalization process focused on positivism and modernity. The critique of positivism was based on the epistemological argument that scientific investigation is a social act, "and in order to understand the significance of facts or of science generally one must possess the key to the historical situation, the right social theory" (Horkheimer 1972, p 159). Positivism supposedly exempted itself from this requirement by separating facts from values. Horkheimer opposed the positivist attempt to separate facts from values on three grounds:

- verification through perception (empiricism) is inadequate as a means to understand reality;
- facts are selected from an infinite number that present themselves more or less arbitrarily; and
- no distinction is made between appearance and essence (Horkheimer 1972).

The critique of positivism by Horkheimer, Adorno, Marcuse, and later pursued by Habermas, attacks one of the key weaknesses of RAP – how ends are selected. Rational choice theorists have difficulties explaining how actors, be they individuals or groups, decide to pursue particular ends (except for the unenlightening suggestion that whatever ends the actor chooses to follow must have been chosen because he or she expected it would increase his or her overall happiness). As a means for helping people make sense out of the world, RAP by itself cannot provide the depth of meaning that is needed.

The early critical theorists of the Frankfurt School questioned the long-held belief that the rationalization of society amounts to human progress in intellectual achievement and in the realization of political orders that ensure the autonomy of the individual through political freedom and popular sovereignty. Living in Germany in the early 1930s during the rise of Nazism, and later in America during and immediately after World War II, these scholars witnessed events that convinced them that rationalization was not one-sidedly good. The persecution of the Jews was a convincing case in which scientific rationality did not prevent the application of its technologies and efficient organizations toward immoral ends. The American development and use of the atomic bomb furthered the concern that science and technology were not always or necessarily linked to the betterment of the human condition. The rise of the culture industry in the United States, and later in Europe after the war, was further evidence of the superficiality and even deteriorating effects of social rationalization.

Critical theorists were zealous critics of modernity, claiming that it was not synonymous with unqualified human progress; and, indeed, that there were clear signs of the opposite effect. Horkheimer and Adorno eventually hypothesized that rationalization was a dialectical process that resulted in a contradiction between increased potential for manipulating the world and the increased likelihood that ill effects of this manipulation would be felt in the realm of everyday life. In their words, rationalization meant "a return to barbarism" that was equated with increasing forms and intensities of domination (Horkheimer and Adorno 1947).

The conclusion drawn was that rationalization brings a loss of freedom, due in part to its inability to make judgments about the *quality* of goals. In other words, we must face the possibility that liberal democracy and a free market – RAP's normative mechanisms for how social actions should best be coordinated – may well produce a society with monstrous features.

How successful was critical theory at addressing the limitations of RAP? Although the Frankfurt School's critiques identified problems associated with RAP, like Weber before them, the theorists were never able to formulate a reasonable alternative that would counteract the problems or protect society against the negative effects of modernization. Critical theory at this stage was wholly critical. The problem was not that critical theorists were unable to see the positive sides of rationalization; they did. The problem was that they were unable to see how the benefits of RAP could be saved, given the corresponding negative consequences.

Critical theory later attempted to provide a more specific diagnosis of the problems of modernity and to offer proposals for a possible way out of these problems. Habermas broke away from one of the main assumptions of his predecessors. Their main error, he argued, was their commitment to preserving the "philosophy of the subject." Horkheimer, Adorno, and Marcuse were particularly interested in reconciling certain difficulties in Marxist theory by adopting a conception of the subject with a broader concept of rationality, one that could account for experiences such as the reality of the holocaust and the failure of the workers' revolution to materialize. But in focusing on the subject as an at least potentially rational actor, they tacitly preserved a fundamental premise of RAP – that rationality is a feature of the individual. Habermas defined the intellectual perspective underlying this approach as "the philosophy of the subject" or "the philosophy of consciousness" and argued that it must be aban-doned.

Habermas has produced a more restrained and sophisticated diagnosis of the role of rationalization and reason in social evolution, by situating himself between the two extremes taken by the proponents of modernity on the one hand and his predecessors in critical theory on the other. Although he engaged in the critique of positivism in his early years, Habermas never adopted the hostility toward science evidenced by Horkheimer and Adorno.

Habermas recognized that rationalization can have amazing liberating effects, both technologically and politically, but that it can also bring horrible consequences. Horkheimer and Adorno saw these as two sides of the same coin. Habermas retrieves the enthusiasm for rationalization, which he equates with the project of modernity, as expressed by Descartes and others. In stark contrast to his mentors at Frankfurt, Habermas explains the black spots of this century, not as signs of fundamental flaws to rationalization, but as one-sided deviations from a more comprehensive process of rationalization.

In Habermas's view, problems of pending nuclear calamities, environmental destabilization, family disintegration, and political delegitimation are consequences of a "lopsided" rationalization process. As we shall see, his argument is basically an argument against the unchecked dominance of the rational actor paradigm. It is, however, an argument that redeems the legitimacy of rationality from one-sided critical theory.

Habermas's contentions about RAP go straight to the core of what it means to be rational. In a move that further marks his departure from the traditions of critical theory, he asserts that rationality has social origins. As such, it is not possible to detect if a person is acting rationally without knowing her social situation. We only know that we are acting rationally when our past experiences, our cultural knowledge, and other people confirm it. Habermas explains his departure from earlier critical theorists as a departure from the philosophy of consciousness (the stalwart of modernist philosophy). Such a transition could hardly take him to a position further away from RAP. Rather than work from a framework that assigns reason to the individual and works upward to society, Habermas believes that it is only through social interaction (namely communication) that standards for reason develop.

If we wish to understand how people learn to make sense out of the world, then we must look at social interaction, especially at speech. This is exactly what Habermas undertakes in his two-volume *Theory of Communicative Action* (1984; 1987). The theory of communicative action is an attempt to provide a theory based on a conception of rationality that encompasses the selection of ends, or collective norms.

Faithful to his roots in the Frankfurt School, Habermas sets out to construct an argument that exclusive reliance on purposive rationality is not merely deficient, but also dangerous. His connections to Weber and the Frankfurt School quickly become clear. In *Dialectic of Enlightenment*,

Horkheimer and Adorno (1947) spelled out their version of Weber's "iron cage" thesis. The unparalleled triumph of science and technology, mediated through a purposive rational outlook, has not fulfilled the Hobbesian task of constructing mutually beneficial political organizations, at least not without simultaneously releasing a horrible new barbaric force. For Horkheimer and Adorno this force was not a theoretical specter, but the reality of Nazi Germany in the 1930s. Later they saw a different, but still dangerous, form of barbarism in the culture industry of the post-war West. They argued that rationalization, rather than a progression of social learning and human development, was a dialectical process. Advances in human dignity through science and technology were matched by an equally powerful undercurrent that eroded the net effect of advancement and drove the insanity of powerful new forms of suppression and domination. Habermas agrees with this part of the diagnosis.

According to Marx and Weber, purposive rationality was the key to social evolution (or social development). Marx even claimed that the level of social development is directly related to society's ability to harness nature for human needs, via the tools provided by science. Weber's dark conclusion was that, while he saw refinement of purposive rationality as essential for progression from a world ruled by myth and magic to one ruled by reason, he saw no way to prevent the loss of freedom resulting from technocratic social, economic, and political institutions. Habermas recognizes both the progress and the nightmares science can bring to society, but he breaks with tradition in German philosophy by diagnosing the principal problems of society as a consequence of "one-sided rationalization".

6.5.2 Communicative Action

Habermas bases his reconstruction of the meaning of rationality on Weber's concept of value spheres, which in turn goes back to Platonic philosophy. Value spheres refer to delimited domains of action, each with its own characteristic rationality, or standard for evaluating action. He distinguishes three spheres: science and technology; morality and law; and art and art criticism (Habermas 1984, pp 163–165; Habermas 1987, pp 236–240).

Habermas perceives the potential in Weber's work for explaining the emergence of modernity, but he also believes that Weber's typifications and conclusions were inaccurate or incomplete. Weber confuses different kinds of rationality in his analysis. However, the confusion can be remedied by distinguishing between the form of rationality that is implied in the process of disenchantment and the form implied in the process of rationalizing economic, political, and social systems. Habermas concludes that the former is redeemed as a beneficial force and the catalyst for modernity, while the latter, when unchecked, leads to Weber's "iron cage."

With this explanation Habermas redeems Marx's argument about nature and social reproduction, while also claiming that Weber's pessimism was unjustified. Habermas supports the contention that social development comes through social learning of more refined forms of rationality, but he argues that learning must be a balanced process. Scientific rationality (purposive rationality) must be developed, but at a pace that matches growth in practical and aesthetic rationality. Balancing social learning along these lines is the key to successful, i.e. non-regressive, social change.

Habermas realizes that his claim stands or falls on his ability to provide a convincing explanation of how these three spheres of rationality can be balanced in a way that answers the problems associated with coordinating actions. It is here that the first main element of Habermas's argument – that rationality is socially constructed – and his second main theme – that the three rationalities must be coordinated – come together.

Precisely where they come together is marked by a clarification of the social character of rationality. Whereas others have tried to explain action, as such – Parsons's and Shils's (1951) action theory, Goffman's (1959) dramaturgical actions, Homans's (1961) exchange approach – Habermas invokes communicative action. Things people do in any one of the three value spheres, he argues, are oriented to three different worlds: the physical, the social, or the personal. In discourse, people relate to these worlds through speech acts. Inherent to each speech act is a validity claim. Each kind of validity claim has its origins in one of the three value spheres. For example, speech acts claiming to characterize existing states of affairs in the material world reside in the sphere of science and technology.

Most importantly, however, the overriding purpose for speech is not represented by any one of these three action orientations, but by a larger purpose that encompasses all three. This is the universal goal of constructing non-coerced mutual understandings and agreements. Habermas calls this communicative action, and considers it as the highest form of rationality, since all other forms rely on it as their basis. Correspondingly, communicative rationality is the mode of making sense of things that engenders successful communicative actions.

In Habermas's terminology, *instrumental rationality* is the form of reasoning used to build expectations about likely consequences of action in cases devoid of other teleological actors. *Strategic rationality* is the reasoning used in cases where the attainment of one's objectives depends in part on the choices that other purposive actors make. For example, a lumberjack attempting to fell a tree in a specific direction uses instrumental reason to decide where, when, and how deep to cut. When that same lumberjack tries to convince his co-worker to run down to camp to fetch him a cup of coffee, he employs strategic reasoning. Even the seemingly straightforward calculations of instrumental rationality, however, cannot be attributed to isolated individuals, for the facts which enter into those calculations, as well as the choices themselves, are the products of a pre-established social consensus. The lumberjack has learned how to judge what kind of cut to use in each situation to bring about the desired ends. Of course, one might point out that the lumberjack could have learned the skill after years of working alone in the Alaska backcountry. Does that mean instrumental reason is not a social product? Not at all, for although individuals can learn new skills in isolation, they must first acquire the conceptual tools for learning. These are provided only via socialization and social reproduction processes. In sum, our understandings of existing states of affairs, society, and ourselves are the products of socialization, culture, and our cumulative life experiences. Thus, Habermas offers a truly social concept of action, substituting approaches grounded in autonomous and responsible social interaction for those based on atomized individuals calculating costs and benefits of available choices.

Communicative action (the application of communicative rationality), like all action, is inherently purposive; its aim is to reach uncoercive consensus. Still, Habermas distinguishes it sharply from instrumental and strategic action, guided by *Zweckrationalität*, on the basis that the latter represent the pursuit of realizing egocentric, as opposed to collective, preferences. Communicative action requires that one be able to predict certain outcomes, as in instrumental and strategic action, but this is not enough. It also requires that one be able to act in ways consistent with social norms, and furthermore, that one be empathetic by placing oneself in other people's shoes in order to understand their understandings and preferences.

In conclusion, Habermas's concept of communicative action as a means to reach collective, uncoercive definitions of ends provides an analysis of rationality that RAP is unable to supply. Habermas does not claim to know which ends are best, he only provides a description of the rational means that free individuals who are inclined toward cooperating with others can use for selecting and defining common ends.

Critical theory takes seriously the fact that individuals living together must intentionally discuss preferences, interests, norms, and values in a rational way. Habermas's "discourse ethics" is his attempt to explain the principles that guide the construction of a social agreement on norms. It provides a vision of how normative disputes can be worked out in a form of social interaction that goes beyond strategic action. Habermas sees strategic action as inherently manipulative and fundamentally inconsistent with relying on consensus as a means of seeking normative agreement. At the same time, however, his argument should not be misinterpreted as a commitment to Durkheim's (1895/1982) "collective conscious" holistic view of society. It is closer to Mead's (1934) ideas of how the concept of "the generalized other" facilitates self reflection, as well as to American pragmatist thought on how society can be understood through language, and to the phenomenalist concept of the lifeworld (Schuetz and Luckmann 1994).

Habermas argues that if instrumental reason is not to enjoy an unacceptably dominant status, one must explain how reasoning about norms and values is similar to reasoning about facts. By making an analytical distinction between theoretical discourse (about states of affairs in the world) and practical discourse (about preferred patterns of social relations) he argues that a single pattern of reasoning underlies both forms. Claims to validity are essentially different forms in these two kinds of discourse, i.e. assertions about facts are different from assertions about norms, and different standards for redemption are appropriate for each.

At the center of both theoretical and practical discourse is the way in which individuals who are linked by structures of social reproduction and who share a culture and a background of symbolic meanings go about posing and redeeming assertions. In other words, practical discourse is not simply muddling through, or simply "politics". There are mutually agreed upon standards for judging the appropriateness of normative claims, just as there are standards for judging the correctness of facts in theoretical discourse.

The distinction between practical and theoretical discourse is key to Habermas's attempt to redeem instrumental reason and the project of modernity from the dark criticisms and negative dialectics of early critical theory. Habermas wholeheartedly accepts modernity as an extension of the enlightenment, because it marks a leap in social learning of Western society. Society has learned that different kinds of truths are realized and checked in different ways. According to Habermas, society learns in two independent directions: by experiencing cognitive enhancement through refinement of instrumental reason, and by experiencing moral development through refinement of modes for practical discourse. Communicative rationality has the potential to tie these two different learning efforts together into a coordinated process. If communicative rationality is successfully employed, the negative consequences of unbalanced cognitive enhancement can be eliminated.

Habermas presents a vision that can be used to discuss both how collective choices about norms and preferences *are made* and how they *should be made*. This is not the same as saying which preferences should be *accepted*. Preference formation has always been a thorny area for rational action theories, which have difficulty in explaining seemingly irrational behavior, for example risk aversion to some things (nuclear power) and voluntary risk exposure to others (cigarettes) (Slovic 1987; Wilson and Crouch 1987; but see Becker 1996). Habermas is able to develop a view of preference formation because his conception of rationality is much broader than that of RAP.

In Habermas's work, there are two levels of rationality. At one level are three independent forms of rationality associated with the three value spheres Weber described: science and technology, law and morality, and art and art criticism. Within each of these domains, rationality has developed differently. (Habermas claims that this is the defining characteristic of the Enlightenment.) Some problems can be handled with one variant of ratio-nality alone. When dealing publicly with problems in social life that cannot be managed with the normal tools of day-to-day language (for example, where to build a new regional landfill), this is rarely the case, however. In particular, it is often insufficient to apply only one of these three forms of rationality (e.g., to have technical experts decide). Instead, those who will be potentially affected by the problem, and experts alike, must invoke communicative rationality; that is, the ability to use the three specific forms of rationality simultaneously in a consistent and interlacing manner in order to deal with a pressing issue in a cooperative and non-coercive way.

In Habermas's scheme, outcomes are brought into the picture during a discursive process in which participants are free to raise or challenge any proposition. In the setting he imagines, people do this and come to collective agreement on the validity of propositions in a cooperative and non-manipulative way. Habermas believes that such a discourse should be open to anyone – discourse membership cannot be restricted. All must have an equal chance to participate. Of course, this alone cannot guarantee that all interest positions will be equally represented or defended. The best guarantee he offers is that normative claims be resolved only by decisions fulfilling the following rule: "*All* affected can accept the consequences and the side effects its *general* observance can be anticipated to have for the satisfaction of *everyone's* interests". (Habermas 1991, p 65).

Critical theory is neither a unified theory, nor a synthesis of theories, but a collection of critiques of society and notions of rationality. It is based on the observation that blind pursuit of instrumentally rational means has led to disastrous results. Jane Braaten (1991) identifies the following normative elements of these critiques. They: (a) attempt to induce self-clarification by encouraging individuals to reflect upon whether and how their actions may be inconsistent with their interests; (b) seek to identify crisis potentials in society; and (c) want to bring about the emancipation of individuals from arbitrary power relations.

Habermas takes the position that the RAP concept of rationality is a subordinate form of a broader concept of communicative rationality. Viewing society through the medium of communication leads him to observe that strategic forms of action as described by RAP are "parasitic" on communicative action. This is because communicative action relies on the normative force of consensual agreement to resolve differences according to preordained rules for speech.

Explanations of the behavior of individuals within well-specified contexts for action are RAP's principal strength. Explaining how actors select and design these contexts for future action lies outside the purview of RAP.[86] Habermas's conception of communicative rationality has the potential for expressing a version of rationality that attributes to rational choice an important, if subordinate, position. However, it would be naive to believe that critical theorists have completed this task. At the present stage, this is an opportunity, no more, no less.

Although Habermas has proposed a sophisticated philosophy of rationality that goes a long way towards developing a robust and well-rounded concept of social rationality, it is less clear that his typology of action accurately portrays how people employ strategic actions. He does not provide a good explanation for how rhetoric can be used to build understandings and to coordinate the actions of autonomous individuals, for example. An important piece of his normative argument is that only the force of better argument should prevail, rather than coercion, manipulation, threats, and so on. But this begs the question of what constitutes an argument. And it opens the door to the suggestion that strategic action, while it does have its limitations, might also have a valid role to play in non-coercive communication. This is relevant to the objection that Habermas has not provided a convincing explanation for why rational actors would choose to cooperate (Johnson 1991).

It is also relevant to raise the question of whether critical theory can help to replace existing understandings of markets. For example, Habermas's critical theory can tell us quite a bit about how people reach shared agreements on norms. In studying how economic agents develop shared expectations of future events, his arguments can be helpful. For the study of markets, however, the idea of communicative rationality is a challenge for future research, not a consolidated theory.

Communicative rationality has many ramifications for understanding human choice and action in a rational age of irrational ecological

[86] With remarkable clarity, Becker (1996) argues that while RAP can account for tastes (and more generally, preferences) and their formation in a wide range of circumstances, it has a hard time in accounting for how people choose social networks and how they deal with the uncertainties involved in such choices.

destruction and social stress – e.g., in families and rapidly changing workplace environments. Central to the critical theory of both Habermas and the early Frankfurt School is the idea that people can and must reflect upon their interests and their actions if they are to develop authentic personal and collective needs. One belief that has been preserved in critical theory from Horkheimer to Habermas is that blind pursuit of effective means without due reflection and justification for the validity of chosen ends is potentially exploitative and morally irresponsible. People must ask: Does what we do contradict what we want and need?

This question is a way to escape from the relativism of post-modern discourse and the fundamentalism of the new conservatism and to construct consensual agreements on values, interests, and beliefs. Critical self-reflection may produce ways out of the economic, ecological, and political problems facing us today. Critical theory offers a major philosophical contribution to the debate about what political procedures are appropriate for a society characterized by human-made risks. It goes a long way towards developing a coherent concept of social rationality. Using such a concept to better understand and influence the economic reality of our times, to provide a valid successor to RAP in theory and in practice, however, is still unfinished business.

7. FROM MONOLOGICAL TO DIALOGICAL RATIONALITY

7.1 The Nature of the Challenge

7.1.1 Some Problems with the Idea of Rational Action

In the introduction to this book, the idea of rational action was introduced in three variants. In its broadest (worldview) form it presupposes that human beings act by individually linking decisions with outcomes. Humans are goal-oriented; they have options for action available and they select options that they consider appropriate to reach their goals. The rational action worldview is atomistic, assuming that all social actions can be reduced to individual choices and that rationality is preeminently a property of human individuals. It is also mechanistic, assuming that the world can be understood on the basis of the interaction of separate material bodies, including human beings.

In Section 2.1 we discussed some philosophical aspects of this worldview, and in Chapter 6 we looked at several attempts to challenge it. Post-modernism and various fundamentalist approaches share a strenuous opposition to the mechanistic worldview. There may be an important lesson here, but in the form proposed by these two strands of thought it is far from acceptable. As we have argued in Chapter 6, these approaches undermine the very possibility of critical risk assessment. For this purpose, at least, they are self-defeating.

Critical theory proposes an alternative to mechanistic thinking in the social world, as it relies on a relational social philosophy. Avoiding both the Scylla of a solipsistic "I" and the Charybdis of an absolutist "We", it orients itself on the relation between "I" and "Thou".[87] It has done little, however, to overcome the mechanistic view of nature which is so influential in modern society – and which may actually be an important cause of the pervasiveness of technological risks (Primas, 1992). For our present purposes, we simply take notice of the fact that in the social science literature on risk, mechanistic thinking has received limited attention. Therefore, in the present setting we will not pursue the matter of worldviews

[87] Recently, Brandom (1994) has offered a careful analysis of human rationality along these lines. Habermas (1999) discusses relations between Brandom's argument and critical theory. He quotes (p. 176f) a letter of Brandom's which conveys the thrust of a relational social philosophy with an inspiring image: "I have in mind thinking of conversation as somewhat like Fred Astaire and Ginger Rogers dancing: they are doing very different things – at least moving in different ways – but are coordinating, adjusting, and making up one dance. The dance is all they share, and it is not independent of and antecedent to what they are doing."

much further (see Section 7.2.1). What we certainly do maintain, is the view that human beings are capable of rational action in the sense of defining and pursuing projects, of justifying and criticizing them, and of assessing the risks these projects engender.

The second, more elaborate version of the idea of rational action, the general theory of human action which we have designated as RAP, is a different story. As outlined in Chapters 3 and 4, many special theories of risk and uncertainty rely on RAP and its presuppositions. These presuppositions refer to human actions based on individual decisions. Among the most important are the following:

- analytical separability of means and ends (people as well as institutions can in principle distinguish between ends and means to achieve them);
- goal-attainment motivation (individuals are motivated to pursue self-chosen goals when selecting decision options);
- existence of knowledge about potential outcomes (people who face a decision can make judgments about the potential consequences of their choices and their likelihood);
- existence of consistent and stable human preferences (people have preferences about decision outcomes based on tastes, values, expected benefits, and the like);
- maximization or optimization of individual utility (preference structures can be represented by utility functions in such a way that most preferred options show the highest utility); and
- predictability of human actions if preferences and subjective knowledge are known (rational actor theory is not only a normative model of how people should decide, but also a descriptive model of how people select options and justify their actions).

This set of fundamental assumptions linked to individual behavior is also transferred to situations of collective decision making or collective impacts of individual decisions. These situations refer to two classes of phenomena:

- actions that are designed and/or implemented by more than one actor (organizations or social collectivities). In this case, RAP leads to the *treatment of organizations or social groups as "virtual" individuals* – as *personae fictae*; organizations act like individuals, and they select the most efficient means to reach pre-determined goals; and
- a multitude of individual actions create social realities ("social facts" in the language of Durkheim). In this case, RAP leads to *the extension of individual preferences to aggregate preference structures* – institutions that aggregate individual preferences such as markets or

political decision making bodies do not simply reproduce the set of individual preferences, but may define combined collective interests of these individuals.

The overall architecture of RAP has two well-known implications for social theory:

- indifference to the genesis and promulgation of values and preferences (values are seen as pre-existent or exogenous; RAP can only make predictions on the premise that preferences are given, not created in the decision process),[88] and
- independence of allocation of resources and their distributional effects (it is rational for societies to assign priority to the most efficient allocation of resources, regardless of distributional effects, before re-distributing the wealth that is produced among the members according to preconceived principles of social justice).[89]

This suite of presuppositions and implications restricts the scope of RAP to a limited set of objectives. Underlying the individual as well as the collective interpretation of RAP is the basic assumption that all relevant actions can be described as problems of maximization or optimization. The social world consists of countless decision problems, each requiring the generation of options for future actions and an algorithm to choose among them. This algorithm is meant to represent an individual or collective strategy to optimize one's own benefits. This assumption may be well founded in many social situations such as certain market interactions or specific negotiations about the selection of policies. It may, however, be totally inappropriate if we think of other social goals and processes. *RAP is handicapped in creating mutual trust among actors, building individual and social identity, achieving ontological security, or constructing solidarity among people.*

It makes little sense to think of trust, identity, or solidarity as resources to be maximized or optimized as defined in RAP. These social phenomena

[88] Two qualifications are in order with regard to sophisticated versions of RAP including endogenous preference dynamics, as in Becker (1996). First, in a specific decision process, preferences which are relevant for other processes may be generated. Second, preference dynamics begin with a bedrock of basic preferences which are not amenable to further analysis within RAP. They may be analyzed, however, as a result of biological evolution.

[89] In some *laissez-faire* versions of RAP, a Darwinistic selection rule for distributing wealth is assumed as most appropriate, since the most successful entrepreneur should also reap most of the benefits. Modern versions inspired by von Hayek emphasize the distinction between allocation of resources and equitable distribution. RAP models are most appropriate for ensuring the most effective and efficient way of production and exchange of goods and services, but need auxiliary ethical norms to distribute the added value among society's members.

are products of communication and mutual understanding, elements of social life that require mutuality, not simply strategic action to satisfy individual actors alone. They also require constant feedback in order to reach stability. They are endangered by strategic or disruptive social actions and must be fueled by reciprocal actions and exchange of symbols that reaffirm shared values and convictions. Were these to be calculated according to RAP principles, decision-makers would be faced with uncertainties concerning outcomes, their own preferences, and expectations. Thus, the optimization process of RAP cannot account for these processes of offering and sustaining social interactions.

Hence, our main point is that the realm of the monarch RAP has been extended beyond its scope to areas that must not be regarded solely in utility optimization terms. Social reality loses its meaning and vitality if all actions have only the goal of maximizing or optimizing one's utility according to well-defined preferences. Balancing social relations, finding meaning within a culture, showing sympathy and empathy to others as well as being accepted or even loved by other individuals, belong to a class of social phenomena that does not fit neatly into the iron rule of RAP. Many activities of social life deny the RAP expectation that each individual consciously or unconsciously behaves according to a means-ends opti-mization process. RAP presumes that individuals continuously pursue the three steps of decision making: option generation, evaluation of consequences, and selection of the most beneficial option. Without doubt, many social situations can be described, or at least simulated, in such a fashion. Unlike other more radical critiques of RAP, we accept the fact that RAP offers the greatest explanatory power for certain circumscribed domains of human activity. There are many other situations, however, in which the RAP presumption of optimizing outcomes appears to be a weak descriptor of actual behavior, let alone what the actors perceive to happen.

Post-RAP theories, therefore, are faced with a twofold task. First, they must delineate scope conditions for which the presuppositions of RAP are appropriate and those for which they are not; and second, they must identify and define additional schemata of social actions that are based on intentionality, but which use routes other than optimizing outcomes.

This criticism is not directed towards all uses of RAP. Our criticism rather is of the overextension of RAP to social phenomena that are not suitable for the assumptions of RAP at both the individual and the collective levels. To our knowledge, no RAP theoretician or practitioner has taken up this issue. Rather – as proponents of paradigms often do – they continue to treat empirical deviations from the assumptions of RAP as anomalies or "noise." These anomalies, however, raise a major question: Do optimizing strategies underlie all classes of human and social actions?

This question has not been adequately addressed by proponents of RAP. Optimization alone raises serious concern about the limits of RAP as a general theory of human action. The presuppositions of RAP are targets for other criticisms, as well – questions that in turn have been addressed by RAP adherents.

The main criticisms can be placed into two classes: aspects of RAP as a theory of individual choice and action; and aspects of RAP as a theory for collective choices and social action. In both cases, the claim that RAP provides an adequate theory for human choices is at stake.

7.1.2 RAP and Human Choice

One of the greatest threats to RAP comes from empirical evidence showing that humans systematically violate its rules of rational action. Thus, as a descriptive tool for predicting people's actions RAP has limited validity. This is true even if an analyst has access to the preferences and subjective knowledge of an individual decision maker. Most psychological experiments demonstrate only modest correlation between rationally predicted and intuitively chosen options (Dawes 1988; Tversky and Kahneman 1981; von Winterfeldt and Edwards 1986). Furthermore, by asking people in thinking-out loud experiments how they arrive at their decision, many rationales are articulated, only a few of which have any resemblance to the prescribed procedures of rational actors (Earle and Lindell 1984). In response to this empirical challenge, proponents of RAP in psychology and economics have proposed five modifications that would bring the theory more in line with the actual observations of behavior.

First, they claim that the procedures prescribed by RAP-based theory serve only as analytical reconstructions of intuitions individuals use to make choices. Whether humans follow these prescriptions consciously or not does not matter, as long as the outcome of the decision process is close to what rational theories would predict (Edwards 1954; Phillips 1979).

Second, they argue that people use simplified models of rationality such as:

- the lexicographic approach (choose the option that performs best on the most important attribute);
- elimination by aspects (choose the option that meets most of the aspects deemed important); or
- the satisficing strategy (choose the first option that reaches a satisfactory standard on most decision criteria).

All these strategies are instances of bounded rationality with suboptimal outcomes (Simon 1976; Tversky 1972). These suboptimal outcomes, however, are regarded as sufficient for the person, once the expected additional increase in utility is less than the expected cost of arriving at a better decision. In particular, the time one needs to come to an optimal solution is considered as more valuable than the additional benefits derived from an optimal decision option. Recent experiments have shown that people use more complex and elaborate models of optimization when decision stakes are high (large potential payoffs), while they prefer more simple models when decision stakes are low (see Dawes 1988). Introducing simplified models for suboptimal decision making increased considerably the validity of theoretical predictions.

Third, most applications of utility theory equate utility with an increase in material welfare. However, individuals may feel an increase in satisfaction when they act altruistically or when they enhance their reservoir of symbolic gratification. Although experiments are usually designed to exclude these factors (or to hold them constant), it is not clear whether symbolic connotations (such as accepting money for a trivial task) may play a role in the decision making process (Weimann 1991). Some of the aforementioned "thinking-out-loud experiments" revealed that some subjects felt they should opt for the most cumbersome decision option, because that way they felt they deserved the promised payoff.

Fourth, subjects in such experiments often assume their choices will depend on actions and reactions of others, even in experimental settings that focus on individual actions without any interference from other actors (Frey 1992). By deliberating over how others could interfere with their preferences, they may select the suboptimal solution, because such a solution does not solicit competing claims by others. Game theoretical models, though normatively inappropriate for these conditions, may actually offer better predictions than simple SEU-based predictions.

Fifth, some analysts claim that the artificial situation of a laboratory, as well as the "playful nature" of the subjects (normally undergraduate students), are the main reasons for many observed deviations from the rational actor model (see Heimer 1988). In real life situations, with real stakes, people would be inclined to be more engaged in the decision making. As a result, they are more likely to use rational or at least bounded rational models to select their preferred decision options.

Some of these arguments have been discussed in Chapter 3. Three implications emerge from an overview of these proposed modifications. First, if utility encompasses all things that matter to people, then the model may well become impossible to handle. Humans, we agree, do act intentionally. The question is whether these intentions can be represented by well-behaved utility functions. If altruism, sense of style, and struggling for meaning are all manifestations of utility, it may (at least ex post) become possible to represent each sentence of Joyce's *Ulysses* as a utility maximizing item – but does it make sense to do so? Second, if closeness to real life situations and the inclusion of bounded rationality indeed improve the predictability of decision choices, as many experiments suggest, then the conclusion is justified that many individual decisions can better be explained by a modified approach. The modification would need to include the simplified models actors often prefer. Adoption of these modifications is not tautological (many potential options are excluded from the selection), and the implications for understanding market processes and policy measures are far from clear. Third, even granted that rational choice occurs at the individual level in many circumstances, other circumstances do not resemble decision situations or are not perceived as optimization processes by social actors.

Laboratory experiments frame the choice situation as a decision making context. Many human actions, however, are motivated by cultural imperatives, most notably habituation, imitation, conditional learning, emotional responses, and subconscious reactions. They are not perceived as decision situations; hence, rational strategies are not even taken into consideration. In addition, anecdotal evidence reveals that many actions are functions of both cognitive balancing and emotional attractiveness. The fragmentation of the psychological communities into clinical, cognitive, and analytical psychologists has prevented researchers from testing the relative impor-tance of these factors in motivating human actions.

Nevertheless, it seems quite appropriate to conclude that rational actor theories can describe some aspects of human life that center around decisions – in particular those decisions involving purposive choice and measurable outcomes from a limited set of alternatives. The real issue seems to be what happens to other aspects of social life, and to what degree actions are closely or loosely coupled to decision making in the RAP sense.

So far, we have looked at decision making by isolated individuals. The problem of finding the appropriate limitations and boundaries of RAP is even more relevant for the application of RAP to decision making by interacting agents. Our review of the literature – in particular the discussion of risk theories outside of the RAP camp in Chapter 5 – set off fireworks of conceptual problems for RAP when they were applied to interacting or collective decision making.

At the most fundamental level, all rational actor theories imply that individuals have sufficient knowledge about the consequences of their potential courses of actions to choose an optimum course – clearly a doubtful assumption. As a consequence, this assumption has been strongly contested by sociologists with a systems theory background (see Section 5.6). The prediction of outcomes is problematic already for an isolated actor. In a complex world where other actors interfere with, and shape the outcomes of, one's own actions, it is even more so. The presence of a multitude of other actors who are likewise making strategic choices, and thus able to affect the outcomes of one's own action in a myriad of different directions, makes it almost impossible to predict the potential impacts of one's own actions. The individual decision maker is trapped in a web of contingencies and uncertainties. In this case, rational actor theories do not provide a very meaningful orientation. The belief that rational action is possible and normatively required turns out to be an "ideological" element of those social systems (or subcultures) that would like their members to believe that the world is governed by rational decisions (Rayner 1987).

Furthermore, there clearly are social contexts where RAP seems out of place. It is not uncommon for an individual to be clueless about the impacts of various decision options (such as accepting one of several job offers). Furthermore, information seeking may turn out to be insufficient or too time consuming to produce sufficient clarity about options and potential outcomes. In such cases, RAP would suggest guessing or a conservative "better safe than sorry" strategy. Systems theory suggests, however, that social systems display greater coherence and predictability of individual actions than this random strategy would provide. If many human actions were random occurrences, as in the case of insufficient knowledge, social integration would be jeopardized. In response to this problem, systems theorists claim that conformity and predictability are accomplished through a variety of functionally equivalent procedures, of which rational choice is only one among others. In particular, orientation through reference group judgments and secondary socialization by organizations and subcultures provide selection rules for options independent of the expected outcome for the individual. In addition, these selection rules include other motivational factors such as emotions or social bonds that play little or no role in RAP. Again, as noted above, option selection often does not result from an optimization process, but from social or cultural considerations. Individual choices often are made on the basis of social criteria, such as reference group lifestyles.

Conflicts between expected and experienced outcomes are likely to increase with the extent of social and cultural complexity. This leads, on the one hand, to increased variability of human action, and thus to a growing number of potential choices that are open to each individual. Increased complexity, on the other hand, necessitates increased effort for coordinated actions. This dilemma has been resolved in modern societies with the evolution of semi-autonomous systems that provide a network of orien-tations within each distinct group. Such systems organize and coordinate the necessary exchange of information and services through specialized exchange agents. Depending on the cultural rules and images of these groups, rational expectations may play a large or small role in shaping these orientations. This is why rational decision making is attractive to some systems, but repulsive to others. *Absolute rationality – the strongest form of RAP – lacks integrative power.* It cannot integrate the diverse system rationalities that different groups accept as binding reference points for action and legitimization (Luhmann 1986a).

Complex modern societies are characterized by the coexistence of multiple rationalities that compete with each other for social attention and influence. This trend towards multiple rationalities is reinforced by the disappearance of homogeneous knowledge systems accepted by all social actors. Collectively approved and confirmed social knowledge has given way to heterogeneous knowledge systems. Each system, as our discussion of the work of sociologists of science and knowledge (Section 6.2) revealed, produces its own rules for making knowledge claims. These rules also determine which claims are justified to appear as factual evidence and which as normative, or as ideology or myths. Furthermore, they govern the process of selecting those elements of an abundant reservoir of knowledge claims relevant to system adherents; those elements are then matched with the body of previously acknowledged and accepted claims.

Under these circumstances, it is difficult to predict the factual consequences of decision options. How can individuals make prudent judgments about options if the relevant knowledge is not only uncertain, but also contested by relevant stakeholder groups? This is a core problem in risk management. The result of growing uncertainty and indecisiveness of potential outcomes leads, on one hand, to a larger share of non-rational incentives to ensure conformity. On the other hand it also leads to concerted efforts to create an impression of certainty *in the assessment of uncertainty*, often in the form of RAP-based strategies such as conducting quantitative risk assessments.

An important consequence of this state of affairs is that the State and other political agents have become less able to develop and allocate resources for regulating exchanges of materials and communication, and relations between social groups. Because it is increasingly difficult for the State to provide a commonly accepted and legitimate meta-rationality encompassing the plural rationalities of subsystems, State authority is challenged. As a result, institutional trust is eroded and political legitimacy is seriously jeopardized. Because of these ramifications (and other equally grave problems), and because RAP does not provide the requisite meta-rules required for social coherence among competing claims, the need for theoretical development beyond RAP is compelling.

7.1.3 Labyrinths of Perfection

RAP argues that, under certain conditions, the pursuit of individually rational actions leads to collectively rational outcomes. The common empirical observation that, in reality, quite often the opposite occurs embarrasses this argument. The common RAP response to the problem of individual versus collective rationality is well known: RAP theorists believe that this problem stems from external effects – in other words, from a deficiency in conditions.

Conversion of individual to collective rationality via the "invisible hand" is said to occur under the condition that external effects can be internalized as costs to individual actors. In particular, collective actions must be undertaken if the market fails to allocate social costs to those who cause them. In such cases, market-based instruments are readily available to impose external costs on relevant parties. Among the instruments available are the extension of property rights to all affected parties or modification of market prices to include social costs (Coase property rights or Pigou taxes).

An analogous argument can be made for the political sector. Under the premise that each individual has the same right to influence political outcomes, individual rationality provides a framework of laws and regulations that enhances the welfare of all. In both cases (economics and politics) RAP claims that individual actions by all participants lead to an optimal outcome if the conditions of rationality and a perfect institutional structure are met.

This strategy of "blaming" conditions for impeding the congruence of individual and collective rationality effectively builds a wall of immunization around RAP theories. Since all conditions for a perfect market or a truly rational political system are never met, deviations from theoretically derived predictions can always be explained by imperfect conditions. But requiring perfect conditions also reduces the explanatory power of RAP, however, since it does not explain how markets or political structures arise in the first place, or how people actually behave in an imperfect world.

RAP theorists have carefully looked at some imperfect market situations, such as oligopolies, monopolies, oligopsonies, and monopsonies. Again, however, economic theory generally assumes that actors in imperfect market conditions would act according to the assumptions of rational action as actors under perfect market conditions. This assumption may indeed hold true for situations where suppliers of goods or services monopolize the market (clear self-interest in maximizing profit). But it breaks down if the imperfection affects the consumer side, for example if the government is the only customer for a service. In this situation, prices tend to be higher than in perfect market situations, even though the single consumer, the monopson (in this case the government), has the power to reduce the prices to the theoretical minimum of marginal costs. An unending series of scandals, ranging from toilet seat covers costing hundreds of dollars each in military aircraft, to overcharging communities for constructing schools or municipalities for new infrastructure, illustrate this apparent deviation from rational decision making.

RAP suffers from the fiction that a hidden social tendency produces an equilibrium of perfect market conditions and political structure. If this were true, social systems would tend to orient themselves to the guiding principles of rationality in order to reach the goal of perfect conditions. In reality, however, many systems and institutions profit from imperfection. Even assuming intentions of rational behavior, particularistic interests interfere. Resistance to change results from vested interests. Imperfection creates losers and winners. It is not clear whether or why social systems would tend to enforce a movement towards more "perfection". It is especially unclear when the "winners" enjoy power, not only in the marketplace, but elsewhere as well.

There may even be strong normative reasons for allowing some imperfections in society, as they might enhance social stability, solidarity, or cohesion. Furthermore, structural barriers – such as market segmentation, incomplete knowledge, manipulation, market scale effects, etc. – make it unlikely that perfect market conditions can ever exist. Even political will, were it present and effective in contemporary societies, would not be sufficient to create or sustain the perfect conditions of RAP. Structural barriers are impossible to overcome without sacrificing other highly esteemed social goods. The real world will always consist of economic and social structures that deviate from the vision of perfection in RAP. One may ask whether this kind of perfection in economy and politics would be desirable in any case. Rather than a utopian vision for the future, the RAP equilibrium could turn out to be a nightmare consisting of utility-maximizing robots who have perfect knowledge, pre-programmed preferences, all equal opportunities – but who lack social and cultural rationality.

Of course, we live in an imperfect world, especially with regard to the standards of perfection set by RAP. Many external effects are not internalized and many social costs cannot be imposed on bystanders. Thus, the congruity of individual and social rationality presumed in RAP does not exist. Social rationality has to be generated and sustained by institutional or political action. Analyzing and explaining this process lies at the core of sociological theory and analysis. Assigning to political institutions the function of restoring perfect market conditions, as RAP theories would suggest, is insufficient. We need to analyze and design political and institutional processes that enhance social rationality in a world that will never be perfect in a RAP sense.

In this imperfect real world, claims regarding fairness and social justice often dominate discussions of collective actions, especially those that concern uncertainties and risks. Rather than focusing on optimal allocation (the principle to maximizing payoffs for all or achieving a Pareto optimum) such discussions are aimed towards finding legitimate and equitable ways for distributing wealth and opportunities for gain as well as costs and risks of losses. The body of theoretical and empirical knowledge to deal with the problem of allocation developed by RAP fails to offer significant assistance for coping with the problem of defining and implementing fair distribution rules (other than equating distribution with allocation, or to improving the market conditions). In essence, if we assume realistic market conditions and political structures, RAP cannot explain or predict social behavior in the face of risk adequately.

The treatment of externalities poses a philosophical dilemma for RAP; *de facto*, RAP-oriented analysts serve the dual role of analysts and high priests, claiming to explain and predict economic behavior, but interpreting their analyses in terms of theoretical assumptions that apply to an ideal set of circumstances and conditions never found in the real world. Using ideal types as instruments of analysis is not by itself problematic, quite the opposite. The problem, instead, begins when pursuit of self-interest in a perfect market system is used as the one and only ideal type. It continues when all normative advice is based on establishing or restoring the perfect conditions suggested by this ideal type.

The normative bias towards a world in which reality should be modified so that it behaves in accordance with the analytical models of perfect markets has tremendous political and social power – and considerable danger. Economic advisory boards on the national and international level often advocate social and economic measures that facilitate structures and policies in which RAP-based behavior is either assumed or mandated. RAP analysts tend to see and structure the world in accordance with the RAP model, rather than coping with the real world, including its tensions between normative orientations and practical behavior. They also tend to offer normative advice that neither matches the imperfect world, nor helps to make the imperfect world a better place in which to live.

This inability to look for flexible arrangements within an imperfect world is highlighted in the present debate on environmental risks and damages. Most RAP-oriented analysts insist that environmental problems could be handled successfully if external costs were internalized, i.e. if all social costs associated with the damage to the environment were placed on those whose actions led to the damage in the first place. Nobody really expects that complete internalization is technically, let alone politically, feasible. Nevertheless, societies may opt to use their potential for further internalization up to its limits. But there will always be some remaining imperfections that these analysts can point to as the cause of still unresolved environmental problems. Again, the theory is immune to empirical counter-evidence, since the presumed perfect conditions can never be met.

More importantly, the fixation on perfect conditions obscures a wide range of alternative theoretical and normative strategies. For example, there are strategies of collective action that take imperfect conditions for granted and try to develop policies that improve environmental quality within these structural constraints. RAP-oriented analysts often object to strategies or other measures than those that change market conditions. This is because the rationale of RAP limits legitimate collective actions to agreement on the setting and enforcing of common rules, on assuring equal opportunities for all actors, and on achieving and sustaining perfect market conditions. Policies that (seem to) go beyond this list of legitimate collective actions attract the objections of RAP-oriented analysts, who then tend to act in their role as high priests of individual freedom. They warn policy makers of the dreadful consequences for the economy of alternative strategies and they reaffirm the RAP-based importance of competitiveness and efficiency.

Efficiency, competitiveness, and optimal allocation are certainly important and desirable goals of society. The problem resides in the tendency of many RAP adherents to over-extend RAP-based models to problems and questions that require the development of a broader theoretical approach. Furthermore, the thought that an imperfect world, with all its irrationalities, may be more desirable and worth living in than a perfect world of RAP optimizers should have a legitimate place in shaping policies for markets and politics. Finally, the common RAP yardstick of efficiency and utility maximization may conflict with other desirable objectives and goals in society. Their violation may, therefore, be justified on other grounds.

A good illustration may be the sale of human organs from poor to rich individuals. Both individuals may maximize their individual utility. The poor person prefers the money over the possession of a body part, while the rich person values the needed organ more than the money. This is a perfect transaction in the eyes of many RAP adherents. Most people, however, would object to such a deal and call it "unethical." That people associate ethical norms with this choice situation means the abandonment of the principle for maximizing the utility of each affected party. Instead, solutions demanded go beyond self-interest. This illustrates that RAP alone may be inadequate to account for the difference in rationalities for perceiving and evaluating such social transactions.

7.1.4 RAP Claims in the Light of Competing Social Theories

The struggle for a comprehensive and overarching framework of rationality has been the focus of many recent social theories. Indeed, the need for such a framework represents one of the most pressing problems of contemporary societies (Wildavsky 1990). Many social system variables intervene between the axioms of rational action and the experienced cause-effect relationships in everyday life. These discrepancies weaken the appeal of rational actor theories for members of self-governing subsystems, and its power as a meaningful interpreter and predictor for use in social responses.

High risk technologies pose special problems because the RAP approach neglects a commonly accepted rationale for designing and legitimizing public policies for managing uncertainties and risks. As this weakness has become clearer, it has challenged RAP's hegemony over risk knowledge claims and questioned its legitimacy as a policy and management tool.

Two key assaults on the foundation of RAP are especially relevant: First, in a complex society individuals often are unable to foresee the consequences of their actions. Second, there is a persistent social need for conceptual schemes that establish and maintain commonly accepted *decision rules* without reference to optimal outcomes.

If all actors had complete knowledge about future consequences of a given set of decision options, and if choices were confined to zero-sum games, coherence and social conformity presumably would evolve almost automatically, the collective result of all individuals trying to optimize their utilities. Even in cases of conflicting claims or values, RAP has developed means and instruments – such as compensation – that help conflicting parties to negotiate a fair deal.

The problem, however, is that individuals often accomplish less than they set out to achieve. One social mechanism to cope with this conflict is to re-interpret ("rationalize") outcomes as variants of desired outcome: "Isn't what I got what I really wanted or needed?" Another is to create a system of symbolic incentives that compensate for the experience of conflict. However, these post hoc methods for resolving conflict perform poorly in creating the conditions for both social order and individual happiness.

It is this dilemma that attracts the scrutiny of critical theory (discussed in Section 6.5). Critical theorists suggest that, with the decline of a universal rationality principle, new social norms and values need to be generated that provide collective orientations without conflicting with personal aspirations (Couzens Hoy and McCarthy 1994; Habermas 1969). Similar suggestions have recently been proposed by the "new communitarians" (Etzioni 1991). For systems theory, new norms are part of an autonomous, evolutionary process divorced from individual volition. In contrast, critical theory argues that the integrative potential of norms and values can be created through free and open discourse. In this view, *discourse is an arena for the establishment of commonly accepted social norms or values* rather than an arena for negotiating a compromise between competing claims (as it is practised in conflict resolution models based on RAP). All participants voluntarily agree to participate in a discourse setting, because they perceive it as intuitively valid and socially rewarding.

RAP also faces major problems regarding structural influences on individual behavior. The relevant literature describes this problem in two ways (Green and Shapiro 1994; Münch 1992; Taylor 1996). First, many individual actions occur in a restricted social context in which the variety and quantity of decision options are limited, or are perceived as limited, by individual actors. This is not a serious challenge to RAP, however. The well-known RREEMM model (resourceful, restricted, evaluating, expecting, maximizing man) accounts for the restrictions and barriers people face when making rational decisions (Esser 1991b; Lindenberg 1985). Problems arise, however, if the decision maker feels guided by context variables such as norms, and does not perceive the situation as one of individual choice. This pertains not only to habitual behavior in the form of personal routines performed in everyday life below conscious awareness, but also to cultural routines based on more complex mechanisms.[90] Culturally shaped behavior occurs mainly below conscious awareness and does not imply internal cost-benefit analysis.

Second, the outcomes of individual actions produce side effects for other individuals – so-called interferences. These unintended, and very often unpredictable, side effects not only limit the ability of other actors to anticipate or predict the consequences of their actions, they also establish structural conditions for further collective actions. A major reason for the necessity of having supra-individual institutions in society is to ensure predictability and social orientation even in the presence of unpredictable side effects of individual actions. RAP-based theories do not deny the existence or the relevance of these structural elements, but they regard them as constraints on rational choice (constraints that can be traced back to individual actions of the past) – or as elements of a social learning process that teaches individuals to cope with interferences by predicting interactive effects more accurately. Furthermore, if collective action is relatively homogeneous, RAP treats social aggregates as if they were rational individuals. Organizations and households are rational actors, too, with goals and options, and they make choices via an optimization strategy. Like individual decision makers, they select the most efficient means for reaching predefined goals. Supra-individual actors making choices via RAP, like individuals, produce collective actions. All aggregate phenomena are interpreted as if they were reducible to optimizing choices by single individual actors.

[90] See the remarks of Giddens (1984) on the relevance of routines for sociology.

Combining these observations with our previous discussion, the following conclusion seems warranted: The most pressing problem for RAP is the idea that all human behavior, in particular all decision making under uncertainty, can be modeled as variants of optimization procedures. There is sufficient doubt that such a structure superimposed on all human actions can offer a satisfactory perspective for studying and explaining such phenomena as trust, solidarity, identity, affection, and – of course – risk.

7.2 Elements of a New Paradigm

We have argued that RAP has great merits and useful applications in spite of its deficits and problems. RAP offers a coherent and empirically relevant framework for specifying normative rules for rational decision making, for describing, as well as partially explaining, individual decision processes, and for analyzing – on a limited scale – social choice situations. However, it has become clear that several elements and assumptions of RAP require revision. Among the most challenging are those having to do with human behavior under uncertainty.

Many contesting theories concerning this problem have evolved within the social sciences, claiming to be capable of replacing RAP and overcoming its problems. Systems theory in sociology provides theoretical explanations for behaviors that are not governed by optimizing individual outcomes, but are oriented towards group cohesiveness, reference group judgments, and systems rationality. Critical theory provides theoretical interpretations of the genesis of norms and values in discourse settings. Post-modern theories add to our understanding of the subjective and interest-driven foundations underlying universal claims for knowledge and ethics. Nonetheless, all fail to offer a consistent and quantitatively expressible framework comparable to RAP. None has the ability to incorporate the many valuable insights and achievements of RAP. The monarchy may be shaking, but the contesting princes and princesses are still in their adolescence.

Competing perspectives promise some improvement where RAP is doing badly, but cannot offer a satisfactory (let alone superior) alternative where RAP is at its best – as an analytic structure with internal coherence and broad applicability.

At this point in theory development it seems clear that the quality of a new paradigm will be judged according to standards that RAP set in the first place. To meet these standards, a succeeding theory will need to provide additional explanatory power about social behavior in uncertain situations. It is premature, given the current state of knowledge, to attempt in detail a replacement proposal. This book is meant to unpack the layers of the dominant framework – RAP – and to outline *elements* of a new foundation. Our hope is that this effort will inspire others to search for new ideas and to discover strategies for coping with the problems posed by each theory.[91] Our task has been to delineate the strengths and limitations of each of the theories discussed. Meeting that task should make it easier to design new theoretical approaches. In order to provide additional guidance towards theoretical development, this chapter highlights some essential elements of a new theory of social behavior and action, particularly as it applies to risk behavior. These have emerged as problems and deficits in our critical review of RAP. If they are to succeed, we believe that theoretical alternatives to RAP must address these elements.

7.2.1 Scope Conditions

There is an important similarity between the RAP approach to society and the claim that in the natural sciences biology and chemistry could be reduced to the study of physics, ultimately based on the behavior of elementary particles. *Mechanistic thinking* has shaped both the natural and the social sciences. Newtonian physics and RAP are the twin children of mechanistic thinking.

In the natural sciences, however, advances in quantum mechanics have made it clear that while one can produce elementary particles by fragmenting physical systems, the reverse is not generally true – one cannot produce all kinds of physical systems by assembling elementary particles. This holds both experimentally and conceptually. A physical system may display states that do not and cannot correspond to unambiguous states of its elementary particles. Some states of the system can only be under-stood if one accepts that its particles are in a superposition of several states. And sometimes these particles do not even exist separately in any meaningful sense.[92] In a surprising way, mechanistic thinking has been proven wrong in the natural sciences. As a result, it is possible to indicate scope conditions that delimit the range of situations for which classical mechanics provides adequate descriptions.

In the social sciences, RAP, the other child of mechanistic thinking, still holds sway. For this reason, the need to combine the individual focus of RAP and the structural focus of many macro-sociological theories in a

[91] In this respect, we share the spirit in which Sraffa (1960) invited his readers to work on a better understanding of economic reality than the one provided by what he called "marginalism". With this concept he referred to a general theory built around properties of local optima in the use of scarce resources – RAP.

[92] d'Espagnat (1981) provides a careful introduction to these problems.

rigorous manner is urgent. One such attempt is structuration theory, as proposed by A. Giddens (1979; 1984; 1990; see also Section 6.1).[93] He rejects the idea that as a rule individuals orient themselves towards the expected utilities of various behavioral options. He argues instead that they orient themselves within a complex arrangement of traditions, individual routines and socio-cultural expectations. Each individual actor is part of the forces that shape the future context of actions for others, but at the same time is bound by constraints that were constructed by past actions and choices of others.

Such an open system would be chaotic if society did not develop consistent patterns of behavior that act as guidelines for individuals in choice situations. Rather than a logical product of individual actions, these patterns develop a structural logic of their own. For example, many traditional norms do not promise maximum payoff, or even an improvement of individual satisfaction. Rather, they assure continuity and stability of the system. They protect against social risk of disorganization, for example. Likewise, power structures are often supported even by those who lack power, because these structures maintain ontological security for the members of society. Giddens's main argument is that individuals have choices to orient themselves in different social frames (such as traditions, special institutions, system rationalities), but that the frames themselves are structural forces that go beyond individual actions and their effects on others.

Structuration theory is inadequate as a successor to RAP,[94] but we believe it can be used to delineate scope conditions for the future use of RAP. Since Max Weber, it has been claimed again and again that RAP is adequate to describe economic phenomena, but that other tools are necessary if one wants to study other phenomena – say, the role of religions in social development. This division of labor, however, seems unwarranted for two reasons. First, applications of RAP outside the economic realm – including the sociology of religions – have been sufficiently successful to show that RAP cannot and should not be confined to that realm. Second, applications of RAP are by no means sufficient to cover the economic realm as a whole. In particular, problems of risk and uncertainty in cases lacking futures markets elude RAP (see below). It seems more appropriate, therefore, to identify the scope conditions of RAP by singling out a specific kind of structuration processes, namely, *processes that lead to a stable*

[93] A similar attempt has been advanced by Bourdieu (1990; see also Bourdieu and Coleman 1991). We focus on Giddens because he has been more explicit in addressing the topic of risk, which is our main concern in this book

[94] Structuration theory does not even try to offer an analysis of market mechanisms that could compete with RAP, nor does it propose formal models of individual choice and social systems.

structure of decision makers with mutually consistent expectations, stable preferences, and reasonably well-known decision alternatives.

Experience shows that such structures occur in many areas of social life, e.g., in market economies, a "society of nations," and in kinship systems. Experience also shows that the relevant actors are at times individuals, at other times formal organizations, and sometimes more fluid social aggregates. In all these cases, RAP can and should be applied. When problems of risk arise under such circumstances, RAP can be expected to provide helpful insights to solve them.

However, experience also shows that such structures are by no means the default form of social life. Where they do not arise, RAP will have little to contribute, and may even be misleading. The challenge is to develop a theory that can reconstruct RAP as a special case of more general social patterns of structuration. Such a theory will need to address social phenomena that do not fit the RAP framework explicitly, in economic and other settings. Such phenomena are by no means exceptional in situations of risk.

7.2.2 Ambivalent Preferences and Expectations[95]

The task of developing a successor theory for RAP constitutes a research program that can only be realized if the scientific community engages in a long-term effort. In order to present a first outline of such a research program, it is useful to consider a stylized version of a typical problem in risk management (see Table 7.1): two parties struggle about how to deal with a risky choice, e.g., whether to build a specific landfill for toxic waste.

Each outcome of their combined actions may be represented by a lottery. For example, outcome α may designate the case where the two parties agree to build the landfill; this may lead either to a positive result or to a highly undesirable accident. The two parties may differ, however, in how they evaluate the consequences of this outcome, and they may differ even more in their judgment about the likelihood of the accident or the consequences. For example, the row player may represent the proponents of the landfill, the first possible consequence of outcome α may represent a regular operation of the landfill, and the second a possible accident. The column player, representing the opponents, considers the same possibi-lities, but associates them with different utilities.

To keep the numbers simple, no attempt has been made to provide empirically grounded figures for the outcomes in Table 7.1. One may

[95] This section has been adapted from Jaeger (1998).

imagine that outcome δ represents the case where both parties agree not to build the landfill. This is a risky decision: its consequence may be a corresponding reduction in solid waste, but also pollution from excess waste. Outcomes β and γ may represent different on-going quarrels about the landfill, each with its own uncertain consequences.

The parties may well be insecure about their judgments of likelihood. Such a consideration is excluded by the standard assumptions of RAP. In reality, however, ambivalent expectations are an important feature of many social interactions. In Table 7.1, they are represented by distinguishing two different sets of subjective probabilities (I and II), each of which may be assigned to each player. The proponents of the landfill may tend to think that the possibility of an accident is quite remote, but from time to time they may have doubts. So, they will in general hold probabilities I, but sometimes consider probabilities II. The converse holds for opponents. Even from these arbitrary numbers, the possibility of representing various outcomes of social conflicts about risk with plausible figures for utilities and probabilities is clear.

Each party in such a controversial game may now try to convince the other that a certain set of probabilities is correct. At the same time, each may use the controversy to make up his own mind about what he really wants to believe. Whether outcome α (building the landfill) will be an equilibrium of the resulting game depends on how these processes evolve. If the proponents manage to convince both themselves and their opponents of probabilities I, building the landfill will constitute an equilibrium. Other outcomes may also become equilibria: if both players accept probabilities II, it will be δ; β and γ may become equilibria if the players do not stick to the same probabilities.

So far we have considered a situation where actors are ambivalent about their judgments of likelihood. Technically, one possibility for representing such situations is given by the method of imprecise probabilities, involving not a single probability, but two probabilities designating an upper and a lower bound of the relevant range (Walley 1991). But once the importance of ambivalent expectations is acknowledged, there is no reason to restrict them to probability judgments; judgments of preference may be ambivalent, too (Smelser 1998). Technically, a notion of imprecise utilities, again characterized by an upper and a lower bound, may help to represent such situations.

The crucial point is that in situations of ambivalence agents do not simply try to choose optimal strategies on the basis of their expectations and evaluations. Rather, they define their evaluations, expectations, and strategies *in a shared social process*. In this process the identity of the agents may not be independent of their interactions (see White 1992, for a comprehensive analysis of this possibility).

	α	β	γ	δ
Utilities of row player for consequences of outcome lotteries	10, -100	10, -10	10, -10	5, -200
Utilities of column player for consequences of outcome lotteries	5, -200	10, -10	10, -10	10, -100
Probabilities I for consequences of outcome lotteries	99%, 1%	50%, 50%	50%, 50%	90%, 10%
Probabilities II for consequences of outcome lotteries	90%, 10%	50%, 50%	50%, 50%	99%, 1%
Utilities of row player for outcome lotteries with *probabilities I*	9.9 - 1 = **8.9**	5 - 5 = **0**	5 - 5 = **0**	4.5 - 20 = **-15.5**
Utilities of column player for outcome lotteries with *probabilities II*	4.5 - 20 = **-15.5**	5 - 5 = **0**	5 - 5 = **0**	9.9 - 1 = **8.9**
Utilities of row player for outcome lotteries with *probabilities II*	9 - 10 = **-1**	5 - 5 = **0**	5 - 5 = **0**	4.95 - 2 = **2.95**
Utilities of column player for outcome lotteries with *probabilities I*	4.95 - 2 = **2.95**	5 - 5 = **0**	5 - 5 = **0**	9 - 10 = **-1**

Table 7.1: Utility functions for a controversial game

In cases involving ambivalent judgments of probability and utility, stakeholders may engage in fruitful arguments about them, arguments that would be futile in their absence. How is a reasonable argument about ambivalent utilities and probabilities possible? RAP-based research usually assumes that judgments of utility lie beyond the realm of reason, while arguments about probabilities are restricted to Bayesian learning, combining prior probabilities with knowledge about empirical facts (see Binmore 1993, for a recent discussion). Neither is of much use for arguing about ambivalent utilities and probabilities. To understand such arguments, a different approach is needed.

We consider the case of utility first. Here, a standard assumption is that preference orderings can be represented by functions from the space of conceivable commodity bundles to the set of real numbers. People who orient themselves according to such preference orderings may be said to follow their self-interest, and they may be said to maximize their utilities. Preferences such as these are closely associated with experiences of pain and pleasure. Such experiences are shared by humans and animals, and they may be seen as lying at the biological roots of preference orderings. We call these *hedonic* preferences. Hedonic preferences may be considered as brute facts, without the necessity, or even the possibility, of further justification. In this view, human rationality shrinks to a remarkably simple affair. To be rational means, first, to display a preference ordering which fulfills some requirements of consistency and, second, to follow one's preference ordering in actual decisions.

Hedonic preferences are not very instructive for arguments about ambivalent preferences, for which ethical judgments are more relevant: ambivalence is frequent in ethical matters, and arguing about ethical judgments is widespread. How then are ethical judgments related to preferences? They do generate preference orderings – an action may be judged preferable to another on ethical grounds. Ethical judgment can be criticized and defended, and many human beings spend a great part of their lives doing so. Hedonic preferences may be understood without asking how people account for them in social settings; after all, animals have hedonic preferences, too. Ethical preferences, however, cannot be understood in this way. Accounting for one's preferences is a crucial part of ethical issues.

One might expect that ethical preferences are justified by relating them to some first principles, like Kant's categorical imperative, Rawls's maximin principle for social structures, or Bentham's idea of maximizing utility across persons. These principles certainly deserve mindful consideration. Nevertheless, they miss a crucial feature of moral deliberations in real life. A careful look at the "moralities of everyday life" (Sabini and Silver 1982) will recognize "the primary locus of moral understanding as lying in the recognition of paradigmatic examples of good and evil, right and wrong: the typical cases of, for example, fairness or unfairness, cruelty or kindness, truth-telling or lying, whose merits and shortcomings even a small child 'knows at a glance'. In the history of culture and society, and in the child's own growing up also, this moral discernment has to be applied to new and more complex cases with progressively greater refinement" (Jonsen and Toulmin 1988).

This observation can be elaborated by analysing ethics as rule systems (Jaeger and Rust 1994) in which the relevant rules are to be understood not as a deductive system of imperative statements, but as open lists of paradigmatic examples. As Wittgenstein (1953) has shown, a rule can be conveyed by offering a sequence of examples together with the injunction to try to continue it. Through discussion and attempts at continuation, understanding of the rule is developed. The point of this procedure is that there is no separate criterion behind the series in question; the series is the criterion.

Therefore, “An alternative to constructing ethical theories is to ground moral reasoning on central, paradigm cases. On these paradigm cases rests what coherence our ethical concepts can be said to have, and on these cases converges what ethical agreement we actually find between individuals and across cultures. On this model, rational persuasion replaces logical deduction, and cases replace principles as the focus of ethical reasoning. From paradigmatic cases moral argument proceeds outward, rather than downward from rational first principles and ethical systems“ (Elliot 1991).

According to this view, moral reasoning involves two kinds of arguments: first, arguments about what constitutes a paradigmatic example, e.g. a standard of justice; second, arguments about how different practical alternatives fare in comparison with such standards.

Still, ethical preferences may be represented by utility functions, as long as the notion of utility is used simply as a shortcut for describing preference orderings in a succinct way. The relevant preference orderings, however, differ in two important respects from those usually considered by RAP.

First, if the notions of justice, fairness, and the like are defined in terms of paradigmatic examples, then there are standards of justice that cannot be surpassed. Technically, this can be represented by a utility function which assumes an infinite value at well-specified points. Such a utility function parts company with many RAP specimens concerning non-satiability. In RAP, it is common to assume that people can never be fully satisfied. Although this may be a sad truth about some people, it is a questionable assumption even for hedonic preferences (animals seem to experience moments of plain satisfaction). In the case of ethical preferences, however, the assumption of non-satiability misses the very point of ethical judgment.

Second, ethical preferences are necessarily incomplete, shaped by a past that has to be creatively interpreted in order to master the present. Simply to repeat past actions does not honor ethical standards. In ethical matters, rationality includes the ability to develop preference ordering for novel situations in continuation of preference orderings developed in earlier situations. A case in point are the issues raised by current biomedical research. A physician may have a perfectly clear ethical orientation for the set of situations with which she is faced in her professional life, until some new technology raises new ethical problems. With regard to the new situation, she has no clear-cut ethical preferences. But she is not at a total loss. It may be quite appropriate to represent her situation by a utility function that is unequivocal over certain choice sets, but which is ambivalent over other sets.

Ethical and hedonic preferences may be seen as two extremes of a continuum, with aesthetic preferences lying in between. For problems of risk management, ethical preferences are often important. But in many human choices, e.g. with regard to consumption goods, aesthetic preferences are more salient. Both ethical and aesthetic preferences, however, often involve ambivalence in ways that lead to exchanges of arguments along the lines sketched above.

Similar reasoning applies to judgments of likelihood. There are paradigmatic cases for what counts as a certain fact or as something strictly impossible (Putnam 1987, provides a careful account of how human action relates to paradigmatic examples of certainty and to probability judgments). People talking to each other in a train will usually take their train ride as a fact beyond doubt and will consider a sudden jump of their train over the Pacific Ocean as simply impossible. As with ethical standards, standards of certainty may evolve, too, and their practical use leaves ample room for situations of ambivalence.

So far, we have provided a general argument to the effect that reasoned debate about judgments of utility and of probability is in fact possible. Procedures for participatory risk management offer opportunities for such debates. Studying their dynamics and developing them as contributions to solving practical problems of risk management is a fascinating twofold task. On the one hand are debates starting with ambivalent judgments that end with unequivocal judgments. In these cases, RAP-based risk management is embedded in a broader procedure in such a way that the ambivalence, which in the first stages precluded a straightforward application of RAP assumptions, later disappears. On the other hand, some debates reinforce initial ambivalence. In these cases, the decision rules provided by RAP must be modified, perhaps by generalizing them, perhaps by combining them with other decision rules, so as to solve decision problems in the face of persisting ambivalence. And as with standards of justice or certainty, decision rules also evolve (Burns and Dietz 1992).

Exchanges and arguments about ambivalent expectations and evaluations lie at the core of social rationality. This is especially relevant with regard to the analysis of choices by social actors, i.e. collective actors involving many persons. Triggered by Arrow's (1951) seminal im-possibility theorem, an impressive body of literature struggles with the impossibility of defining a reasonably democratic procedure that would aggregate arbitrarily given individual preferences into a social preference ordering (for a review, see Sen 1995). In conflicts about decisions under risk, widely varying preferences about actions with severe but uncertain consequences must be blended into a socially acceptable decision. But aggregation problems are not unique to society at large. Organizations have a similar problem, as do families and other social bodies. Human beings entertain a multitude of preferences and probability judgments that are consistent only in the frames of well-known choice situations. Novel and even familiar situations confront us with the ambivalence of our own expectations and preferences. These situations trigger discussions about judgments of utility and probability which may lead to social choices by resolving critical ambivalences. Participatory tools of risk management create open spaces for such discussions.

At both individual and social levels, expectations and preferences may evolve in the course of decision making. Sometimes they are static, as assumed by RAP; on other occasions they are not. "The definition of democracy as 'government by discussion' implies that individual values can and do change in the process of decision-making" (Buchanan 1954). Risk managers need to acknowledge that "Values are established or validated and recognized through discussion, an activity which is at once social, intellectual, and creative" (Knight 1947). And what holds for values, holds for expectations, which often evolve by passing through phases of ambivalence.

7.2.3 Rule Evolution

In the preceding section, we have looked at processes in which social agents may assume the shape of rational actors in the sense of RAP – or lose that shape. However, even the existence of social agents cannot simply be assumed. Individuals are born and die, institutions emerge and fade away, organizations are founded and disbanded. A better understanding of the processes by which social agents emerge, evolve, and possibly disappear is needed. An especially promising approach to understanding these processes consists in analyzing social reality in terms of evolving systems of rules (e.g. Burns and Dietz 1992; Jaeger 1994).

This view of social systems evolution is based on a metaphor taken from biology. In biological evolution, genes are seen as experiencing random variations which eventually lead to modifications in organisms. Organisms are subjected to non-random selection by the environment. Over the life span of single organisms, major patterns of the environment usually remain relatively stable. As a result, biological evolution leads to a remarkable fit between species and the environment in which they live. But over long time periods, environments change, favoring some species, which survive, and disfavoring others, which become extinct. By analogy, cultural rules can be seen as experiencing variations which lead to modifications of institutions. Institutions in turn experience selection from their environments. As a result, socio-cultural evolution leads to patterns of mutually interlocking institutions whose social reproduction is structured by cultural rules. (Note the similarity to Giddens's structuration theory.)

The role played by genes in evolutionary processes was discovered long after Darwin; but evolutionary approaches to social systems have a much older lineage than the discovery of genes. Two roots are especially important – German historicism and Spencer's theory of cultural evolution.

Germany has a long, rich intellectual tradition of trying to understand society and its principal components via historical analysis. For many theorists, understanding society requires that first history be understood; and understanding history is, often, understanding conflict between competing ideas and interests. Indeed, the German historical school was fundamentally antipositivist (and *de facto*, anti-RAP) because it objected to the positivist's faith in the omnipotence of the scientific method. Unlike the natural world, the social world was not believed to have any unbending, universal laws.

The German school of economics, despite a tighter focus, was shaped by the same intellectual presuppositions. It "nurtured...the idea that *any* social science had to be historical, [and that this view] did offer a genuine alternative to 'classical' political economy" (Manicas 1987, p 124). German economic thinkers were oriented to understanding economic processes, not by examining the laws of the market, but by examining the history of economic institutions – including the market. They did not examine the laws of the market, nor any economic laws, for the simple reason that they believed, consistent with historicist roots, that there were no universal laws in the social world. Thus, in contrast to the RAP of classical economics, there could be no universal laws of economics. Instead, the world was composed of particularistic histories that could be used to uncover the variety of historical periods leading to the various types of economies.

The historical and, to some extent, romantic emphasis of German scholarship shaped and was reinforced by theoretical method. A long line of German historical thinkers, including such luminaries as Herder, Dilthey, Hegel, von Ranke, and even Nietzsche, "conceived of the state as the integrated system of institutions rather than, as Hobbes and the tradition which followed him had seen it, an aggregation or aggregations of individuals acting in consort to satisfy individual interests" (Manicas 1987, p 94).

With few exceptions, positivist theories, such as the RAP of classical economics, have typically been individualistic – epistemologically, methodologically, and politically. In contrast, anti-positivist historical and institutional theories of thought, such as the German historical school, have tended to be sociolistic or sociologic.

The founding fathers of sociology grew up in this German intellectual tradition. Marx and Weber received their university training in Germany, and Durkheim – though educated in France – was the son of a Rabbi from Alsace, a region of France where strong Germanic traditions prevailed, and "was one of many French intellectuals who made the trip to sit at the feet of German masters" (Collins 1994, p 22).

Marx, Weber, and Durkheim, deeply influenced by the German intellectual tradition, turned toward macro-sociology as the principal orientation in theoretical method. Because of this influence in its formative years in the 19th century, sociology adopted a macrotheoretical method almost exclusively;[96] it was the principal orientation in theoretical method of Marx, Weber and Durkheim, the secular trinity. The adoption of a macrosocio-logical approach was simultaneously a rejection of the methodological individualism of RAP. Only by placing holistic lenses against the social world could one reach an understanding of society, said the leading founders of sociology, not by aggregating the vast and widely differing actions of individuals who were presumed to be rational. This macro-theoretical legacy, though later challenged within American sociology, remains as a crucial distinction between sociology and "individualistic" sciences, such as economics and psychology.

While the German historicist tradition had similarities with evolutionary thinking, Herbert Spencer, British empiricist and social philosopher turned sociologist (1852/1888; 1874-96/1898) took up Darwin's ideas explicitly. Spencer's personal transformation, from philosophy to empirical sociology, was instrumental to and reflective of the larger intellectual ferment that witnessed the independence of sociology as a scientific discipline from social philosophy. Spencer was highly influenced by three big ideas: the notion of evolution, an idea that was clearly in the air and generating great excitement in England and elsewhere; by the utilitarian orientation and *laissez faire* doctrine associated with classical economics; and by the positivism of August Comte. In essence, Spencer hoped to subsume RAP into evolutionary theory scientifically, thereby providing a grand theoretical synthesis. He therefore adopted the utilitarian premises of classic economics based upon RAP, but pushed this line of thinking into a larger mold – that of evolution.

Just as biological evolution in the Darwinian scheme of things showed that organisms systematically changed from simple to complex, so, too, did societies. And just as there were laws of change in biology, there were also laws for society. These laws are the unintended outcomes of the actions of self-interested actors making contracts in order to exchange with one another. Thus, for Spencer it was the combination of individual liberty and the open capitalist market that ensured evolutionary progress

[96] The two key exceptions were the writings of Comte and Spencer, but both were eclipsed by the work of Marx, Weber, and Durkheim.

– and the best type of society. The straightforward policy prescription from this theoretical analysis was *laissez faire:* leave these processes and institutions alone so that the laws of social evolution, unimpeded, would lead to progress.[97]

The idea that laws governed society was the credo of Comtian positivism. It was a straightforward borrowing from the then dominant scientific paradigm – the mechanics of Newtonian physics. As laws of gravity, force, and thermodynamics constituted the physical universe so, too, the laws of the market and evolution constituted the social world. The thoughts and intentions of the social actor, despite its strategic utilitarianism, mattered little after all. Human actors, like units of matter, were the vehicle through which the laws of social life were expressed.

The analogy between biological and cultural evolution established by Spencer has been reincarnated in the cultural rules approach. While a powerful idea, it must be treated with caution. Variations of cultural rules typically are not random processes. Parliaments, for example, do not formulate new laws by rearranging words or meanings in a random process. In rule evolution, the concept of an institution is used with a very broad interpretation that differs greatly from reference to biological organisms. Institutions rarely have well-defined life spans, and there is no analogue of sexual reproduction in the case of institutions. Moreover, the life span of some institutions may well be so long that in many respects they are more stable than their environments. Witness the financial market of the city of London or the Catholic church, for example. Moreover, the environment of institutions includes other institutions as well as physical persons and the biophysical environment. Finally, the use of genetic metaphors can be misleading when, as is often the case, social rules are confused with some kind of programmed statements allegedly stored in the human brain. The law certainly can be described as a set of rule systems, but these rules are usually codified and archived – not simply held by cultural actors. While some specialists will know some of these rules without having to access them in the written word, even most specialists will have to rely on the latter most of the time. The primary repositories of social rules are not individual brains, but social systems, aided, of course, by individuals, libraries, and other resources.

Despite these difficulties, the idea of rule evolution has important advantages over the social systems view of evolution formulated by Parsons (1967) and taken up by Habermas (1984; 1987). For Parsons, societal evolution is a process marked by differentiation of societal subsystems that, when successfully integrated, achieve a higher potential of adaption. Thus, evolution is equated with the ability of social systems to

[97] Spencer's ideas were adopted and promoted in the United States by William Graham Sumner.

deal positively with emergent problems (enhancement of adaptive capacity) (Parsons 1967, pp 21–22). Habermas's idea of communicative rationality, as the ability to integrate the rationalities of subsystems in modern society, develops this idea in a similar way.

Both Parsons and Habermas, however, have difficulty showing how processes of variation and selection interact in the dynamics of socio-cultural evolution. They also have difficulty showing how their concepts of socio-cultural evolution can provide an analysis of market mechanisms superior to RAP. Bluntly put, they are in danger of claiming that human history represents the organic development of a super-organism – human society – towards a teleological state that fits their theoretical desideratum. By leaving the analysis of market processes to RAP, they imply that somehow RAP could be integrated into some broader theoretical scheme without far-reaching modifications of RAP. Both claims are hard to sustain, and might be substantially improved by specifying the general idea of societal evolution in terms of rule evolution.

So far, the idea of rule evolution has not been employed in this way. On the contrary, Burns and Dietz (1992) offer a striking contrast between rule evolution and developmental conceptions of society. Luhmann (1984) also rejects the developmental view of society proposed by critical theory in favor of an open-ended process of variation and selection of social systems patterned by socio-cultural norms. He blends his evolutionary approach with Parsonian ideas into a neo-functionalist scheme which seems antithetical to any sociological critique of modern society.

However, a different approach is possible. Individually and collectively, we often try to realize a desirable state of affairs. Let us call this the pursuit of a *project* (following Jaeger 1994). The pursuit engenders a set of preferences which must be satisfied under given constraints. Suppose someone, starting with a poor education and a boring job, pursues the project of leading an interesting and meaningful professional life. Such a person will prefer jobs that offer opportunities for continuing education over jobs that do not offer such opportunities, and so on. The pursuit of projects gives rise to preference orderings of the sort presupposed by RAP.

Clearly, preferences are rooted in biological properties of the human organism. Babies prefer sweet over bitter food, and they prefer smiling faces over angry faces. They soon discover that they can take action in many different ways and begin to pursue projects, however – from the first attempts to throw a ball to those aimed at becoming professional at the task.

Because goal orientation shapes preference orderings, human preferences are often context-dependent. Individuals typically entertain different projects, which sometimes contradict one another, and their interest in particular projects shifts over time. Moreover, projects are constructed of smaller projects: one wants to spend a nice evening; therefore one wants to go to see a movie; therefore one takes the subway, etc. That is, there are different levels of preferences. Somebody wants to be strong, and as he believes that steak will make him strong, he wants to eat steak. He then learns about BSE, hormones fed to cattle, etc., and comes to believe that eating steaks may actually make him sick. At one level, his preferences are unchanged, while at another level they may have changed dramatically. Preferences may even contradict one another, as when a smoker wants to stop smoking, but also wants to smoke a cigarette.

Human projects are shaped by social rules and, as Giddens (1984) emphasizes, rules simultaneously enable and constrain individual and collective human agents. Without the rules of the soccer game, it would be impossible for a team to design a project to score a goal. At the same time, the rules limit the possibilities of realizing the project, e.g. by excluding the possibility of passing the ball with one's hands.

Mediated by projects, then, the evolution of social rules shapes human preference orderings. This occurs with given individual and social agents, but it also happens as agents emerge and develop. Newborns become persons by being exposed to social rules, including the rules of language. Human beings become social agents under the influence of social rules, as when they team up for a game, a business, a social movement, a family, or any of the myriad of activities enabled and constrained by the evolution of social rule systems.

This body of ideas offers promise for the development of a successor to RAP. An important asset is their amenability to the kind of formal treatment on which the success of RAP-based models relies to a large extent. In bio-logy and ecosystems studies, tools of non-linear dynamics and computer simulation have been combined with large data sets. As a result, formal models of population dynamics have reached a remarkable level of sophistication. Moreover, game theory has been applied to biological evo-lution with remarkable success, yielding the well-known models of evolutionary game theory. Further generalizations of game theory should be undertaken in such a way as to get closer to an understanding of rule evolution in multi-layered networks of individual and social agents.[98]

7.2.4 Prices, Investment, and Risk

Prospects for embedding RAP in a broader theoretical framework are enhanced both by ambivalence regarding preferences and expectations

[98] Work in progress by Tom Burns and colleagues (Burns et al 1997) seems especially promising in this regard.

and by rule evolution. However, any such attempt must accept the challenge of the invisible hand; that is, preservation and possibly improvement of the insights regarding the operation of markets made possible by RAP.[99]

As a case in point, consider Lindenberg's (1989) remarkable attempt to generalize the RAP model of human action. He first contrasts RAP with a sociological theory of action, exemplified by the seminal work of Parsons and Shils (1951). He claims that "no serious integration of the traditions will be possible without first integrating the most prominent *behavioral insights* of both traditions: the insight (mainly from economics) that changes in relative scarcities (i.e., in relative prices) will change behavior accordingly, and the insight (mainly from sociology) that the definition of the situation matters" (pp 175–76).

Lindenberg points to subjective utility theory (SEU) to justify the insight that relative prices matter. In contrast to pure neoclassical price theory, SEU permits explicit incorporation of uncertainty and, therefore, consideration of social psychological influences on the effects of uncertainty. This is an important step toward integration, and for this reason Lindenberg chooses SEU rather than price theory in developing his theory. He views Kahneman and Tversky's (1979) prospect theory as "the most prominent attempt to bring the definition of the situation (or 'framing', as they call it) into scarcity-based action theory" (p 176). Finally, he proposes the improvement of prospect theory with a *discrimination model of stochastic choice*. About the resulting theory, Lindenberg advances the following claim: "Like prospect theory, it has fared well in empirical tests, but unlike prospect theory, it is explicitly geared to sociological rather than psychological problems" (p 187).

Suppose now that empirical evidence would show that the elaboration of RAP with a discrimination model of stochastic choice is clearly superior to the traditional RAP models of human actions. Would this be sufficient to drop the latter models? The answer is a categorical "no". The real importance of RAP lies not in its ability to describe the choices of human agents, but in its treatment of how the invisible hand of the market is able to coordinate these choices in a highly efficient manner.

Proposals like Lindenberg's, therefore, face the challenge of how to account for the operation of markets. Otherwise, the attempt to make the theory of human choices more realistic leads to an unacceptable consequence; namely, that the successes of market economies are rendered unintelligible.

---

[99] "Made possible" both in the sense that the insights are due to RAP, and in the sense that the market economy itself has been shaped by RAP.

We are not arguing that Lindenberg's model should be discarded. We want to emphasize, rather, that attempts to supersede RAP by subsuming it within a larger framework face the challenge of developing an understanding of markets that is superior to that currently available. Another example is given by Macy's (1995) impressive attempt to overcome the limitations of RAP in models of choice. Is it possible to develop a better understanding of markets on such a basis? The answer to this question will be crucial for the future of models like Macy's – and for the future of RAP.

Or, consider the "broadly conceived rational-actor model" used by Opp and Gern (1993). In this model, beliefs, social norms, and interpersonal networks are all considered as somehow affecting an actor's utility function – a perfectly reasonable assumption for the study of individual choices. In particular, it may be quite helpful for research about choices under conditions of risk and uncertainty. It is certainly possible to elaborate meaningful findings about the aggregate outcome of individual choices in a given population. It is much less clear, however, how such a model could explain the workings of markets, a challenge that must be met by any attempt to supersede RAP theoretically.

It is here that the intellectual power of RAP lies. And here lies RAP's Achilles heel. The main claim of RAP is that if a multitude of actors interact in interdependent markets, there exists at least one vector of relative prices for which demand and supply match for all markets. The market achieves equilibrium and the market clears. When no external effects arise, any resulting equilibrium is Pareto optimal; i.e. it cannot be modified without reducing the utility level of at least one actor. The background assumptions needed to obtain these results describe consumers, producers, and markets in a highly idealized manner. For example, they imply that in equilibrium every firm is faced by unlimited demand for its products and that no firm can reduce unit costs by increasing production.

The basic results of RAP are static in nature. Contrary to conventional wisdom, it has never been proved that interdependent markets will converge to an equilibrium if they are not already there, much less that they will do so at a speed that exceeds the incidence of external shocks. If labor markets take ten years to adjust to technological innovations which happen twice in a decade (an eternity in today's rapidly changing technological environment), equilibrium will rarely, if ever, be reached. Even without external shocks, however, the dynamics of markets may well follow chaotic trajectories without ever reaching equilibrium.

Research on chaotic systems started with the study of price changes in competitive markets (Mandelbrot 1963). Moreover, Sonnenschein (1973) demonstrated that demand functions in interdependent markets can literally take on any form. This implies that there will usually be multiple equilibria, and RAP has no way to explain why the system arrives at one equilibrium rather than another (Kreps 1992).

This is especially worrying because of the crucial role played by expectations in the global economy, where decisions about industrial and financial investment occur constantly. In theory, these decisions could be based on futures prices for goods that might be produced later on with the help of investments made today. Such futures prices could result from a stock exchange in which contracts about future transactions were traded today. In practice, such contracts exist only for a limited number of goods, such as cereals and some raw materials, and for a very limited time span – a few years at most. What determines investments, then, are not futures prices, but expectations of future prices.[100] These expectations are the result of sociological processes of opinion formation and "educated guesses" based on imperfect information, not the result of market operations. *Expectations* are a presupposition of a functioning market economy.

If an ideal market economy had one and only one equilibrium, it could be argued that the expectations of economic agents might at least approximate that equilibrium. Given that real markets have many possible equilibria, however, the expectations of economic agents will in fact determine the path along which the economy will develop. The uncertainty involved in the interaction of economic agents is not resolved by market mechanisms, but by sociological processes of opinion formation. Clearly, these processes are strongly shaped by social structures; not every opinion enters them with the same weight.

Market mechanisms can only operate on the basis of these processes. Risk and uncertainty are not peripheral issues for RAP. Rather, they are crucial challenges for the central task which RAP claims to have achieved – explaining the operation of a market economy.

This situation suggests that the familiar explanation of prices in terms of supply and demand is far from satisfactory. A better understanding of markets would no doubt be highly welcome. The need for such a theory has recently been stressed by Drucker (1993), with an argument about innovative economies. The definition of the price for new products is an essential step in product innovation. Prices emerge out of a complex social process involving engineers, marketing specialists, and others involved in product development. Prices cannot be explained by demand functions, because preferences for the products in question will be formed only after they have been developed and marketed. And they can hardly be explained by production costs, because under schemes of target costing, engineers are bound to develop technologies at production costs preset by management.

[100] Morishima (1992) gives a remarkable analysis of this process.

This situation calls for a sociologically enriched analysis of markets. Although some efforts are underway, few social theorists have produced an explicit theory of market processes which could challenge RAP on its own ground. Among the most promising are those elaborated by network theorists that combine, albeit somewhat idiosyncratically, elements of very different theoretical traditions. Given the crucial relevance of market processes for RAP, it may well be that a successor for RAP will only emerge if this kind of theoretically hybrid analysis of markets is integrated into an overarching framework.

One of the most advanced attempts to develop network theory along these lines is the sustained effort of American sociologist Harrison White. White's sociological analysis of market processes (White 1981) has resulted in a sophisticated and promising alternative to the RAP model of economic man (White 1992). Uncertainty plays a crucial role in his analysis of social reality in general and of markets in particular. "No firm can reliably assess relative qualities of other firms, and every firm knows that its position could be affected by choices made by any one or more of its competitors" (White 1981, p 519). Uncertainty about the choices of competitors conforms to patterns familiar from game theoretical analyses of markets, where the outcome of an actor's choices depends not simply on the working of an anonymous market, but rather on the choices of identifiable actors operating in the same market. Uncertainty about *relative qualities*, however, breaks new ground. In White's view, the main problem refers not to the choices of individual consumers with regard to identifiable producers, but to relative qualities reflecting aggregate evaluations of consumers about the complete array of products offered in a specific market. From the point of view of the single producer, the total market reflects a social environment which cannot be known with certainty.

In White's approach, markets are seen as social networks that process the uncertainty associated with conditions of demand. It is quite plausible that a single organization (a monopolist) is not very efficient in handling the inevitable fluctuations of demand for a specific kind of product. A market consisting of several firms will usually be more flexible in this respect. White's argument, however, implies that so-called perfect competition is rarely the best way of achieving the flexibility required for market success.

More often than not, a network of little more than a dozen firms seems to be appropriate for the analysis of a given market. These firms observe each other more than they observe consumers, because together they are able to establish relatively stable expectations for the operations of a given market – expectations that none could establish alone. These expectations include specifications of prices for products of different qualities. They are not, however, simply expectations about some state of the world, but mutually interlocked expectations which define a role structure. The expectations then crystallize into social norms.

The notion that social networks may be able to process specific risks by creating social norms that organize behavior is a powerful idea with great potential. Thus far, however, its scope has been limited to isolated markets. White has not been able to offer an analysis of systems of interdependent markets. Nevertheless, network studies of isolated markets show remarkable capability to challenge RAP on its home ground. Other studies strengthen the expectation that the study of markets as social networks may lead to an approach which could supersede RAP,[101] including works by Granovetter (1973; 1984; 1985) on the importance of specific social relations in specific markets (e.g., the role of weak ties for job searches on labor markets). Similar ideas are being explored by Burt (1992), who considers competitive advantages as being related to the ability to bridge structural holes in specific social networks.

One promising feature of network theory is a more "realistic" conceptualization of the individual agent than that provided by RAP. Both Burt (1992) and White (1992) offer an explicit conceptualization of human agency. While these two approaches differ in substantial respects, they share one very important feature – an emphasis on processes of identity formation in social settings. One key shortcoming of RAP, as often noted, is its presupposition that values, tastes, preferences, and other social elements exist independently of the decision context. The idea of identity formation, familiar to sociologists at least since the work of Mead (1934) and the symbolic interactionists, can be used to overcome this unsatisfactory feature of RAP.

Personal identity is formed via social processes. This leads inevitably to dynamically evolving preferences, as when actors require certain kinds of products to acquire or reinforce a desired identity. Whether or not such preferences will acquire the kind of overall consistency presupposed by RAP is another question. In any case, relating preferences to processes of identity formation leads to another theme of considerable relevance for attempts to develop a successor to RAP – the theme of human reflexivity.

7.2.5 Reflexivity

Classical social theory distinguishes social behavior from social action. Behavior refers to what humans do (including doing nothing), while social action refers to those behavioral elements to which social actors assign a specific meaning or *intention*. Intention constitutes a central tenet of RAP,

[101] For an attempt to tackle the problem of interdependent markets with a sociological notion of normal prices, see Jaeger (1994).

expressed as goal-expectancy. However, as proponents of symbolic interactionism and ethnomethodology have pointed out, meaning cannot be reduced to intention alone. Meaning, in part, includes the impact of one's actions on things that one values; but it also refers to many other aspects of interpretation and social interaction. Individuals often act on intuitive social theories. That is, self-made social theories tell them why and with what effect others act as they do. It is part of the human inference system to generalize on the basis of a limited number of observations and draw conclusions about the general nature of situations and the persons. Without understanding the intuitive social theories of actors, it is hard to understand why they behave as they do.

Only if we know the subjective stories and/or the guiding social constructions of the actors involved, are we in a position to explain present behavior and, under special circumstances, predict future behavior. Giddens (1984) speaks of the double hermeneutics of understanding social action. Social scientists inevitably must interpret the actions, symbols, rituals, etc., which are the subject of their study. Without interpretation, it would be impossible to identify a body movement as a form of greeting, or a piece of paper as a contract. But this hermeneutic process refers to material which has already been interpreted by the social actors involved – people greeting each other, organizations bound by a contract. Social scientists develop interpretations of a social actor's interpretation of his or her social context and hence construct theories about the construction of their subject's theories, under the premise that the latter determine social behavior to a large extent.

Behaviorism, and more recently sociobiology, have questioned (or ignored) the importance of subjective interpretations of actors for explaining and predicting behavior. Similarly, systems theories, world system theorists and other macro-oriented sociologists leave little theoretical room for the individual – let alone her subjective rationale for action. By contrast, micro-oriented scientists are convinced that human behavior cannot be understood without reference to the subjective expectations and meaning that each actor associates with action. RAP is based on the same conviction, but limits the range of meaning to a calculus of costs and rewards associated with decision outcomes. In doing so, it subsumes a whole class of motivations, including emotion, bonding, trust, duty, honor, and others as black boxes without investigation.

This problem is especially relevant for the study of organizations, including private businesses competing in the marketplace. Organizational theory has evolved from perceiving organizations as monolithic entities with fixed goals, rules, and borders, to perceiving them as reflexive bodies that observe the performance of other organizations and learn to adjust to changing social conditions (Scott 1995). Recent organizational sociology has demonstrated that Weber's characterization of organizations as prototypes of instrumental rationality provides only a narrow perspective on how organizations are structured and what they do.

Organizations offer niches for informal social learning and often provide rationales of their own, including commonly shared knowledge claims, values, and convictions, as well as common views of the outside world. Many organizational cultures have evolved not only functionally; but as providers of social security, solidarity, and emotional support, as well.

Furthermore, many successful organizations have established subsections responsible for studying the outside world and translating their observations into internal organizational language. RAP oriented theories provide a conventional explanation of these empirical results, interpreting such efforts as indirect means to accomplish better performance and to realize instrumental rationality. This interpretation misses, however, the whole world of symbolic interactions and their relevance for social learning. The more outcomes are uncertain in a complex world, the more important it is for organizations to provide additional incentives for consistent and predictable actions by individual members.

Because individuals must search for meaning in a complex and sometimes confusing world, social institutions develop special agencies to collect, process, and integrate social feedback as a means to redirect their own behavior and to cultivate their social status in the outside world. This is more than just clever marketing. The point is that predictions about outcomes have become so uncertain that purely internal deliberations about future actions are unlikely to lead to reasonable decisions. To make decision making feasible, then, institutions define their tasks in accordance with how they are perceived by the outside world. They generate options in a trial and error process and assess outcomes by pre-testing for social acceptance. At each stage of decision making, institutions reflect about their goals and strategies by looking at their performance through the lenses of outside observers. This type of self-reflective behavior is particularly relevant for institutions that need public support, such as economic and political institutions.

In striking affinity with the results of new organizational studies, Beck, Giddens and Lash have recently emphasized a new perspective for social theory: reflexivity (Beck et al 1994). The idea of reflexivity represents an important departure from the traditional approach of RAP (and other theoretical approaches in the social sciences). Because this approach has attempted to construct special theories independent of the subjects to which the theories apply, this perspective may require thorough revision. The old hermeneutic claim that social actions can only be understood if the investigator understands the actors is reversed by the claim for reflexive social science: social research can only be fruitful for explaining social actions if the actors on whom the research is performed can understand its theoretical reasoning. Incongruity between theories explaining behavior and the theory held by those behaving may well be an indication of serious flaws in the analyst's understanding, rather than of a lack of theoretical literacy of the objects under study.

Assumptions and abstractions that form the structure of a theory must be tested not only on statistical grounds, as traditional social science theory demands. They must also be tested against the beliefs of the subjects of social theory. Do theoretical results fit the actors' interpretation of their behavior? Investigators who claim to have understood the motives or shared meanings of social actions must demonstrate that they have not superimposed their own meanings on sets of social actions. They need to show that their interpretations fit what the actors meant when acting. Assumptions and theoretical explanations need to be constantly tested against the actual meaning that actors construct in the course of social interaction.

An example may clarify this proposition. In many value studies, value clusters have been identified by statistical factor analysis. Factor analysis is a statistical method for combining individual items into one conceptual dimension. The items are then combined in one factor labeled by the analyst with what he or she felt was the common denominator of all these items. As a measure of the subjective meaning of actors, factor analysis in isolation has two serious limitations. First, factor analysis does not automatically identify a content-related shared element among items that form a statistical dimension. Second, the process of labeling is itself governed by the inter-pretation of the analyst. A reflexive researcher would not leave factor results unchallenged. Instead, he would go back to the subjects and ask them whether they felt that these items had something in common. He might also ask them to provide a meaningful "label" for the elements as a reflection of their own values. Practically, this can only be done with selected indi-viduals or institutions. But it represents a fundamental change in method. The reflexive procedure extends the traditional rule of falsification beyond the testing of theoretical predictions (which has been weak in the social sciences, anyway). Thus, we have an additional rule for validating social science studies. A study is considered valid to the extent the subjects involved can see "themselves" in the results.

How does this affect RAP? In a reflexive use of social science, RAP assumptions and perspectives can be used as hypotheses of why actors behave in a particular way. But such hypotheses would not (only) be tested by comparing predictions and actual outcomes, but by giving actors the opportunity to respond to the findings. Only then could the investigator determine whether those who responded acted in accordance with RAP assumptions. If actors disagree, alternative interpretations may be proposed until actors and analysts agree that they have found a non-trivial level of mutual understanding. In some cases, the resulting interpretation may be consistent with RAP; in other cases, alternative interpretations may be required. It is always possible, of course, that theorists and actors cannot arrive at a non-trivial level of mutual understanding.

Clearly, there is a danger here that critical research becomes impossible by definition. If a study of CEOs yields results that are highly critical of their behavior, it will be hard to secure agreement with the researchers' interpretations of their behavior. It would be futile to handle such situations by searching for a common interpretation; rather, the task will be one of finding ways of agreeing to disagree. This may be even harder in cases of open, possibly violent conflict. But in such cases it is especially important to stick to the general rule that whatever macro-explanation is being offered can be related to the subjective definition of the situation by the actors involved (Esser 1991a).

This is especially important because otherwise a crucial option for future action will be missed; namely, redefining the situation. As Wittgenstein (1953) once asked: Can we not imagine a group of people playing on a lawn with a ball and making up the rules as they go along? Clearly, this means that preferences and expectations evolve through phases of ambivalence as described in Section 7.2.2, and that rules as well as social agents – like teams – evolve as discussed in Section 7.2.3. Similar patterns arise in situations of considerable social and historical weight, as when a political constitution is drafted or amended, or when a labor market is formed or transformed. Taking the potential for reflexivity into account does not deny the adequacy of RAP to describe and interpret many important social situations. But it does enlarge the options available beyond the limits implied by RAP assumptions. It expands opportunities for explaining social change and designing solutions for practical problems.

7.3 The Practical Context

In the preceding section, we have looked at four elements of a possible successor for RAP: ambivalence, rule evolution, markets as networks, and reflexivity. All are necessary, but by no means sufficient to the task. Considerable effort by many researchers will be required. Conceptual analysis, empirical investigation and formal modelling will all be required. Does this mean that that endeavor lacks practical relevance for the present? Quite the opposite. Just as RAP evolved in close connection with practical efforts to design economic policy in the times of industrialization (especially in the U.K.), a successor is likely to evolve in a practical context, as well. This is even more likely as dealing with uncertainty is a core challenge for policy analysis in many fields (Morgan and Henrion 1990).

7.3.1 Implications for Political Behavior

The metaphor of the invisible hand is used by RAP to explain how satisfactory collective outcomes emerge from an array of independent individual choices. At the economic level, the invisible hand supposedly guarantees that social welfare is maximized by market mechanisms as individuals pursue their private rational economic choices. The corollary at the political level is that social welfare is maximized through institutions of liberal democracy, where individuals pursue their private rational political choices. This is the fundamental presupposition of political theory based upon RAP – rational choice theory. One of its leading proponents, Buchanan (1988), stated the problem this way: How are differing individual preferences to be reconciled in reaching results that must, by definition, be shared jointly by all members of the community? The answer, according to RAP, lies in a competitive process for majority voting. The politicians' rational quest for votes, which citizens rationally cast, will ensure that the use of state power serves the interests of the general population as well as is possible to achieve.

The political argument of RAP is closely linked to the economic perspective in Joseph Schumpeter's economic theory of democracy. Schumpeter is credited with transforming classical democratic theory into a descriptive theory of contemporary democracy. By re-defining democracy as simply a method for achieving desired political outcomes, he likened politics to markets. Pursuing this line of thought in an analysis of voting, Anthony Downs (1957) viewed politicians as similar to entrepreneurs offering goods in the market, and voters as similar to consumers. Downs suggested that voters act to attain their preferences by "spending" their votes on politicians who offer the most appealing profile. The assumption is that politicians serve as a liaison between the public and the goods-providing arm of government to provide returns that individual voters desire. Social welfare is presumed to be maximized by voting, even though not everyone's candidate is elected.

This is no doubt an elegant but very simple picture of how the democratic process works. Recent experiences in public management of technological and environmental risks, however, strongly suggest that the mechanisms so described are insufficient to deal with current problems of risk and uncertainty. The institutional machinery of representative democracy has again and again been paralyzed by risk controversies. In some cases, such controversies have effectively precluded practical solutions to pressing problems. This negative result is another motivation for the critical assessment of RAP presented in this book.

Representative democracy is remarkably resilient. It has proven to be capable of spawning innovative mechanisms that can overcome many inherent structural limitations. Forms of public participation have recently been one of those innovations.[102] Public participation offers a workable setting for the solution of many – but certainly not all – controversies concerning environmental and technological risks. A stylized version of such procedures is described in four steps, as follows (also see Section 5.1).

First, one or more decision making bodies which are legally entitled to make decisions about some specific risk commit themselves to a public participation procedure. They agree to accept the outcome of participation in a well-defined decision range. The question for public participation might be where to site a new landfill, while the question of whether the landfill will be sited at all remains the province of the decision makers.

Delimiting the scope of a public participation procedure is of critical importance. It may be impossible to reach consensus if the scope of the procedure is too small or if it is too large. Surprisingly, perhaps, experience shows that it is often possible to define a workable decision range for the public procedure to work. This usually implies that the procedure starts with a finite set of alternative options among which a choice must be made.

Second, people from affected social groups and organizations are invited to join a forum for open discussion of the decision. Typically, the forum is composed of a series of working groups. It is important that these groups not be too large or too small. Here, too, however, experience shows that a workable size can often be arranged. The crucial move is to constitute the forum as a jury system by inviting laypersons, not specialists. Laypersons are usually not formal representatives of the groups from which they come. They are instructed to engage in debate as individuals committed to the search for a collective decision. Their motivation may range from curiosity to a clear sense of some collective interest. In particular, positive and negative incentives may be used by the original decision makers to get the relevant groups and institutions to participate in the procedure. The time invested in the procedure will usually be paid for, while groups who do not participate may face decisions taken at their disadvantage.

Third, the forum so established has the right and the resources to hold briefings with professional experts from fields relevant to the inquiry. These experts will be instructed not to suggest how to solve the overall problem, but rather to answer factual questions raised in the forum. Experts need not agree with each other, nor do they need to convey uncontested knowledge. The crucial element here is the division of labor by which the forum decides which questions shall be asked, while the experts decide

[102] They may well be combined with forms of semi-direct democracy, as the referenda and initiatives used in the Swiss and, to a lesser extent, in many other political systems.

which answers to give. This enables the forum to focus on both the factual knowledge and on unavoidable uncertainties that are relevant to the decision process.[103]

Fourth, the forum will work towards a decision by following a well-defined sequence. With the help of a facilitator trained in public participation procedures, a list of criteria to evaluate the consequences of given options is elaborated, possible consequences of each option are assessed, and the criteria are then applied to these consequences. In some ways this resembles the RAP view of rational decision making. The crucial difference, however, is in the emphasis on open dialogue and collective debate as the procedure by which the sequence is actually performed. No algorithm can guarantee a consensus on an optimal decision. Differing value orientations, strategic interests, expectations of consequences, and interpretations of arguments may or may not result in consensus. The key point where a consensus needs to be established is in the choice among options.

The remarkable fact about public participation in risk management is that, if the procedure is properly followed, a consensus can often be established. The procedure must permit the possibility of not reaching a consensus. In this case the decision falls back on the decision makers who started the public participation procedure. Experience demonstrates, however, that a consensus is much more likely to be reached when these procedures are followed than is the case when similar problems are tackled using only the institutional means advocated by RAP.

The experience of successful public participation strongly suggests that the most committed post-modernists push their critique of rational procedures too far. By contrast, important features of public participation can be understood by using the concept of communicative action as advanced by critical theory. It must be noted, however, that settings of public participation are far from the ideal speech communities of critical theory. Public participation relies much more directly on power relations and strategic rationality than purists of critical theory might be willing to envisage. In the practical development of public participation procedures, RAP will not be simply dismissed; instead they will be integrated in a more comprehensive understanding of public deliberations and decisions.

The form of public participation described above is by no means the only one; nor is public participation the only form of institutional innovation which may lead beyond RAP-based institutions of democratic decision making. The example chosen, however, may convey a sense of practical steps toward the development of a theoretical successor of RAP.

[103] Advances in computer technology will make it increasingly attractive to use computer models as an interface between expert knowledge and laypersons' debates. This possibility is currently explored by research in Integrated Assessment (Jaeger et al 1999).

7.3.2 Implications for Economic Behavior

If RAP is not to be the final word of the social sciences on economic institutions, the implications for risk management in business are not yet clear. Until just a few years ago, the debate about the future of capitalism was shaped by the global struggle with planned economies. Now that this struggle seems settled, the prospect of an economy planned by political authority lies in the past, not in the future. The future seems to belong to a global market economy. How this economy will evolve, however, also is far from clear. Some authors argue that the contemporary global economy has already outgrown many of the concepts we use to describe it (Drucker 1993). Among the factors shaping its future, the challenges of risk management loom large.

This holds first of all for the risks of financial markets (Feldstein 1991), the institutional setting for which has been in continuous evolution since the beginnings of modern financial operations during the Renaissance. One of the driving forces of rule evolution in this instance has been the need for increasingly powerful tools for pooling and managing financial risks. It would be surprising, indeed, if this evolution were to suddenly come to an end. Further changes are to be expected, especially in view of the increasing role of institutional investors and problems related to large speculative bubbles that may trigger new initiatives in financial risk management.

Financial markets will be increasingly important in dealing with environmental risks, as well (Schmidheiny and Zorraquín 1996). The current division of labor between re-insurance firms and banks is by no means fixed forever. New arrangements are being explored, and international environmental policy is gradually approaching the point where negotiations about, say, risks related to climate change, also will involve large financial operators. Redefining the relation between capital markets and insurance markets is likely to affect the management of technological risks as well.

Other risks are certain to arise in the coming century, e.g., the risks of war, including nuclear wars on a local as well as a global scale. Related, but different, are the risks of endemic violence, exceptional in some parts of the world, terribly normal in others. Still another class are health risks from illness, including risks for the affluent, ranging from heart attack and cancer to Alzheimer and yet unknown diseases; and risks to the poor, stemming from polluted water, malnutrition, lack of medical infrastructure, etc. They also include the risks of epidemics fostered by population density and the evolutionary selection of micro-organisms that thrive in the high-tech environment created by humankind.

There is no point in trying to provide a comprehensive list of risks; what matters is to understand that nearly all risks are closely related to economic dynamics and technological developments.

The RAP view of risk has a simple answer to this challenge. First, let government deal with risks to public goods – as with the risks of war. Second, let government internalize external effects, as those of climate change, by suitable instruments – like a carbon tax. Third, let government modify the allocation of resources wherever this seems appropriate on grounds of fairness and feasible in terms of efficiency. Then, let the markets take care of whatever other risks arise.

This answer is flawed for three reasons. First, it presupposes the existence of efficiently operating futures markets for all relevant goods over all relevant time spans. Creating new futures markets will no doubt be an important contribution to the management of many risks. But market economies are not based on an infinite suite of futures markets, as implied by such seminal economic models as that advanced by Debreu (1959). Instead, they are based on expectations formed by interacting social agents, including both industrial and financial investors (Morishima 1992; see also Section 7.24). These investors, in turn, are and must be sensitive to the expectations of the public at large.

The second flaw is the presupposition that people act on the basis of stable and consistent preferences and expectations which take into account all available information. Ambivalent preferences and expectations, as well as the many processes by which people interpret the meaning of information, are unduly neglected.

Third, the RAP approach to economic risk management presupposes that the distribution of resources and income among economic agents can either be modified without interfering with the operation of market mechanisms or simply presents no problem. In a world where billions of human beings are deprived of elementary resources and capabilities, and where hundreds of millions live in conditions of misery which are hard to imagine for the more fortunate, this presupposition presents a very serious problem. The current inequality in global distribution of income and wealth obviously involves an extremely unequal distribution of many risks and creates additional ones.

The goal of sustainable development sets these risks in perspective. It challenges decision makers to find a new balance between rightness of procedures and goodness of outcomes (Sen 1995). RAP claimed that the market mechanism would automatically establish such a balance: market equilibria are optima when all concerned parties interact according to the rules of free enterprise. The good outcome is guaranteed by the right procedure. The very need to declare a goal such as sustainable develop-ment shows that things are not that simple.

A more comprehensive understanding of economic institutions looks at the economy as a complex adaptive system (Anderson et al 1988), in which free markets can lead to very different outcomes depending on the expectations and preferences of investors. The system can follow different paths which from time to time offer branching points. Once a given path has been taken, the system may remain locked into that path for quite some time (Arthur 1990). Under some circumstances, to switch from one path to another may be extremely difficult; under other circumstances, such a switch may be relatively easy and require only some triggering action. In some respects, the future course of the system may be well predictable, while in other respects it cannot be predicted because it is actually indeterminate (Jaeger and Kasemir 1996). Such a system may best be handled with a strategy of robust action – pursuing short-term goals in such a way as to keep long-term options open (Eccles and Nohria 1992).

There can be little doubt that the world economy presently is locked into a path involving a large array of quite serious risks. An essential feature of this path is its dependency on increasing energy use (Imboden and Jaeger 1999). Risks like those of global warming and nuclear waste are geared into it. Less directly, risks of ecosystem destruction and traffic accidents are related to the same path. Increasing global population and energy use per capita increasingly narrow long-term options and dependency on fossil fuels gets tighter, leading to additional economic and geopolitical risks.

A strategy of robust action might require that steps be taken to enhance the feasibility of long-term low-energy consumption on a global scale. In RAP terms, such an option is usually seen as involving a tradeoff between achieving high economic growth and reducing environmental and technological risks. In a broader perspective, different paths of socio-technical development may be equally feasible. The array of options open can be widened, and the goal of sustainable development specified so as to enhance the possibility of a society with considerably reduced risks.

At a related but somewhat less general level, consider the risks of automobile traffic. Besides the global risks engendered by carbon dioxide emissions and local health risks engendered by air pollution, there are the risks of accidents. As is well known, in many countries the death rate from auto accidents is much larger than the death rate from wars. Many of these accidents are caused by young males enjoying the risks of driving. At the same time, the automobile has so many advantages that it is very likely to dominate traffic systems worldwide in the foreseeable future.

Suppose now that autos were used mainly on a leasing basis while remaining the property of auto manufacturers. Most likely, many would use the same car for everyday purposes but switch to other models when the need arises. This would facilitate new forms of design that would better fit specific uses. Liability in case of accidents would lie with the manufacturers so long as the driver behaves lawfully. Clearly, an insurance premium would then be a core component of the leasing contract. But now there would be a strong incentive for companies to design cars so as to avoid accidents. This is a very different goal from the one of letting the car user survive an accident, and one can be curious of what technologies this new goal would lead to. Moreover, car companies now would have an active interest in traffic-related laws that reduce the likelihood of accidents. Again, it would be interesting to observe what consequences this would have for law making in the traffic sector. Eventually, the cultural meanings of the automobile – a major symbol in contemporary society – would change along with new patterns of use.

Whether such a scheme will ever be implemented or whether superior alternatives to deal with the risks of auto traffic should be developed is not the issue. The crucial point is the fact that market mechanisms can be transformed in order to improve the societal capability of risk management. For this purpose, the considerable potential of RAP for institution design can and should be further enhanced by embedding it in a broader understanding of economic institutions. There are many ways of implementing market mechanisms, and learning to take advantage of this variety will be a major task for risk management in the coming century.

7.3.3 Social Theory in Historical Context

Having examined the domains of politics and economics, we end by setting the quest for a successor to RAP in historical perspective. This will enable us to reinforce some core points of our argument by putting them in a new setting. From a discussion framed mainly by the internal logic of scientific investigation, we move to a view of social science as integral to society at large. The evolution of social theories quite often seems to reflect and sometimes even shape the evolution of society's self-image and spirit – its "*Zeitgeist*". Social theories often run parallel to major social developments.

The evolution of rationality was intimately tied to the evolution of modernity. RAP models fit nicely into the historical dynamics of modernization. In this view, rational actors advance the welfare of society by taking options most beneficial for themselves. The invisible hand of all rational actions assures that the overall effect is positive for society and that progress is maintained over time.

The competitors of RAP express serious doubts about the progressive nature of modernization. Since the beginnings of the industrial revolution, social experiences of cultural and social alienation have contributed to deep distrust in the promises of modernity. Furthermore, the link between individual rationality and social progress was difficult to maintain *vis-à-vis* obvious social costs associated with individual optimization processes. At the same time that individuals were acting rationally in their own terms, societies were experiencing social misery, destructive wars, geopolitical upheavals, and serious environmental degradation. RAP theorists blame imperfect markets and ill-designed political institutions for these social costs; more radical as well as more conservative theorists blame the RAP prophets themselves. Disappointment about the broken promises of RAP has been the driving engine of criticism of the belief that rational action is the key to progress.

In the social sciences, this criticism takes mainly two forms. On the one hand, Marx and his followers provided a vision of capitalist society as a transitory historical phase on the road to socialism. On the other hand, the conservative tradition of social theory has led to the holistic view of society formulated by Durkheim as the basis of sociology. As sociology also came to include Marxist-inspired research, it became the home of the two major theoretical alternatives to RAP.[104]

In the decades following World War II, however, opposition to Marxism united both RAP and the more holistic thinking of mainstream sociology. In the West, these were decades of unchallenged modernization and a near-unanimous belief in technological progress. Within sociology, structural functionalism was, by far, the dominant theory. Social actions were interpreted as functional elements for the advancement of society, reflecting technological optimism, and in this respect constituting an adequate description of social reality. At the same time, this line of thought implicitly accepted RAP as an adequate description of economic markets.

In part because it failed to capture emerging social processes of the late 1960s and early 1970s, the influence of structural functionalism within sociology declined. The decline was not due to theoretical or logical deficiencies (which were known from the beginning), but because of its seeming inability to account for many forms of social change – especially rapid change – as exemplified by the social upheaval of this period. The evolution of new social movements evolving and increasing calls for the emancipation of oppressed minorities, demonstrated the theoretical weakness of functionalism as a description of social reality.

In the face of these challenges, some versions of conflict theory became associated with functionalism. Conflicts were interpreted as functional for the goals of emancipation and changing power structures (Coser 1956; Dahrendorf 1992). While conflict theory was partially based on the classic

[104] Sociology also became home of a third tradition, the tradition of symbolic interactionism, inaugurated by Mead (1934). This tradition provides important seeds for an alternative to RAP by its awareness for the intricacies of human reflexivity. So far, however, it has been used only to a very limited extent to analyze, model, and design economic institutions, the stronghold of RAP.

analyses of Marx and Weber, its new incarnation was organized around agents of social change and their functional (or dysfunctional) roles in the creation of social cohesion. Conflict theories were also paramount in promoting Marxist views of society, although neo-Marxists were very skeptical about the premises and assumptions of conflict theories. At the same time, however, conflict theories laid the foundations for new philosophies of relativism and social constructivism, which were to gather a strong following in sociology.

The demise of functionalism left a deep scar in American sociology for almost two decades. Social theory was in constant motion and often in disarray. Macro-theories lost favor; small scale theoretical approaches were in high demand. Researchers focused on areas of sociology that permitted inferences of causal relationships. The focus on sub-fields and topical areas increased specialization and further fragmented the field. In the U.S., some questioned the viability of sociology, arguing that sub-fields of sociology should either develop independent disciplinary status or merge with existing fields like economics or psychology.

Responses in Europe were different. The fragments of structural-functional analysis were collected piece by piece by several leading social theorists. Niklas Luhmann, one of Parsons's last students, was committed to a systems approach with functional underpinnings. Luhmann and the Bielefeld school adopted the idea of evolutionary functionalism and transferred the concept into a constructivist framework in which social systems develop functional autonomy by promoting internal change, internal structure building, and adaptation at the expense of cross-systems integration (Luhmann 1984).

To some extent, this view of society reflected the post-war development of Germany. The traumatic experience of Nazism left most Germans skeptical about common visions of society and made them eager to invest their energy in subsystems congenial to themselves. They became devoted players within particular subsystems – economic, political, or whatever – but they rejected a common vision for society as a whole. Being active in optimizing one's own subsystem without building bridges to neighboring systems reflected the reluctance of most Germans to get emotionally involved in overarching new ideologies. This disposition also contributed to constructing a new (desired) image of Germans as busy, efficient, but non-ambitious cosmopolitan citizens of the world. Luhmann's analysis of self-organizing subsystems with as few communication pipelines to other neighboring systems as required to fulfill their internal mandate matched the predominant social identity of many people in Germany. Another RAP competitor, systems theory, provided an elegant explanation for the discrepancies between RAP promises and reality, but offered no counter-model that could serve as a new guide post for societal development.

For some time, the main challenger to such "bourgeois" ideas all over Europe was still Marxism. Since the 1950s and 1960s, Marxists were divided between the traditional and the anti-authoritarian left. The first was highly influenced by thoughts and worldviews developed in the then communist countries. The second, also anti-authoritarian, was inspired by the Marxist critique of capitalism and by an anarchic vision of a new society. Both schools fought endless battles against structural-functionalism in all its variants. Much social momentum for the rise of neo-Marxist views was provided by the student rebellions of the 1960s. These rebellions, in turn, were a reflection of concerns about social injustice and disappointment with the promises of modernity.

These Marxist schools were not opposed to modernization, social risk taking, or to implementing modern technologies. Their main criticisms were with the unequal distribution of resources that occurred in the process of modernization, the allegedly illegitimate basis for economic and political control of resources, and the ruthless manner in which modern lifestyles overtook traditional value patterns in society. Marxist thoughts elaborated by formal and informal think-tanks were widely adopted by different movements on the political left, as well as by some new ecological movements (Mauss 1975).

One of the most prominent of these neo-Marxist think-tank versions was critical theory as elaborated in the so-called Frankfurt School. Originally developed as a neo-Marxist theory, critical theory adopted the anti-capitalist viewpoint of traditional Marxism, but rejected the teleological philosophy of continuous progress towards a class-free communist end-society. Rationality as a yardstick for evaluating social actions was retained but the traditional means-ends (instrumental) rationality was supplemented by additional forms of rationality: social and communication. This amendment to rational action was derived from analyses of Max Weber and his vision of a bureaucratic society and, from a normative view, the need for emancipatory change.

Critical theory became a major element of social theory at a crucial moment in history, when:

- traditional Marxist views were difficult to defend in view of their authoritarian implementation in communist countries;
- dark sides of the capitalist mode of production (such as external effects on the environment and on socially disadvantaged persons) became obvious to intellectuals and laypersons alike;
- a quest for self-realization served as the dominant individual vision for one's own life, particularly for affluent members of societies; and
- a sufficient level of affluence was reached to give rise to increased concerns of quality of life rather than economic standards of survival and to pluralistic visions of future societal development.

In Germany, critical theory was a counterpart to both RAP and social systems theory. It retained the idea of rationality, but aligned this idea with a prescription for substantive change. Because they had been victims of the Nazi regime, many proponents deeply distrusted the "rationality of instrumental rationality." They noted that concentration camps could be interpreted as excellent examples of instrumental rationality. Even the holocaust could be reconciled with RAP theories if instrumental rationality were left to individual preferences. This form of rationality thus appeared to be amoral at best, often immoral in its consequences. As concern about the infamous role of technological elites during the Nazi regime mounted, critical theory reflected the growing skepticism of many German intellectuals with respect to instrumental rationality.

Against this background, critical theory developed as a theory of emancipation, attempting to reconcile subjective preferences and objective rules of rationality with respect to moral norms as well as factual knowledge. Increasing pluralism of values and knowledge systems provided a major challenge to the quest for theoretical reconciliation. One response was the development of the "theory of communicative action" in which emancipation meant that all actors are free to define their common norms of behavior within a specified discourse setting. The setting needs to be free of coercive powers and status differences. In this view, the classic conflict between reliance on substantive reasoning and voluntary agreement among all actors is resolved by structuring the necessary conditions for an "ideal" discourse. The theory of "communicative action" emerged from critical review of RAP's instrumental rationality and from internal responses to problems inherent in the early work of critical theory. Similar to systems analysis, this development reflects an ongoing trend in modern society, toward more open public deliberation. As powerful as this development is, it nevertheless lacks the ability to dethrone the monarch of RAP and become its successor.

The lack of attractiveness of any kind of Marxism among workers led to the decline of many Marxist-inspired movements, including the student rebellions. Two decades later, the breakdown of communism created an intellectual black hole into which the whole cultural fabric shaped by Marxism collapsed within a very short period. One wonders whether this was a case of healthy intellectual progress or whether the social sciences have simply recoiled from the badly needed task of developing a thorough understanding of the historical experience of the labor movement, including the strengths and limits of Marxism. The collapse of Marxism did not, in fact, lead to a new consensus in post-functional sociology, but to increased divergence of approaches.

The inability to settle theoretical disputes in the social sciences paved the path for the rise of social relativism. Quickly, French post-modernism became its herald. The provision in all post-functional theories for plurality and lifestyle changes reinforced this development. It was further reinforced by the inability of society to create concrete visions of the social world and by its fragmentation into pluralistic structures and partial belief systems without overall coherence. The interest in deconstructing social constructions and building up constructions upon small fields of expertise flourished as long as society followed a similar path. Post-modernism was a radical response to the RAP crisis by defying the concept of rationality altogether. To post-modernists, rationality is an ideological disguise for legitimating power.

Progress was removed from sociological language and replaced by social change or transformation, a change that was not merely cosmetic. Relativism does not recognize any privileged direction of change. At the same time the breakdown of modernity and an increasingly cynical attitude towards common values (stemming from dogmatic individualism) fueled the belief that social actions are only bound to the consent of those involved. No other moral or social order could claim compliance or redemption. Overall, the central philosophy was that anything goes. Methodological constraints are superfluous, while a critical stance is needed to unmask the hidden interests in all social activities that lay claim to truth or common morale.

Post-modernism does not strive to replace RAP with another general theory. To the contrary, it denies the necessity for theoretical integration. In Chapter 6 we criticized this image as neither intellectually convincing nor practically useful in the social sciences. Notwithstanding its intellectual faults and merits, however, post-modernism has been a mirror of important aspects of social reality in the 1970s and 1980s. From the arts to the sciences, from the humanities to the social sciences, from the architecture of buildings to the architecture of thought, relativism was *en vogue*. In almost all economically advanced societies people felt insecure about moral and aesthetic standards, expressed skepticism about claims for objective truth, treated "authenticity" as the main yardstick for judging validity, and adjusted their values to the changing demands of external situations. In post-modern society it appeared that one faced no objective constraints other than those imposed by oneself or by power-abusing agents.

It soon became clear – but only understood in its full dimensions somewhat later – that society was not liberated from involuntary constraints. Such constraints had always been present. The latent promise, however, was that these constraints could be overcome with the conventional methods of modernity (the instrumental rationality powered by RAP) or by the diversity of post-modernity (let each individual do his or her thing). Early in the last decade of the twentieth century it was painfully clear that these recipes would not work. New constraints had emerged, creating a demand for new concepts of society.

Among the new constraints are:

- *the ecological crisis:* Billions of individuals optimizing their individual choices disrupt and degrade the global environment. As a result, large segments of humankind are victimized, with little hope for better lives in the future. Although external effects have long been recognized by the RAP-dominated field of economics, in practice neither massive social costs nor environment costs provide sufficient basis for the assignment of new property rights or shadow prices for natural resources. Even if such measures were implemented on a large scale, it is doubtful that they would be sufficient to solve such problems.
- *the reemergence of nationalism and tribalism*: Recent outbreaks of violence and tribalism are powerful indicators that people the world over look for some measure of direction and stability in their social environments. When either is threatened, scapegoats are created to channel frustrations, a mechanism as old as the existence of social systems. The post-modern promise that society can live without social cohesion and that stable orientations in everyday life are not necessary elements of a functioning society has been seriously challenged by reality.
- *increased social costs of poorly integrated subsystems:* A number of global trends have contributed to growing difficulties with the autonomy claimed and defended by representatives of various subsystems of society: globalization of markets, the relative importance of political and social factors for sustaining international competitiveness, the dependence of social security and political stability on economic performance, and the importance of cultural identity for building and sustaining a flexible but predictable response to external challenges. Weak economic performance has repercussions on social security, on the support for culture and education, and on the self-image of society. Political problems give rise to economic and other subsystems. The fact that everything is connected to everything is a truism that has long been known. The direct correlation between the performance of one system and the constraints of other systems, however, has only recently caught the attention of social scientists and social agents. This runs counter to the theoretical assumptions of systems theory as well as post-modernism in which everything goes – but only at tremendous social costs. With its emphasis on individual utility, RAP is poorly equipped to provide a meaningful theoretical framework for understanding and managing these important interdependencies. Even critical theory has little more to offer than the stereotypical demand for more discourse. *Defining rationality for collective action without sacrificing individual freedom and cultural pluralism will be one of the most serious social challenges in the next century.*

Dealing with such constraints requires collective social action and shared values. Neither the singular pursuit of instrumental rationality nor the application of relativism will be able to deal with these constraints, and will very likely aggravate them. Some of these constraints seem to call for RAP-based approaches: in particular, cases in which collective solutions for global problems such as pollution or climate change are needed. However, RAP theories will surely fail to produce the intellectual base for producing the solidarity and trust necessary to produce instrumentally sound solutions that have a realistic chance of being implemented.

Solutions must be embodied in culture as shared meaning. Agreements about normative visions and guidelines are required for consensus on collective actions necessary if these goals are to be achieved. Discourse theories may help to provide better knowledge about collective policy making, and reflexive social science may aid in promoting increased reflection among leaders and other social actors as they struggle to cope with constraints such as those noted above.

RAP, the monarch, thus faces severe new challenges that risk the kingdom. At the same time, a new, functionally equivalent and equally potent paradigm has not yet emerged. The new paradigm must include the RAP perspective and it must be as rigorous and decisive as RAP with regard to computational structure and mathematical articulations of its assumptions. It must also be receptive to insights from sources as diverse as discourse theory, systems theory and post-modern thinking. None of these princesses or princes is yet ready to take over the kingdom, but by analogy to modern democratic principles, they may try jointly to reign over the emerging kingdom as best as they can. Each may be incapable of taking over the kingdom, but by working together they may come closer to a social theory that provides an adequate response to the challenges of our rapidly changing world. They may even achieve a more viable version of RAP that is properly restricted in scope.

Risk is a paradigmatic issue for this challenge. Initially, the inclusion of risk in the mainstream social sciences was an attempt to use instrumental rationality to cope with uncertain but increasingly dangerous technologies. We can now see the types of serious problems that are associated with over-reliance on one type of rationality. Criteria are needed for social rationality that encompasses collective values and goals. On this basis we may strive for an understanding of society that is congruent with new constraints rather than working against them or ignoring them.

Understanding of this sort will no longer view rationality as a property of isolated agents. Clearly, rationality is a desirable, possible, but sometimes lacking property of interactions between social agents, both at the level of the global changes that are already shaping the coming century and at the level of the personal encounters that continually shape our lives.

The study of risk and uncertainty requires a debate among the proponents of RAP and competing approaches. This book is written in the hope that such a debate might serve as a common platform for developing a consistent but pluralistic view in the era of increasing uncertainty and risks. The next century – indeed, the next millennium – requires a research process that will produce shared meanings and suitable institutional settings for living with both.7

REFERENCES

Adams, J. 1995. *Risk*. London: UCL Press.

Adorno, T. W., E. Frenkel-Brunswik, D.J. Levinson, and N. Sanford, 1950/1969. *The Authoritarian Personality*. New York: W. W. Norton & Company Inc.

Alexander, J. C. 1994. "Modern, Anti, Post and Neo: How Social Theories Have Tried to Understand the 'New World' of 'Our Time'." *Zeitschrift für Soziologie* 23:165–197.

Alexander, J.C., and B. Giesen. 1987. "From Reduction to Linkage: the Long View of the Micro-Macro Link." Pp 1–44 in *The Micro-Macro Link*, edited by J.C. Alexander, B. Giesen, R. Münch, and N.J. Smelser. Berkeley and Los Angeles: University of California Press.

Allen, F.W. 1987. "Towards a Holistic Appreciation of Risk: The Challenge for Communicators and Policymakers." *Science, Technology, and Human Values* 12:138–143.

Allport, G.W. 1935. "Attitudes." Pp 798–844 in *Handbook of Social Psychology*. Worcester, MA: Clark University Press.

Anderson, P.W. , K.J. Arrow, and D. Pines. 1988. *The Economy as an Evolving Complex System, Santa Fe Institute Studies in the Sciences of Complexity.*

Arms, K. 1990. *Environmental Science*. Philadelphia: Saunders College Publishing.

Arrow, K., and G. Debreu. 1954. "Existence of an Equilibrium for a Competitive Economy." *Econometrica* 22:265–290.

Arrow, K.J. 1951. *Social Choice and Individual Values*. New York: Wiley.

Arrow, K.J. 1970. *Essays in the Theory of Risk Bearing*. Amsterdam: North Holland.

Arrow, K.J. 1987. "Oral History I: An Interview." Pp 191–242 in *Arrow and the Ascent of Modern Economic Theory*, edited by G.R. Feiwel. Basingstoke: Macmillan.

Arthur, W.B. 1990. "Positive Feedbacks in the Economy." *Scientific American*:80–85.

Axelrod, R.M. 1984. *The Evolution of Cooperation*. New York: Basic Books.

Baird, B.N.R., T.C. Earle, and G. Cvetkovich. 1985. "Public Judgment of an Environmental Health Hazard: Two Case Studies of the ASARCO Smelter." Pp 383–398 in *Risk Assessment and Management*, edited by L.B. Lave. New York: Plenum Press.

Banse, G. 1996. "Herkunft und Anspruch der Risikoforschung." Pp 15–72 in *Risikoforschung zwischen Disziplinarität und Interdisziplinarität*, edited by G. Banse. Berlin: Edition Sigma.

Barber, B.R. 1995. *Jihad vs. McWorld. How the planet is both falling apart and coming together and what this means for democracy.* New York: Times Books.

Barnard, C. 1968. *The Functions of the Executive*. Cambridge, MA: Harvard University Press.

Beach, L.R., and R. Lipshits. 1993. "Why Classical Decision Theory Is an Inappropriate Standard for Evaluating and Aiding Most Human Decision Making." Pp 21–35 in *Decision Making in Action: Models and Methods*, edited by G.A. Klein, J. Orasnu, R. Calderwood, and C.E. Zsambok. Norwood, NJ: Ablex.

Beck, U. 1986/1992. *Risk Society: Towards a New Modernity*. London and Newbury Park, CA: Sage Publications.

Beck, U. 1991/1995. *Ecological Enlightenment: Essays on the Politics of the Risk Society*. Atlantic Highlands, NJ: Humanities Press.

Beck, U. 1994. "The Reinvention of Politics: Towards a Theory of Reflexive Modernization." Pp 1–55 in *Reflexive Modernization: Politics, Tradition and Aesthetics in the Modern Social Order*, edited by U. Beck, A. Giddens, and S. Lash. Cambridge: Polity Press.

Beck, U., A. Giddens, and S. Lash. 1994. *Reflexive Modernization: Politics, Tradition and Aesthetics in the Modern Social Order*. Cambridge: Polity Press.

Becker, D. 1991. *Womit handeln Banken? Eine Untersuchung zur Risikoverarbeitung in der Wirtschaft*. Frankfurt/M: Suhrkamp.

Becker, G.S. 1976. *The Economic Approach to Human Behavior*. Chicago: University of Chicago Press.

Becker, G.S. 1996. *Accounting for Tastes*. Cambridge, MA: Harvard University Press.

Bentham, J. 1789/1970. *An Introduction to the Principles of Morals and Legislation*. London: Athlone Press.

Binmore, K. 1993. "De-Bayesing Game Theory." in *Frontiers of Game Theory*, edited by K. Binmore, A. Kirman, and P. Tani. Cambridge, MA: MIT Press.

Blau, P. 1964. *Exchange and Power in Social Life*. New York: Wiley.

Bonss, W. 1996. "Die Rückkkehr der Unsicherheit. Zur gesellschaftstheoretischen Bedeutung des Risikobegriffes." Pp 166–185 in *Risikoforschung zwischen Disziplinarität und Interdisziplinarität*, edited by G. Banse. Berlin: Edition Sigma.

Borcherding, K., and B. Rohrmann. 1990. "An Analysis of Multi-attribute Utility Models Using Longitudinal Field Data." Pp 223–244 in *Contemporary Issues in Decision Making*, edited by K. Borcherding, O.I. Larichev, and D.M. Messick. Amsterdam: North-Holland.

Bourdieu, P. 1990. *The Logic of Practice*. Stanford: Stanford University Press.

Bourdieu, P., and J.S. Coleman. 1991. *Social Theory for a Changing Society*. New York, NY: Russell Sage Foundation.

Braaten, J. 1991. *Habermas's Critical Theory of Society*. Albany, NY: State University of New York Press.

Brandom, Robert B. 1994. *Making It Explicit. Reasoning, Representing, and Discursive Commitment*. Cambridge, Mass.: Harvard University Press.

Brehmer, B. 1987. "Risk and Decicions." Pp 25–39 in *The Psychology of Risk*, edited by W.T. Singleton and J. Howden. New York: Wiley.

Buchanan, J. 1954. "Social Choice, Democracy, and Free Markets." *Journal of Political Economy* 62:114–123.

Buchanan, J.M., and G. Tullock. 1962. *The Calculus of Consent. Logical Foundations of Constitutional Democracy.* Ann Arbor, MI: University of Michigan Press.

Buchanan, J.R. 1988. "The Economic Theory of Politics Reborn." *Challenge* 1988:4–10.

Burke, E. 1790/1987. *Reflections on the Revolution in France.* Indianapolis: Hackett.

Burns, T.R., and T. Dietz. 1992. "Cultural Evolution: Social Rule Systems, Selection and Human Agency." *International Sociology* 7:259–283.

Burns, W., R. E. Kasperson, J. X. Kasperson, O. Renn, S. Emani, and P. Slovic. 1990. *Social Amplification of Risk: An Empirical Study.* Carson City, NV: Yucca Mountain Socioeconomic Project, State of Nevada Nuclear Waste Project Office.

Burt, R.S. 1992. *Structural Holes. The Social Structure of Competition.* Cambridge, MA: Harvard University Press.

Burton, I., R. Kates, and G. White. 1978. *The Environment as Hazard.* New York: Oxford University Press.

Camerer, C. 1992. "Recent tests of Generalizations of Expected Utility Theory." Pp 207–251 in *Utility Theories: Measurements and Applications,* edited by W. Edwards. Boston: Kluwer.

Chase, V.M., R. Hertwig, and G. Gigerenzer. 1998. "Visions of Rationality." *Trends in Cognitive Sciences* 2:206–214.

Clarke, L. 1989. *Acceptable Risk? Making Choices in a Toxic Environment.* Berkeley, CA: University of California Press.

Clarke, L., and J.F. Short. 1993. "Social Organization and Risk: Some Current Controversies." *Annual Review of Sociology* 19:375–399.

Coase, R.H. 1937. "The Nature of the Firm." *Economica* 4:386–405.

Cole, B. and Gealt, A. 1989. *Art of the Western World: from Ancient Greece to Post-Modernism.* New York: Summit Books.

Coleman, J.S. 1990. *Foundations of Social Theory.* Cambridge, MA: Harvard University Press.

Collins, R. 1988. *Theoretical Sociology.* New York: Harcourt, Brace, Jovanovich.

Collins, R. 1994. *Four Sociological Traditions.* New York: Oxford University Press.

Commons, J.R. 1924. *Legal Foundations of Capitalism.* New York: Macmillan.

Coppock, R. 1986. *Regulating Chemical Hazards in Japan, West Germany, France, the United Kingdom, and the European Community: A Comparative Examination.* Washington, D.C.: National Academy Press.

Coser, L.A. 1956. *The Function of Social Conflict.* New York: The Free Press.

Couzens Hoy, D., and T. McCarthy. 1994. *Critical Theory.* London: Routledge.

Covello, V., P. Sandman, and P. Slovic. 1989. "Risk Communication, Risk Statistics and Risk Comparisons: A Manual for Plant Managers." Pp 297–359 in *Effective Risk Communication: The Role and Responsibility of Government and Nongovernment Organizations,* edited by V.T. Covello, D.B. McCallum, and M.T. Pavlova. New York: Plenum Press.

Covello, V.T. 1983. "The Perception of Technological Risks: A Literature Review." *Technological Forecasting and Social Change* 23:285–297.

Covello, V.T. 1991. "Risk Comparisons and Risk Communication: Issues and Problems in Comparing Health and Environmental Risks." Pp 79–124 in *Communicating Risk to the Public,* edited by R.E. Kasperson and P.M. Stallen. Dordrecht: Kluwer.

Covello, V.T., J.L. Mumpower, P.J.M. Stallen, and V.R.R. Uppuluri (Eds.). 1985. *Environmental impact assessment, technology assessment and risk analysis.* Berlin, Heidelberg/Germany: Springer Verlag.

Covello, V.T., P. Slovic, and D. von Winterfeldt. 1986. "Risk Communication – A review of literature." *Risk Abstracts* 3:172–182.

Covello, V.T., P. Slovic, and D. von Winterfeldt. 1988. "Disaster and Crisis Communications. Findings and Implications for Research and Policy." in *Risk Communication,* edited by H. Jungermann, R.E. Kasperson, and P.M. Wiedemann. Jülich: Research Center Jülich.

Crawford, M., and R. Wilson. 1996. "Low-Dose Linearity: The Rule or the Exception?" *Human and Ecological Risk Assessment* 2:305–330.

Crouch, E.A.C., and R. Wilson. 1982. *Risk/Benefit Analysis.* Cambridge, MA: Ballinger Publishing Company.

Cvetkovich, G., C. Vlek, and T.C. Tearle. 1989. "Designing Technological Hazard Information Programs: Towards a Model of Risk-adaptive Decision Making." in *Social Decision Methodology for Technological Projects,* edited by C. Vlek and G. Cvetkovich. Dordrecht: Kluwer.

Dahrendorf, R. 1992. *Der moderne soziale Konflikt.* Stuttgart: Deutsche Verlagsanstalt.

Dake, K. 1991. "Orienting Dispositions in the Perceptions of Risk: An Analysis of Contem-porary Worldviews and Cultural Biases." *Journal of Cross-Cultural Psychology* 22:61–82.

Dake, K. 1992. "Myths of nature, culture and the social construction of risk." *Journal of Social Issues* 48:21–37.

Davies, J. 1962. "Toward a Theory of Revolution." *American Sociological Review* 27:5–19.

Dawes, R.M. 1988. *Rational Choice in an Uncertain World.* New York: Harcourt, Brace, Jovanovich.

Debreu, G. 1959. *Theory of Value: an Axiomatic Analysis of Economic Equilibrium.* New York: Wiley.

DeFleur, M.L., and S. Ball-Rokeach. 1982. *Theories of Mass Communication.* New York: Longman.

d'Espagnat, B. 1981/1983. *In Search of Reality.* New York: Springer.

Di Maggio, P.J., and W.W. Powell. 1983. "The Iron Cage Revisted: Institutional Isomorphism and Collective Rationality in Organizational Fields." *American Sociological Review* 48:147–160.

Dietz, T., R.S. Frey, and E.A. Rosa. In press. "Risk, Technology, and Society." In *Handbook of Environmental Sociology*, edited by R.E. Dunlap and W. Michelson. Westport, CT: Greenwood Press.

Dietz, T., and R. Rycroft. 1987. *The Risk Professionals*. New York: Russell Sage Foundation.

Dietz, T., and P. Stern. 1995. "Toward a Theory of Choice: Socially Embedded Preference Construction." *Journal of Socio-Economics* 24:261–279.

Dijksterhuis, E.J. 1961. *The Mechanization of the World Picture: Pythagoras to Newton*. Oxford: Clarendon Press.

Douglas, M. 1966. *Purity and Danger*. London: Routledge & Kegan Paul.

Douglas, M. 1985. *Risk Acceptability According to the Social Sciences*. New York: Russell Sage Foundation.

Douglas, M., and A. Wildavsky. 1982. *Risk and Culture: The Selection of Technological and Environmental Dangers*. Berkeley, CA: University of California Press.

Downs, A. 1957. *An Economic Theory of Democracy*. New York, NY: Harper and Row.

Drabek, T.E. 1986. *Human System Response to Disaster. An Inventory of Sociological Findings*. Berlin and New York: Springer.

Drottz-Sjöberg, B.-M. 1991. *Perception of Risk. Studies of Risk Attitudes, Perceptions, and Definitions*. Stockholm: Center for Risk Research.

Drucker, P.F. 1993. *Post-Capitalist Society*. New York: HarperCollins.

Duesenberry, J. 1960. "Comment on 'An Economic Analysis of Fertility'." Pp 231–234 in *Demographic and Economic Change in Developed Countries: A Conference of the Universities—National Bureau Committee for Economic Research*, edited by Universities—National Bureau Committee for Economic Research. Princeton, NJ: Princeton University Press.

Dulaney, D.E. 1968. "Awareness, Rules, and Propositional Control. A Confrontation with S-.R Behavior theory." in *Verbal Behavior and General Behavior Theory*, edited by T.R. Dixon and D.R. Horton. Englewood Cliffs: Prentice-Hall.

Dunlap, R.E., G.H. Gallup, Jr., and A.M. Gallup. 1993. *Health of the Planet: Results of a 1992 International Environmental Opinion Survey*. Princeton, NJ: George H. Gallup International Institute.

Dunlap, R.E., and A.G. Mertig (Eds.). 1992. *American Environmentalism*. Washington, D.C.: Taylor and Francis.

Durkheim, E. 1893/1984. *The Division of Labor in Society*. New York: The Free Press.

Durkheim, E. 1895/1982. *The Rules of Sociological Method*. New York: The Free Press.

Durkheim, E. 1915/1965. *The Elementary Forms of the Religious Life*. New York: The Free Press.

Earle, T.C. , and M.K. Lindell. 1984. "Public Perceptions of Industrial Risks: A Free-Response Approach." Pp 531–550 in *Low-probability/High-Consequence Risk Analysis*, edited by R.A. Waller and V.T. Covello. New York: Plenum.

Eccles, R.G., and N. Nohria, with J.D. Berkley. 1992. *Beyond the Hype: rediscovering the essence of management*. Cambridge, MA: Harvard Business School Press.

Edenhofer, O., and C.C. Jaeger. 1998. "Power Shifts. The Dynamics of Energy Efficiency." *Energy Economics* 20:513–538.

Edgeworth, F.Y. 1881. *Mathematical Psychics*. London: C. Kegan Paul & Co.

Edwards, W. 1954. "The Theory of Decision Making." *Psychological Bulletin* 51:380–417.

Edwards, W. 1977. "How to Use Multiattribute Utility Measurement for Social Decision Making." *IEEE Transactions on Systems, Man, and Cybernetics* SMC-7:326–340.

Eibl-Eibesfeldt, I. 1974. *Grundriss der vergleichenden Verhaltensforschung*. München and Zürich: Piper.

Eibl-Eibesfeldt, I. 1979. "Human Ecology: concepts and Implications for the Science of Man." *Behavioral and Brain Sciences* 1:1–57.

Einstein, A., and L. Infeld. 1966. *The evolution of physics from early concepts to relativity and quanta*. New York: Simon and Schuster.

Elster, J. 1979. *Ulysses and the Sirens: Studies in Rationality and Irrationality*. New York: Cambridge University Press.

Elster, J. 1983. *Sour Grapes*. New York: Cambridge University Press.

Elster, J. 1989a. *Nuts and Bolts for the Social Sciences*. New York: Cambridge University Press.

Elster, J. 1989b. "Wage Bargaining and Social Norms." *Acta Sociologica* 32:113–136.

Emel, J, and R. Peet. 1989. "Resource Management and Natural Hazards." Pp 49–76 in *New Models in Geography*, edited by R. Peet and N. Thrift. London: Unwin Hyman.

Emerson, R.M. 1972. "Exchange Theory, Part I: A Psychological Basis for Exchange. Part II: Exchange Relations and Network Structures." in *Sociological Theories in Progress*, edited by J. Berger, M. Zelditch, Jr., and B. Anderson. New York: Houghton Mifflin.

Esser, H. 1991. *Soziologie. Allgemeine Grundlagen*. Frankfurt: Campus.

Esser, H. 1991b. *Alltagshandeln und Verstehen. Zum Verhältnis von erklärender und verstehender Soziologie am Beispiel von Alfred Schuetz und "rational choice"*. Tübingen: Mohr.

Etzioni, A. 1961. *A Comparative Analysis of Complex Organizations*. New York: The Free Press.

Etzioni, A. 1991. *A responsive society. Collected essays on guiding deliberate social change*. San Francisco: Jossey-Bass Publishers.

Evers, A., and H. Nowotny. 1987. *Über den Umgang mit Unsicherheit. Die Entdeckung der Gestaltbarkeit von Gesellschaft.* Frankfurt/Main: Suhrkamp.
Fairchild, H.P. 1955. *Dictionary of Sociology.* Amos: Littlefield Adams & Co.
Farmer, F.R. 1967a. "Reactor Safety and Siting: A Proposed Risk Criterion." *Nuclear Safety* 8:11–12.
Farmer, F.R. 1967b. "Siting Criteria – A New Approach." in *International Atomic Energy Agency Symposium.* Vienna.
Fattorelli, S., M. Borga, and D. Da Ros. 1995. "Integrated Systems for Real-time Flood Forecasting." Pp 191–212 in *Natural Risk and Civil Protection*, edited by T. Horlick-Jones, A. Amendola, and R. Casale. London: Chapman and Hall.
Feldstein, M. (Ed.). 1991. *The risk of economic crisis.* Chicago: University of Chicago Press.
Fessenden-Raden, J., J.M. Fitchen, and J.S. Heath. 1987. "Providing Risk Information in Communities: Factors Influencing What is Heard and Accepted." *Science, Technology, and Human Values* 12:94–101.
Festinger, L. 1957. *A Theory of Cognitive Dissonance.* Stanford : Stanford University Press.
Fischer, H.R. 1987. *Sprache und Lebensform: Wittgenstein über Freud und die Geisteskrankheit.* Frankfurt/Main: Athenäum.
Fischhoff, B. 1996. "Public Values in Risk Research." Pp 75–84 in *Challenges in Risk Assessment and Risk Management. Annals of the American Academy of Political and Social Science, Special Issue,* edited by H. Kunreuther and P. Slovic. Thousand Oaks: Sage Publications.
Fischhoff, B., B. Goitein, and Z. Shapiro. 1982. "The Experienced Utility of Expected Utility Approaches." Pp 315–340 in *Expectations and Actions: Expectancy-Value Models in Psychology*, edited by N.T. Feather. Hillsdale, NJ: Lawrence Erlbaum Associates.
Fischhoff, B., S. Lichtenstein, P. Slovic, S.L. Derby, and R.L. Keeney. 1981. *Acceptable Risk.* Cambridge: Cambridge University Press.
Fischhoff, B., P. Slovic, S. Lichtenstein, S. Read, and B. Combs. 1978. "How Safe is Safe Enough? A Psychometric Study of Attitudes Toward Technological Risks and Benefits." *Policy Sciences* 9:127–152.
Fischhoff, B., S.R. Watson, and C. Hope. 1984. "Defining Risk." *Policy Sciences* 17:123–139.
Fishbein, M., and I. Ajzen. 1975. *Belief, Attitude, Intention and Behavior: An Introduction to Theory and Research.* Reading, MA: Addison-Wesley.
Foucault, M. 1971. *The order of things: an archaeology of the human sciences.* New York: Pantheon Books.
Freese, L. (Ed.). 1980. *Theoretical Methods in Sociology: Seven Essays.* Pittsburgh: University of Pittsburgh Press.
Freese, L. 1997a. *Evolutionary Connections.* Advances in Human Ecology, Supplement 1 (Part A). Greenwich, CT: JAI Press.
Freese, L. 1997b. *Environmental Connections.* Advances in Human Ecology, Supplement 1 (Part B). Greenwich, CT: JAI Press.
Freud, S. 1917/1946. *Totem and Taboo: Resemblances Between the Psychic Lives of Savages and Neurotics.* New York: Vintage Books.
Freudenburg, W.R. 1988. "Perceived Risk, Real Risk: Social Science and the Art of Probabilistic Risk Assessment." *Science* 239:44–49.
Freudenburg, W.R. 1989. "The Organizational Attenuation of Risk Estimates." in *Annual Meeting of the Society for Risk Analysis.* San Francisco.
Freudenburg, W.R. 1992. "Nothing Recedes Like Success? Risk Analysis and the Organizational Amplification of Risk." *Risk Issues in Health and Safety* 3:1–35.
Freudenthal, G. 1982. *Atom und Individuum im Zeitalter Newtons: zur Genese der mechanistischen Natur- und Sozialphilosophie.* Frankfurt/Main: Suhrkamp.
Frey, B.S. 1992. *Umweltökonomie.* Göttingen: Vandenhoek & Ruprecht.
Fröbel, F., J. Heinrichs, and O. Kreye. 1980. *The new international division of labour: structural unemployment in industrialised countries and industrialisation in developing countries.* Cambridge, New York and Paris: Cambridge University Press; Editions de la Maison des Sciences de l'Homme.
Funtovicz, S.O., and J.R. Ravetz. 1992a. "Three Types of Risk Assessment and the Emergence of Post-Normal Science." Pp 251–297 in *Social Theories of Risk*, edited by S. Krimsky and D. Golding. Westport, CT: Praeger.
Funtovicz, S.O., and J.R. Ravetz. 1992b. "Risk Management as a Postnormal Science." *Risk Analysis* 12:95–97.
Gamson, W. H., and B. Fireman. 1979. "Utilitarian Logic in the Resource Mobilization Perspective." Pp 8–45 in *The Dynamics of Social Movements*, edited by M.N. Zald and J.M. McCarthy. Cambridge, MA: Winthrop.
Gamson, W. H., B. Fireman, and B. Rytina. 1982. *Encounters with Unjust Authority.* Homewood, IL: Dorsey.
Gardner, G.T., and P.C. Stern. 1995. *Environmental Problems and Human Behavior.* Needham Heights, MA: Allyn and Bacon.
Gesellschaft für Reaktorsicherheit. 1979. *Deutsche Risikostudie Kernkraftwerke Phase A.* Köln: Verlag TÜV Rheinland.
Gesellschaft für Reaktorsicherheit. 1989. *Deutsche Risikostudie Kernkraftwerke Phase B.* Köln: Verlag TÜV Rheinland.

Gethmann, C.F. 1993. "Zur Ethik des Handelns unter Risiko im Umweltstaat." Pp 1–54 in *Handeln unter Risiko im Umweltstaat*, edited by C.F. Gethmann and M. Kloepfer. Berlin: Springer-Verlag.
Giddens, A. 1979. *Central Problems in Social Theory: Action, Structure and Contradiction in Social Analysis.* Berkeley, CA: University of California Press.
Giddens, A. 1984. *The Constitution of Society. Outline of the Theory of Structuration.* Berkeley, CA and Los Angeles: University of California Press.
Giddens, A. 1990. *The Consequences of Modernity.* Stanford, CA: Stanford University Press.
Giddens, A. 1991. *Modernity and Self-Identity. Self and Society in the Late Modern Age.* Cambridge: Polity Press.
Gigerenzer, G., and R. Selten (Eds.). In press. *Bounded Rationality: The Adaptive Toolbox.* Cambridge, MA: MIT Press.
Gigerenzer, G., and P.M. Todd. In press. "Ecological Rationality: Introduction." in *Handbook of experimental economics results*, edited by C.R. Plott and V.L. Smith. Amsterdam: North Holland/Elsevier.
Goffman, E. 1959. *The Presentation of Self in Everyday Life.* New York: The Free Press.
Granovetter, M. 1973. "The strength of weak ties." *American Sociological Review* 78:1360–1380.
Granovetter, M. 1984. "Small Is Bountyful: Labor Markets and Establishment Size." *American Sociological Review* 49:323–334.
Granovetter, M. 1985. "Economic Action and Social Structure: The Problem of Embeddedness." *American Journal of Sociology* 91:481–510.
Graumann, C.-F., and L. Kruse. 1990. "The Environment: Social Constructions and Psychological Problems." Pp 212–229 in *Societal Psychology*, edited by H.T. Himmelweit and G. Gaskell. London: Sage.
Green, D.P., and J. Shapiro. 1994. *Pathologies of Rational Choice Theory. A Critique of Applications in Political Science.* New Haven: Yale University Press.
Guagnano, G.A., P.C. Stern, and T. Dietz. 1994. "Willingness to Pay for Public Goods: A Test of the Contribution Model." *Psychological Science* 5:411–415.
Habermas, J. 1969. *Strukturwandel der Öffentlichkeit. Untersuchungen zu einer Kategorie der bürgerlichen Gesellschaft.* Neuwied: Luchterhand.
Habermas, J. 1984. *Reason and the Rationalization of Society.* Boston: Beacon Press.
Habermas, J. 1987. *System and Lifeworld.* Boston: Beacon Press.
Habermas, J. 1991. *Moral Consciousness and Communicative Action.* Cambridge, MA: MIT Press.
Habermas, Jürgen. 1999. *Wahrheit und Rechtfertigung.* Frankfurt: Suhrkamp.
Hacking, I. 1975. *The Emergence of Probability: a philosophical study of early ideas about probability, induction and statistical inference.* London and New York: Cambridge University Press.
Häfele, W. (Ed.). 1990. *Energiesysteme im Übergang – Unter den Bedingungen der Zukunft.* Landsberg/Lech: Poller.
Häfele, W., O. Renn, and G. Erdmann. 1990. "Risiko, Unsicherheit und Undeut-lichkeit." Pp 373–423 in *Energiesysteme im Übergang – Unter den Bedingungen der Zukunft*, edited by W. Häfele. Landsberg/Lech: Poller.
Hamilton, W.D. 1964. "The Genetical Theory of Social Behavior." *Journal of Theoretical Biology* 7:1–52.
Harcourt, G.C. 1972. *Some Cambridge Controversies in the Theory of Capital.* Cambridge, UK: Cambridge UP.
Harding, Sandra. 1992. "After the Neutrality Ideal: Science, Politics and 'Strong Objectivity'." *Social Research* 59:567–587.
Harding, S. (Ed.). 1993. *The Racial Economy of Science. Toward a Democratic Future.* Bloomington, IN: Indiana University Press.
Harless, D.W., and C. Camerer. 1994. "The Predictive Utility of Generalized Expected Utility Theories." *Econometrica* 62:1251–1289.
Harlow, R.E. 1986. "The EPA and Biotechnology Regulation: Coping with Scientific Uncertainty." *Yale Law Journal* 95:553–576.
Harvey, D. 1996. *Justice, Nature and the Geography of Difference.* Cambridge, Mass.: Blackwell.
Hauptmanns, U., M. Herttrich, and W. Werner. 1987. *Technische Risiken: Ermittlung und Beurteilung.* Berlin: Springer-Verlag.
Heap, S.H., M. Hollis, B. Lyons, R. Sugden, and A. Weale. 1992. *The Theory of Choice. A critical guide.* Oxford: Blackwell.
Hechter, M. (Ed.). 1983. *The Microfoundations of Macrosociology.* Philadelphia: Temple University Press.
Hechter, M. 1987. *Principles of Group Solidarity.* Berkeley, CA: University of California Press.
Heimer, C. 1988. "Social Structure, Psychology, and the Estimation of Risk." *Annual Review of Sociology* 14:491–519.
Heimer, C., and L.R. Staffen. 1998. *For the Sake of the Children. The Social Organization of Responsibility in the Hospital and the Home.* Chicago: University of Chicago Press.
Herrnstein, R.J., and J.E. Mazur. 1987. "Making Up Our Minds." *The Sciences* 27:40–47.
Hicks, J.R. 1940. "The valuation of social income." *Economica* 7:105–124.
Hilgartner, S., and C.L. Bosk. 1988. "The Rise and Fall of Social Problems: A Public Arenas Model." *American Journal of Sociology* 94:53–78.

Hobbes, T. 1651/1968. *Leviathan.* Harmondsworth: Penguin.

Hogarth, R.M., and M.W. Reder (Eds.). 1987. *Rational Choice: The Contrast between Economics and Psychology.* Chicago: University of Chicago Press.

Hohenemser, C., R.W. Kates, and P. Slovic. 1983. "The Nature of Technological Hazard." *Science* 220:378–384.

Hollis, M., and R. Sugden. 1993. "Rationality in Action." *Mind* 102:1–35.

Holzkamp, K. 1983. *Grundlegung der Psychologie.* Frankfurt/Main: Campus Verlag.

Homans, G.C. 1950. *The Human Group.* New York: Harcourt, Brace.

Homans, G.C. 1961. *Social Behavior: Its Elementary Forms.* New York: Harcourt, Brace.

Hoos, I. 1980. "Risk Assessment in Social Perspective." Pp 57–85 in *Perceptions of Risk*, edited by National Council on Radiation Protection and Measurements. Washington, D.C.: NCRP.

Horkheimer, M. 1972. *Critical Theory: Selected Essays.* New York: Herder and Herder.

Horkheimer, M., and Th.W. Adorno. 1947. *Dialektik der Aufklärung: Philosophische Fragmente.* Frankfurt/Main: Fischer Verlag.

Hughes, H.S. 1958. *Consciousness and Society: The Reorientation of European Social Thought: 1890–1930.* New York: Vintage Books.

Hume, D. 1740/1978. *A Treatise of Human Nature.* Oxford: Clarendon Press.

Hume, D. 1997. *An Enquiry Concerning the Principles of Morals.* Oxford: Oxford University Press.

Husserl, E. 1936/1970. *The Crisis of European Sciences and Transcendental Phenomenology: An Introduction to Phenomenological Philosophy.* Evanston, IL: Northwestern University Press.

IAEA. 1995. *Guidelines for Integrated Risk Assessment and Management in Large Industrial Areas.* Vienna: International Atomic Energy Agency.

Imboden, D.M., and C.C. Jaeger. 1999. "Towards a Sustainable Energy Future." in *Energy: The Next Fifty Years*, edited by OECD. Paris: OECD.

Ingham, G. 1984. *Capitalism Divided.* London: Macmillan.

Jackson, D., and E.A. Rosa. 1994. "Organization, Rationality, and Public Safety: A Socio-Economic Model of Regulatory Compliance." in *Sixth Annual Conference on Socio-Economics.* Jouy-en-Josas, F.

Jaeger, C.C. 1994. *Taming the Dragon. Transforming Economic Institutions in the Face of Global Change.* Yverdon: Gordon and Breach Science Publishers.

Jaeger, C.C. 1998. "Risk Management and Integrated Assessment." *Environmental Modeling and Assessment* 3:211–225.

Jaeger, C.C., and B. Kasemir. 1996. "Climatic Risks and Rational Actors." *Global Environmental Change* 6:23–36.

Jaeger, C.C., O. Renn, E.A. Rosa, T. Webler, R. Cantor, G. McDonell, O. Edenhofer, S. Funtovicz, S. Rayner, J. Ravetz, and G. Sergen. 1998. "Decision Analysis and Rational Action." Pp 141–215 in *The Tools for Policy Analysis*, edited by S. Rayner and E. Malone. Columbus, OH: Battelle Press.

Jaeger, C.C., and A.J. Rust. 1994. "Ethics as Rule Systems: The Case of Genetically Engineered Organisms." *Inquiry* 37:65–84.

Jaeger, C.C., R. Schüle, and B. Kasemir. 1999. "Focus Groups in Integrated Assessment: A Microcosmos for Reflexive Modernization." *Innovation* 12:195–219.

Janowitz, M. 1978. *The Last Half Century.* Chicago: University of Chicago Press.

Jasanoff, S. 1982. "Science and the Limits of Administrative Rule-Making: Lessons from the OSHA Cancer Policy." *Osgood Hall Law Journal* 20:536–561.

Jasanoff, S. 1986. *Risk Management and Political Culture.* New York: Russell Sage Foundation.

Jenkins, J. C. 1983. "Resource Mobilization Theory and the Study of Social Movements." *Annual Review of Sociology* 9:527–553.

Jevons, W.S. 1871/1970. *The Theory of Political Economy.* Harmondsworth: Penguin.

Johnson, B.B. 1991. "Risk and Culture Research. Some Cautions." *Journal of Cross-Cultural Psychology* 22:141–149.

Jonsen, A., and S. Toulmin. 1988. *The Abuse of Casuistry.* Berkeley: University of California Press.

Jungermann, H. 1986. "The Two Camps of Rationality." Pp 627–641 in *Judgment and Decision Making. An Interdisciplinary Reader*, edited by H.R. Arkes and K.R. Hammond. Cambridge, MA: Cambridge University Press.

Jungermann, H., H.-R. Pfister, and K. Fischer. 1998. *Die Psychologie der Entscheidung. Eine Einführung.* Heidelberg and Berlin: Spektrum Verlag.

Jungermann, H., and P. Slovic. 1993a. "Die Psychologie der Kognition und Evaluation von Risiko." Pp 167–207 in *Risiko und Gesellschaft. Grundlagen und Ergebnisse interdisziplinärer Risikoforschung*, edited by G. Bechmann. Opladen: Westdeutscher Verlag.

Jungermann, H., and P. Slovic. 1993b. "Charakteristika individueller Risikowahrnehmung." Pp 79–100 in *Riskante Technologien: Reflexion und Regulation. Einführung in die sozialwissenschaftliche Risikoforschung*, edited by W. Krohn and G. Krücken. Frankfurt/Main: Suhrkamp.

Kahneman, D., P. Slovic, and A. Tversky (Eds.). 1982. *Judgment under uncertainty: heuristics and biases.* Cambridge and New York: Cambridge University Press.

Kahneman, D., and A. Tversky. 1979. "Prospect Theory: An Analysis of Decision Under Risk." *Econometrica* 47:263–291.

Kaldor, N. 1939/1969. "Welfare Propositions in Economics and Intertemporal Comparisons of Utility." Pp 387–389 in *Readings in Welfare Economics*, edited by K.J. Arrow and T. Scitovsky. Homewood, IL: Irwin.

Kaminski, G. 1983. "Probleme einer ökopsychologischen Handlungstheorie." Pp 35–53 in *Kognition und Handeln*, edited by L. Montada, K. Reusser, and G. Steiner. Stuttgart: Klett.

Kant, I. 1785/1959. *Foundations of a Metaphysics of Morals*. Indianapolis, IN: Bobbs-Merrill.

Kasperson, R.E. 1992. "The Social Amplification of Risk: Progress in Developing an Integrative Framework." Pp 153–178 in *Social Theories of Risk*, edited by S. Krimsky and D. Golding. Westport: Praeger.

Kasperson, R.E., and J.X Kasperson. 1983. "Determining the Acceptability of Risk: Ethical and Policy Issues." Pp 135–155 in *Assessment and Perception of Risk to Human Health*, edited by J.T. Rogers and D.V. Bates. Ottawa: Royal Society of Canada.

Kasperson, R.E., and I. Palmlund. 1988. "Evaluating Risk Communication." Pp 143–158 in *Effective Risk Communication. The Role and Responsibility of Government and Nongovernment Organizations*, edited by V.T. Covello, D.B. McCallum, and M.T. Pavlova. New York: Plenum Press.

Kasperson, R E., O. Renn, P. Slovic, H.S. Brown, J. Emel, R. Goble, J.X. Kasperson, and S. Ratick. 1988. "The Social Amplification of Risk. A Conceptual Framework." *Risk Analysis* 8:177–187.

Kasperson, R.E., O. Renn, P. Slovic, J.X. Kasperson, and S. Emani. 1989. "The Social Amplification of Risk: Media and Public Response." Pp 131–135 in *Waste Management '89*, edited by R.G. Post. Tucson: Arizona Board of Regents.

Kasperson, R.E., and P.M. Stallen. 1991. "Introduction." Pp 1–11 in *Communicating Risk to the Public*, edited by R.E. Kasperson and P.M. Stallen. Dordrecht: Kluwer Academic Press.

Kates, R.W. 1976. "Experiencing the Environment as Hazard." Pp 133–156 in *Experiencing the Environment*, edited by S. Wappner, S.B. Cohen, and B. Kaplan. New York and London: Plenum Press

Keeney, R.L. 1992. *Value-Focused Thinking. A Path to Creative Decision Making*. Cambridge, MA: Harvard University Press.

Keeney, R.L., and H. Raiffa. 1976. *Decision with Multiple Objectives. Preferences and Value Tradeoffs*. New York: Wiley.

Keeney, R.L., O. Renn, D. von Winterfeldt, and U. Kotte. 1984. *Die Wertbaumanalyse. Entscheidungshilfe für die Politik*. München: HTV.

Keeney, R.L., and D. von Winterfeldt. 1986. "Improving Risk Communication." *Risk Analysis* 6:417–424.

Kellner, D. 1990. "The Postmodern Turn: Positions, Problems and Prospects." in *Frontiers of Social Theory. The New Syntheses*, edited by G. Ritzer. New York: Columbia University Press.

Kemp, R. 1985. "Planning, Political Hearings, and the Politics of Discourse." Pp 177–201 in *Critical Theory and Public Life*, edited by J. Forester. Cambridge, MA: MIT Press.

Kempton, W., J. Boster, and J.A. Hartley. 1995. *Environmental Values in American Culture*. Cambridge, MA: MIT Press.

Kinghorn, S. 1984. "Corporate Harm: A Structural Analysis of the Criminogenic Elements of the Corporation." Ann Arbor: University of Michigan Press.

Kitschelt, H. 1980. *Kernenergiepolitik. Arena eines gesellschaftlichen Konflikts*. Frankfurt/Main and New York: Campus Verlag.

Klandermanns, B. 1984. "Mobilization and Participation: Social Psychological Expansion of Resource Mobilization Theory." *American Sociological Review* 49:583–600.

Kleinhesselink, R.R., and E.A. Rosa. 1991. "Cognitive Representation of Risk Perceptions: A Comparison of Japan and the United States." *Journal of Cross-Cultural Psychology* 22:11–28.

Kleinhesselink, R.R., and E.A. Rosa. 1994. "Nuclear Trees in a Forest of Hazards: A Comparison of Risk Perceptions Between American and Japanese Students." Pp 101–119 in *Nuclear Power at the Crossroads: Challenges and Prospects for the Twenty-First Century*, edited by T.C. Lowinger and G.W. Hinman. Boulder, CO: International Research Center for Energy and Economic Development (ICEED).

Knight, F. 1947. *Freedom and Reform. Essays in Economic and Social Philosophy*. New York: Harper.

Kolluru, R.V. 1995. "Risk Assessment and Management: A Unified Approach." Pp 1.3–1.41 in *Risk Assessment and Management Handbook. For Environmental, Health, and Safety Professionals*, edited by R. Kolluru, S. Bartell, R. Pitblade, and S. Stricoff. New York: McGraw-Hill.

Kornhauser, W. 1959. *The Politics of Mass Society*. Glencoe, IL: The Free Press.

Kreps, D. 1988. *Notes on the Theory of Choice*. Boulder, CO: Westview Press.

Kreps, D.M. 1992. *A Course in Microeconomic Theory*. New York: Harvester Wheatsheaf.

Krimsky, S. 1992. "The Role of Theory in Risk Studies." in *Social Theories of Risk*, edited by S. Krimsky and D. Golding. Westport, CT: Praeger.

Kropotkin, P. 1902/1914. *Mutual Aid: A Factor in Evolution*. Boston: Extending Horizons Press.

Kuhn, T.S. 1962/1970. *The Structure of Scientific Revolutions*. Chicago: University of Chicago Press.

Kunreuther, H. 1992. "A Conceptual Framework for Managing Low-Probability Events." Pp 301–320 in *Social Theories of Risk*, edited by S. Krimsky and D. Golding. Westport, CT: Praeger.

Kunreuther, H. 1995. "Voluntary Siting of Noxious Facilities: The Role of Compensation." Pp 283–295 in *Fairness and Competence in Citizen Participation. Evaluating New Models for Environmental Discourse*, edited by O. Renn, T. Webler, and P. Wiedemann. Dordrecht: Kluwer.

Kunreuther, H., L. Ginsberg, P. Miller, P. Sagi, P. Slovic, B. Borkan, and N. Katz. 1978. *Disaster Insurance Protection: Public Policy Lessons*. New York: Wiley.
Kunreuther, H., and P. Slovic. 1996. *Science, Values, and Risk*. Thousand Oaks: Sage.
Laird, F. 1993. "Participatory Analysis: Democracy and Technological Decision Making." *Science, Technology, and Human Values* 18:341–361.
LaPorte, T., and P.M. Consolini. 1991. "Working in Practice but Not in Theory: Theoretical Challenge of High Reliability." *Journal of Public Administration Research and Theory* 1:19–47.
Lasswell, H.D. 1948. "The Structure and Function of Communication in Society." Pp 32–51 in *The Communication of Ideas*, edited by L. Brison. New York: Institute of Religious & Social Studies.
Lave, L.B. 1987. "Health and Safety Risk Analyses: Information for Better Decisions." *Science* 236:291–295.
LeBon, G. 1960. *The Crowd*. New York: Viking Press.
Lee, T.R. 1981. "The Public Perception of Risk and the Question of Irrationality." Pp 5–16 in *Risk Perception*, edited by Royal Society of Great Britain. London: Royal Society.
Lee, T.R. 1986. "Effective Communication of Information about Chemical Hazards." *The Science of the Total Environment* 51:149–183.
Leifer, E.M. 1983. *Robust Action: Generating Joint Outcomes in Social Relationships*. Cambridge, MA: Harvard University (Dissertation).
Lemert, C. 1990. "The Uses of French Structuralism in Sociology." Pp 230–254 in *Frontiers of Social Theory. The New Syntheses*, edited by G. Ritzer. New York: Columbia University Press.
Levi, M., K.S. Cook, J.A. O'Brien, and H. Faye. 1990. "Introduction: The Limits of Rationality." Pp 1–16 in *The Limits of Rationality*, edited by K. Schweers Cook and M. Levi. Chicago: University of Chicago Press.
Lewis, H.W. 1978. "Risk Assessment Review Group. Report to the U.S. Nuclear Regulatory Commission." Washington, D.C.: NUREG/CR-0400.
Lindenberg, S. 1985. "An Assessment of the New Political Economy: Its Potential for the Social Sciences and for Sociology in Particular." *Sociological Theory* 3:99–114.
Lindenberg, S. 1989. "Choice and Culture: The behavioral basis of cultural impact on transactions." Pp 175–200 in *Social Structure and Culture*, edited by H. Haferkamp. Berlin: Walter de Gruyter.
Lipset, S.M., and W. Schneider. 1983. *The Confidence Gap. Business, Labor, and Government, in the Public Mind*. New York: The Free Press.
Lopes, L.L. 1983. "Some Thoughts on the Psychological Concept of Risk." *Journal of Experimental Psychology: Human Perception and Performance* 9:137–144.
Lowrance, W.W. 1976. *Of Acceptable Risk: Science and the Determination of Safety*. Los Altos: William Kaufman.
Luce, R.D., and E.U. Weber. 1986. "An Axiomatic Theory of Conjoint, Expected Risk." *Journal of Mathematical Psychology* 30:188–205.
Luhmann, N. 1968. *Vertrauen. Ein Mechanismus der Reduktion sozialer Komplexität*. Stuttgart: Enke.
Luhmann, N. 1971. "Sinn und Kontingenz." In *Theorie der Gesellschaft oder Sozialtechnologie. Was leistet die Systemforschung?*, edited by J. Habermas and N. Luhmann. Frankfurt: Suhrkamp.
Luhmann, N. 1982. *Aufsätze zur Theorie der Gesellschaft*. Opladen: Westdeutscher Verlag.
Luhmann, N. 1984. *Soziale Systeme. Grundriss einer allgemeinen Theorie*. Frankfurt/Main: Suhrkamp.
Luhmann, N. 1986a. *Ökologische Kommunikation. Kann die moderne Gesellschaft sich auf ökologische Gefährdungen einstellen?* Opladen: Westdeutscher Verlag.
Luhmann, N. 1986b. "The Autopoiesis of Social Systems." Pp 172–192 in *Sociocybernetic Paradoxes: Observation, Control and Evolution of Self-Steering Systems*, edited by R.F. Geyer and J. van der Zouven. London: Sage Publications.
Luhmann, N. 1990. "Technology, Environment, and Social Risk: A Systems Perspective." *Industrial Crisis Quarterly* 4:223–231.
Luhmann, N. 1993. *Risk: A Sociological Theory*. New York: Aldine de Gruyter.
Luhmann, N. 1997. "Grenzwerte der ökologischen Politik. Eine Form von Risikomanagement." Pp 195–221 in *Risiko und Regulierung*, edited by P. Hiller and G. Krücken. Frankfurt/Main: Suhrkamp.
Lyotard, J. 1984. *The Postmodern Condition*. Minneapolis: University of Minnesota Press.
Machina, M.J. 1987a. "Decision Making in the Presence of Risk." *Science* 236:537–543.
Machina, M.J. 1987b. "Choice Under Uncertainty: Problems Solved and Unsolved." *Journal of Economic Perspectives* 1:121–154.
Machlis, G.E., and E.A. Rosa. 1990. "Desired Risk: Broadening the Social Amplification of Risk Framework." *Risk Analysis* 10:161–168.
MacIntyre, A.C. 1984. *After Virtue: a Study in Moral Theory*. Notre Dame, Ind.: University of Notre Dame Press.
MacIntyre, A.C. 1988. *Whose Justice? Which Rationality?* Notre Dame, Ind.: University of Notre Dame Press.
Macy, M.W., and A. Flache. 1995. "Beyond Rationality in Models of Choice." *Annual Rev. Sociol.* 21:73–91.
Mandelbrot, B.B. 1963. "The variation of certain speculative prices." *Journal of Business*: 394–419.
Manicas, P.R. 1987. *A History and Philosophy of the Social Sciences*. Oxford, UK: Basil Blackwell.
March, J.G. 1978. "Bounded rationality, ambiguity, and the engineering of choice." *Bell Journal of Economics and Management Science* 9:587–608.

March, J.G. 1991a. "How Decisions Happen in Organizations." *Human – Computer Interaction* 6:95–117.
March, J.G. 1991b. "Social Science and the Myth of Rationality. Invited presentation to the School of Social Sciences, University of California." In *Invited as part of the 25th anniversary celebration of UCI.* University of California, School of Social Sciences.
March, J.G., and H.A. Simon. 1958. *Organizations.* New York: Wiley.
Marcus, A.D. 1988. "Risk, Uncertainty, and Scientific Judgment." *Minerva* 26:138–152.
Margolis, H. 1996. *Dealing with Risk. Why the Public and the Experts Disagree on Environmental Issues.* Chicago: University of Chicago Press.
Markart, M. 1990. "Kritische Psychologie." Pp 119–123 in *Oekologische Psychologie*, edited by L. Kruse, C.-F. Graumann, and E.-D. Lantermann. München: Psychologie Verlags-Union.
Markowitz, J. 1990. "Kommunikation über Risiken – Eine Theorie-Skizze." *Schweizerische Zeitschrift für Soziologie* 3:385–420.
Marris, C., I. Langford, T. Saunderson, and T. O'Riordan. 1997. "Exploring the Psychometric Paradigm: Comparisons between aggregate and individual analyses." *Risk Analysis* 17:303–312.
Marx, K. 1852/1963. *The Eighteenth Brumaire of Louis Bonaparte.* New York: International Publishers.
Maslow, A.H. 1954/1970. *Motivation and Personality.* New York: Harper and Row.
Mauss, A.L. 1975. *Social Problems as Social Movements.* Philadelphia: Lippincott.
Mazur, A. 1984. "The Journalists and Technology: Reporting about Love Canal and Three Mile Island." *Minerva* 22.
Mazur, A. 1985. "Bias in Risk-Benefit Analysis." *Technology in Society* 7:25–30.
Mazur, A. 1987. "Does Public Perception of Risk Explain the Social Response to Potential Hazard?" *Quarterly Journal of Ideology* 11:41–45.
McAdam, D. 1982. *Political Process and the Development of Black Insurgency, 1930–1970.* Chicago: Chicago University Press.
McAdam, D., J. McCarthy, and M.N. Zald. 1988. "Social Movements." Pp 695–737 in *Handbook of Sociology*, edited by N. Smelser. Newbury Park, CA: Sage Publications.
McCallum, D. 1987. "Risk Factors for Cardiovascular Disease: Cholesterol, Salt, and High Blood Pressure." Pp 67–70 in *Risk Communication*, edited by J.C. Davies, V.T. Covello, and F.W. Allen. Washington, D.C.: The Conservation Foundation.
McCarthy, J.D., and M.N. Zald. 1987. "Resource Mobilization and Social Movements: A Partial Theory." In *Social Movements in an organizational society: collected essays*, edited by M.N. Zald, J.D. McCarthy, W.A. Gamson, R.A. Garned, P. Denton, M.A. Berger, J.R. Wood, and B. Useem. New Brunswick, NJ: Transaction Books.
McDougall, H. 1920. *The Group Mind.* Cambridge: Cambridge University Press.
Mead, G.H. 1934. *Mind, Self and Society.* Chicago: University of Chicago Press.
Meeks, J.G.T. 1984. "Utility in Economics: A Survey of the Literature." in *Surveying Subjective Phenomena*, edited by C.F. Turnerand and E. Martin. New York: Russell Sage Foundation.
Meijers, J.M.M., G.M.H. Swaen, and L.J.N. BLoemen. 1997. "The predictive value of animal data in human cancer risk assessement." *Regulatory Toxicology and Pharmacology* 25:94–102.
Merkhofer, L.W. 1984. "Comparative Analysis of Formal Decision-Making Approaches." Pp 183–220 in *Risk Evaluation and Managment*, edited by V.T. Covello, J. Menkes, and J. Mumpower. New York: Plenum Press.
Merkhofer, M.W. 1987. *Decision science and societal risk management.* Boston, MA: Reidel.
Merton, R.K. 1968. *Social theory and social structure.* New York: The Free Press.
Midden, C. 1988. "Credibility and Risk Communication." In *International Workshop on Risk Communication.* Jülich: Research Center.
Mileti, D., Th.E. Drabek, and J.E. Hass. 1975. *Human Systems in Extreme Situations.* Boulder, CO: Institute of Behavioral Science, University of Colorado.
Milgram, S. 1963. "Behavior Study of Obedience." *Journal of Abnormal and Social Psychology* 67:371–378.
Miller, G.A., E. Galanter, and K.H. Pribam. 1960. *Plans and Structure of Behavior.* New York: Holt, Rinehart & Wilson.
Mitchell, R.C. 1979. "National Environmental Lobbies and the Apparent Illogic of Collective Action." Pp 87–121 in *Collective Decision Making Applications from Public Choice Theory*, edited by C.S. Russell. Baltimore: Johns Hopkins University Press.
Morgan, G., and M. Henrion. 1990. *Uncertainty. A guide to dealing with uncertainty in quantitative risk and policy analysis.* Cambridge: Cambridge University Press.
Morgan, M.G. 1990. "Choosing and Managing Technology-Induced Risks." Pp 5–15 in *Readings in Risk*, edited by T.S. Clickman and M. Gough. Washington, D.C.: Resources for the Future, Inc.
Morishima, M. 1992. *Capital & Credit. A new formulation of general equilibrium theory.* Cambridge: Cambridge University Press.
Münch, R. 1982. *Basale Soziologie: Soziologie der Politik.* Opladen: Westdeutscher Verlag.
Münch, R. 1992. "Rational Choice Theory: A Critical assessment of Its Explanatory Power." Pp 121–137 in *Rational Choice Theory. Advocacy and Critique*, edited by J.S. Coleman and T.J. Farago. Newbury Park: Sage.

Nash, J.F. 1950. "Equilibrium points in n-person games." *Proceedings of the National Academy of Sciences of the USA* 36:48–49.

National Research Council. 1991. *Environmental epidemiology: Public health and hazardous wastes.* Washington, D.C.: National Academy Press.

National Research Council, Committee on Methods for the In Vivo Toxicity Testing of Complex Mixtures. 1988. *Complex Mixtures. Methods for In Vivo Toxicity Testing.* Washington: National Academy Press.

National Research Council, Committee on the Institutional Means for Assessment of Risks to Public Health. 1983. *Risk Assessment in the Federal Government: Managing the Process.* Washington, D.C.: National Academy Press.

National Research Council, Committee on Risk Perception and Communication. 1989. *Improving Risk Communication.* Washington, D.C.: National Academy Press.

Neisser, U. 1967. *Cognitive Psychology.* New York: Meredith.

Nelkin, D. 1984. *Controversy: Politics of Technical Decisions.* Beverly Hills: Sage Publications.

Nelkin, D. , and S. Gilmam. 1988. "Placing Blame for Devastating Disease." *Social Research* 55:378–384.

Nelkin, D., and M. Pollak. 1979. "Public Participation in Technological Decisions: Reality or Grand Illusion." *Technology Review* 9:55–64.

Nigg, J.M. 1995. "Risk Communication and Warning Systems." Pp 369–382 in *Natural Risk and Civil Protection,* edited by T. Horlick-Jones, A. Amendola, and R. Casale. London: Chapman and Hall.

Nisbett, R.E., and L. Ross. 1980. *Human Inference: Strategies and Shortcomings in Social Judgment.* Englewood Cliffs, N.J.: Prentice-Hall.

Nordhaus, W.D. 1994. *Managing the Global Commons. The Economics of Climate Change.* Cambridge, MA.: MIT Press.

Nowotny, H. 1976. "Social Aspects of the Nuclear Power Controversy." Laxenburg: Internationales Institut für Angewandte Systemanalyse (IIASA).

Nowotny, H. 1979. *Kernenergie: Gefahr oder Notwendigkeit? Anatomie eines Konfliktes. Einleitung: Peter Weingart.* Frankfurt/Main: Suhrkamp.

Nowotny, H., and R. Eisikovic. 1990. *Entstehung, Wahrnehmung und Umgang mit Risiken.* Bern: Schweizerischer Wissenschaftsrat.

Oberschall, A. 1973. *Social Conflict and Social Movements.* Englewood Cliffs, N.J.: Prentice-Hall.

Okrent, D. 1996. "Risk Perception Research Program and Applications: Have They Received Enough Peer Review?" Pp 1255–1259 in *Probabilistic Safety Assessment and Management '96 ESREL '96 – PSAM '96,* edited by C. Cacciabue and I.A. Papazoglu. Berlin: Springer-Verlag.

Oliver, P. 1984. "If You Don't Do It, Nobody Will. Active and Token Contributors to Local Collective Action." *American Sociological Review* 49:601–610.

Olson, M. 1965. *The Logic of Collective Action.* Cambridge, MA: Harvard University Press.

Opp, K.-D. , and C. Gern. 1993. "Dissident Groups, Personal Networks, and Spontaneous Cooperation: The East German Revolution of 1989." *American Sociological Review* 58:659–680.

O'Riordan, T. 1982. "Risk perception studies and policy priorities." *Risk Analysis* 2:95–100.

O'Riordan, T., and B. Wynne. 1987. "Regulating Environmental Risks: A Comparative Perspective." Pp 389–410 in *Insuring and Managing Hazardous Risks: From Seveso to Bhopal and Beyond,* edited by P.R. Kleindorfer and H.C. Kunreuther. Berlin: Springer-Verlag.

Otway, H., and K. Thomas. 1982. "Reflections on Risk Perception and Policy." *Risk Analysis* 2:69–82.

Otway, H., and B. Wynne. 1989. "Risk Communication: Paradigm and Paradox." *Risk Analysis* 9:141–145.

Palmlund, I. 1992. "Social Drama and Risk Evaluation." Pp 197–212 in *Social Theories of Risk,* edited by S. Krimsky and D. Golding. Westport: Praeger.

Pareto, V. 1916/1935. *The Mind and Society.* New York: Harcourt, Brace, and Co.

Pareto, V. 1927. *Manual of political economy.* New York: Kelley.

Parsons, T.E. 1937. *The Structure of Social Action: A Study in Social Theory With Special Reference to a Group of Recent European Writers.* New York: The Free Press.

Parsons, T.E. 1951. *The Social System.* Glencoe, IL: The Free Press.

Parsons, T.E. 1963. "On the Concept of Political Power." *Proceedings of the American Philosophical Society* 17:352–403.

Parsons, T.E. 1967. *Sociological Theory and Modern Society.* New York: The Free Press.

Parsons, T.E., and E.A. Shils. 1951. *Toward a General Theory of Action.* Cambridge: Cambridge University Press.

Parsons, T.E., and N.J. Smelser. 1956. *Economy and Society. A Study in the Integration of Economic and Social Theory.* Glencoe, IL: The Free Press.

Peltu, M. 1989. "Media Reporting of Risk Information: Uncertainties and the Future." Pp 11–32 in *Risk Communication,* edited by H. Jungermann, R.E. Kasperson, and P.M. Wiedemann. Jülich: Research Center Jülich.

Perrow, C. 1984. *Normal Accidents. Living with High-Risk Technologies.* New York: Basic Books.

Perrow, C. 1986a. "The Habit of Courting Disaster." Pp 1 in *The Nation.*

Perrow, C. 1986b. *Complex Organizations: A Critical Essay.* New York: Random House.

Perrow, C. 1991. "A Society of Organizations." *Theory and Society* 20:725–762.

Perry, R.W., M.R. Greene, and M.K. Lindell. 1980. "Enhancing Evacuation Warning Compliance: Suggestions for Emergency Planning." *Disasters* 4:433–449.

Petak, W.J., and A.A. Atkinson. 1982. *Natural Hazards: Risk Assessment and Public Policy*. Berlin: Springer.

Peters, H.P. 1984. "Entstehung, Verarbeitung und Verbreitung von Wissenschaftsnachrichten am Beispiel von 20 Forschungseinrichtungen." Jülich: Research Center.

Peters, H.P. 1990a. "Warner oder Angstmacher? Thema Risikokommunikation." *Funkkolleg, 'Medien und Kommunikation'*, edited by D. Mertens. Frankfurt/Main: Fischer.

Petersen, E.L., and N.O. Jensen. 1995. "Storms: Statistics, Predictability and Effects." Pp 147–177 in *Natural Risk and Civil Protection*, edited by T. Horlick-Jones, A. Amendola, and R. Casale. London: Chapman and Hall.

Petty, R.E., and E. Cacioppo. 1986. "The Elaboration Likelihood Model of Persuasion." *Advanced Experimental Social Psychology* 19:123–205.

Phillips, L.D. 1979. "Introduction to Decision Analysis." London: London School of Economics and Political Science.

Pinkau, K., and O. Renn (Eds.). 1998. *Environmental Standards. Scientific Foundations and Rational Procedures of Regulation with Emphasis on Radiological Risk Management*. Dordrecht and Boston: Kluwer.

Pinsdorf, M.K. 1987. *Communicating When Your Company Is Under Siege*. Lexington: Lexington Books.

Piore, M.J., and C.F. Sabel. 1984. *The second industrial divide. Possibilities for prosperity*. New York: Basic Books.

Plough, A., and S. Krimsky. 1987. "The Emergence of Risk Communication Studies: Social and Political Context." *Science, Technology, and Human Values* 12:4–10.

Pollatsek, A., and A. Tversky. 1970. "A Theory of Risk." *Journal of Mathematical Psychology* 7:540–553.

Powell, W.W., and P. DiMaggio (Eds.). 1991. *The New Institutionalism in Organizational Analysis*. Chicago: Chicago University Press.

Priddat, B.P. 1996. "Risiko, Ungewissheit und Neues: Epistemiologische Probleme ökonomischer Entscheidungsbildung." Pp 105–124 in *Risikoforschung zwischen Disziplinarität und Interdisziplinarität. Von der Illusion der Sicherheit zum Umgang mit Unsicherheit*, edited by G. Banse. Berlin: Edition Sigma.

Primas, H. 1992. "Umdenken in der Naturwissenschaft." *GAIA* 1:5–15.

Quarantelli, E.L. 1988. "Disaster Crisis Management. A Summary of Redearch Findings." *Journal of Management Studies* 25:373–385.

Rapoport, A., and A.M. Chammah. 1965. *Prisoner's Dilemma*. Ann Arbor, MI: University of Michigan Press.

Rappaport, R.A. 1971. "Nature, Culture and Ecological Anthropology." Pp 237–267 in *Man, Culture, and Society*, edited by H.C. Shapiro. Oxford and London: Oxford University Press.

Rayner, S. 1986. "Management of Radiation Hazards in Hospitals: Plural Rationalities in a Single Institution." *Social Studies of Science* 16:573–591.

Rayner, S. 1987. "Risk and Relativism in Science for Policy." Pp 5–23 in *The Social and Cultural Construction of Risk*, edited by B.B. Johnson and V.T. Covello. Dordrecht: Reidel.

Rayner, S. 1988. "Muddling Through Metaphors to Maturity: A Commentary on Kasperson et al., The Social Amplification of Risk." *Risk Analysis* 8:201–204.

Rayner, S. 1990. *Risk in cultural perspective: acting under uncertainty*. Norwell: Kluwer.

Rayner, S. 1992. "Cultural Theory and Risk Analysis." Pp 83–115 in *Social Theories of Risk*, edited by S Krimsky and D Golding. Westport, CT: Praeger.

Rayner, S., and R. Cantor. 1987. "How Fair is Safe Enough? The Cultural Approach to Societal Technology Choice." *Risk Analysis* 7:3–13.

Reiss, A. 1992. "The Institutionalization of Risk." Pp 299–308 in *Organizations, Uncertainties, and Risk*, edited by J.F. Short and L. Clarke. Westview: Boulder.

Renn, O. 1981. "Man, Technology, and Risk." Jülich: Research Center Jülich.

Renn, O. 1984. *Risikowahrnehmung der Kernenergie*. Frankfurt/Main and New York: Campus Verlag.

Renn, O. 1985. "Risk Analysis – Scope and Limitations." Pp 111–127 in *Regulating Industrial Risks: Science, Hazards and Public Protection*, edited by H. and M. Peltu Otway. London: Butterworth.

Renn, O. 1988. "Evaluation of Risk Communication: Concepts, Strategies, and Guidelines." Pp 99–117 in *Managing Environmental Risks. Proceedings of an APCA International Speciality Conference in Washington D.C., October 1987*. Washington, D.C.: APCA.

Renn, O. 1989. "Risikowahrnehmung – Psychologische Determinanten bei der intuitiven Erfassung und Bewertung von technischen Risiken." Pp 167–192 in *Risiko in der Industriegesellschaft*, edited by G. Hosemann. Nürnberg: Universitätsbibliothek.

Renn, O. 1990a. "Die Psychologie des Risikos. Die intuitive Erfassung technischer Risiken." *Energiewirtschaftliche Tagesfragen* 8:558–567.

Renn, O. 1990b. "Risk Perception and Risk Management" *Risk Abstracts* 7:1–9.

Renn, O. 1991b. "Risk Communication and the Social Amplification of Risk." Pp 287–324 in *Communicating Risks to the Public: International Perspectives*, edited by R.E. Kasperson and P.J.M. Stallen. Dordrecht: Kluwer Academic Publishers.

Renn, O. 1992a. "Concepts of Risk: A Classification." Pp 53–79 in *Social Theories of Risk*, edited by S. Krimsky and D. Golding. Westport, CT: Praeger.

Renn, O. 1992b. "Risk Communication: Towards a Rational Dialogue with the Public." *Journal of Hazardous Materials* 29:465–519.

Renn, O. 1998a. "The Role of Risk Communication and Public Dialogue for Improving Risk Management." *Risk Decision and Policy* 3:5–30.

Renn, O. 1998b. "The Need to Integrate Risk Assessment and Perception." *Reliability Engineering and System Safety* 59:49–62.

Renn, O., W. Burns, R.E. Kasperson, J.X. Kasperson, and P. Slovic. 1992. "The social amplification of risk: theoretical foundations and empirical application." *Social Issues* 48:137–160.

Renn, O., and D. Levine. 1991. "Trust and credibility in risk communication." Pp 175–218 in *Communicating risk to the public*, edited by R. Kasperson and P.J. Stallen. Dordrecht: Kluwer.

Renn, O., and E. Swaton. 1984. "Psychological and Sociological Approaches to Study Risk Perception." *Environment International* 10:557–575.

Renn, O., and T. Webler. 1992. "Anticipating Conflicts: Public Participation in the Solid Waste Crisis." *GAIA* 1:84–94.

Renn, O., T. Webler, H. Rakel, P.C. Dienel, and B. Johnson. 1993. "Public Participation in Decision Making: A Three-Step-Procedure." *Policy Sciences* 26:189–214.

Rohrmann, B. 1990. "Perception and evaluation of risk: a cross-cultural comparison." In *Comparative Risk Perception*, edited by B. Rohrmann and O. Renn. Dordrecht: Kluwer.

Rokeach, M. 1969. *Beliefs, Attitudes, and Values.* Berkeley/San Francisco, CA: California University Press.

Ronge, V. 1980. "Theoretical Concepts of Political Decision-Making Processes." Pp 209–240 in *Society, Technology and Risk Assessment*, edited by J. Conrad. London: Academic Press.

Rosa, E.A. 1994. "Mirrors and Lenses: Toward Theoretical Method in the Study of the Nature-Culture Dialectic." in *La Société au Naturel: Functions de la Nature (Nature's Society: Social Functions of Nature)*, edited by D. Duclos. Paris, France: L'Harmattas.

Rosa, E.A. 1998. "Metatheoretical Foundations for Post-Normal Risk." *Journal of Risk Research* 1:15–44.

Rosa, E.A., and S.K. Wong. 1992. "Weaving the Social Fabric of Risk Perceptions: The Cultural Context." Paper presented at the *Annual Meeting of the American Sociological Association.* Pittsburgh, PA.

Rosenbaum, W. 1979. "Elitism and Citizen Participation." In *Citizen Participation Perspectives. Proceedings of the National Conference on Citizen Participation, Washington D.C., Sept. 28 – Oct. 1, 1978*, edited by S. Langton. Medford: Tufts University Lincoln Filene Center for Citizenship and Public Affairs.

Ross, L.D. 1977. "The Intuitive Psychologist and His Shortcomings: Distortions in the Attribution Process." Pp 173–220 in *Advances in Experimental Social Psychology*, edited by L. Berkowitz. New York: Random House.

Ross, R.J.S., and K.C. Trachte. 1990. *Global Capitalism: the New Leviathan.* Albany, NY: State University of New York Press.

Rostow, W.W. 1971. *Politics and the Stages of Growth.* Cambridge: Cambridge University Press.

Rowe, W.D. 1977. *An Anatomy of Risk.* New York: Wiley.

Royal Society. 1992. *Risk: Analysis, perception and management.* London, UK: Royal Society.

Rust, A. 1987. *Die organismische Kosmologie von Alfred N. Whitehead.* Frankfurt/Main: Athenäum.

Sabini, J., and M. Silver. 1982. *Moralities of Everyday Life.* Oxford: Oxford University Press.

Salcedo, R.N., H. Read, J.F. Evans, and A.C. Kong. 1974. "A Successful Information Campaign on Pesticides." *Journalistic Quarterly* 51:91–95.

Samuelson, P.A. 1950/1969. "Evaluation of real national income." Pp 402–433 in *Readings in Welfare Economics*, edited by K.J. Arrow and T. Scitovsky. Homewood, IL: Irwin.

Sandman, P. 1987. "Risk communication: Facing public outrage." *Environmental Protection Journal* November:21–22.

Sandman, P.M., N.D. Weinstein, and M.L. Klotz. 1987. "Public Response to the Risk from Geological Radon." *Journal of Communication* 37:93–108.

Schelling, T. 1960/1980. *The Strategy of Conflict.* Cambridge, MA: Harvard University Press.

Schmidheiny, S., and F. Zorraquín. 1996. *Financing Change. The Financial Community, Eco-efficiency, and Sustainable Development.* Cambridge, Mass: The MIT Press.

Schoemaker, P.J. 1982. "The Expected Utility Model: Its Variants, Purposes, Evidence, and Limitations." *Journal of Economic Literature* 30:529–563.

Schoemaker, P.J. 1987. "Mass Communication by the Book: A Review of 31 Texts." *Communication* 37:109–133.

Schuetz, A., and T. Luckmann. 1994. *Strukturen der Lebenswelt.* Frankfurt: Suhrkamp.

Scott, R.W. 1995. *Institutions and Organizations.* Thousand Oaks: Sage.

Seamon, D.R. 1987. "Phenomenology and Environment-Behavior Research." Pp 3–27 in *Advances in Environment, Behavior, and Design*, edited by E.H. Zube and G.T. Morre. New York.

Sen, A. 1995. "Rationality and Social Choice." *American Economic Review* 85:1–24.

Shannon, C.E., and W. Weaver. 1949. *The mathematical theory of communication.* Urbana: University of Illinois Press.

Short, J.F. 1984. "The Social Fabric at Risk: Toward the Social Transformation of Risk Analysis." *American Sociological Review* 49:711–725.

Short, J.F. 1989a. "On Defining, Describing, and Explaining Elephants (and Reactions to Them): Hazards, Disasters, and Risk Analysis." *International Journal of Mass Emergencies and Disasters* 7:397–418.
Short, J.F. 1989b. "Toward a Sociolegal Paradigm of Risk." *Laws and Policy* 11:242–252.
Short, J .F. 1997. "The Place of Rational Choice in Criminology and Risk Analysis." *The American Sociologist* 28:61–72.
Short, J.F., Jr., and L. Clarke (Eds.). 1992a. *Organizations, Uncertainties, and Risk.* Boulder, CO: Westview Press.
Short, J.F., and L. Clarke. 1992b. "Social Organization and Risk." Pp 309–321 in *Organizations, Uncertainties, and Risk*, edited by J.F. Short and L. Clarke. Boulder, CO: Westview Press.
Shrader-Frechette, K. 1985. *Risk Analysis and Scientific Method: Methodological and Ethical Problems with Evaluating Societal Risks.* Boston, MA: Reidel.
Shrader-Frechette, K.S. 1991. *Risk and Rationality: Philosophical Foundations for Populist Reforms.* Berkeley, CA, and Los Angeles: University of California Press.
Siegel, L.J. 1989. *Criminology.* St. Paul, MN: West Publishing Co.
Simon, H.A. 1955. "A Behavioral Model of Rational Choice." *Quarterly Journal of Economics* 69:174–183.
Simon, H.A. 1976. *Administrative Behavior: A Study of Decision-Making Processes in Administrative Organizations.* New York: Basic Books.
Simon, H.A. 1987. "Rationality in Psychology and Economics." Pp 25–40 in *Rational Choice. The Contrast between Economics and Psychology*, edited by R.M. Hogarth and M.W. Reder. Chicago and London: The University of Chicago Press.
Sjöberg, L. 1996. "A Discussion of the Limitations of the Psychometric and Cultural Theory Approaches to Risk Perception." *Radiation Protection Dosimetry* 68:219–225.
Sjöberg, L. 1997. "Risk Sensivity, Attitude and Fear as Factors in Risk Perception. Manuscript." in *Annual Meeting of the Society for Risk Analysis-europe. Manuscript.* Stockholm.
Skinner, B.F. 1971. *Beyond Freedom and Dignity.* New York: Knopf.
Slovic, P. 1987. "Perception of Risk." *Science* 236:280–285.
Slovic, P. 1992. "Perception of Risk: Reflections on the Psychometric Paradigm." Pp 117–152 in *Social Theories of Risk*, edited by S. Krimsky and D. Golding. Westport, CT: Praeger.
Slovic, P., B. Fischhoff, and S. Lichtenstein. 1980. "Facts and Fears: Understanding Perceived Risk." Pp 181–216 in *Societal Risk Assessment: How Safe is Safe Enough?*, edited by R. Schwing and W. Albers. New York: Plenum Press.
Slovic, P., and E. Peters. 1996. "The role of affect and worldviews as orienting dispositions in the perception and acceptance of nuclear power." *Journal of Applied Social Psychology* 26:1427–1453.
Smelser, N.J. 1998. "The Rational and the Ambivalent in the Social Sciences." *American Sociological Review* 63:1–16.
Smith, A. 1776/1976. *An Inquiry into the Nature and Causes of the Wealth of Nations.* Oxford: Clarendon Press.
Somers, E. 1995. "Perspectives on risk management." *Risk Analysis* 15:677–684.
Sonnenschein, H. 1973. "Do Walras' Identity and Continuity Characterize the Class of Community Excess Demand Functions?" *Journal of Economic Theory* 6:345–354.
Spaemann, R. 1980. "Technische Eingriffe in die Natur als Problem der politischen Ethik." Pp 192–204 in *Ökologie und Ethik*, edited by D. Birnbacher. Stuttgart: Reclam.
Spencer, H. 1852/1888. *Social Statics; Or the Conditions Essential to Human Happiness Specified, and First of Them Developed.* New York: Appelton.
Spencer, H. 1874–96/1898. *Principles of Sociology.* New York: Appelton.
Sraffa, P. 1960. *Production of Commodities by Means of Commodities. Prelude to a Critique of Economic Theory.* Cambridge: Cambridge University Press.
Starr, C. 1969. "Social Benefit Versus Technological Risk: What is Our Society Willing to Pay for Safety." *Science* 165:1232–1238.
Starr, C., and C. Whipple. 1980. "Risks of risk decisions." *Science* 208:1114–1119.
Stern, P.C. 1992. "Psychological Dimensions of Global Environmental Change." *Psychological Review* 43:269–302.
Stern, P.C., and H.V. Fineberg (Eds.). 1996. *Understanding Risk: Informing Decisions in a Democratic Society.* Washington, D.C.: National Academy Press.
Steward, J. 1955. *Theory of Cultural Change.* Urbana: University of Illinois Press.
Strange, S. 1986. *Casino Capitalism.* Oxford: Blackwell.
Sugden, R. 1991. "Rational Choice: A Survey of Contributions from Economics and Philosophy." *The Economic Journal* 101:751–785.
Sugden, R. 1993. "Thinking as a Team: Towards an Explanation of Nonselfish Behavior." *Social Philosophy & Policy* 10: 69–89.
Susam, P., P. O'Kefe, and B. Wisner. 1983. "Global Disasters – A Radical Interpretation." Pp 263–283 in *Interpretations of Calamity*, edited by K. Hewitt. Boston: Allen and Unwin.
Tännsjö, T. 1990. *Conservatism for Our Time.* London: Routledge.

Tarnas, R. 1990. *The Passion of the Western Mind.* New York: Ballantine Books.

Taylor, M. 1996. "When Rationality Fails." Pp 223–234 in *The Rational Choice Controversy. Economic Models of Politics Reconsidered*, edited by J. Friedman. New Haven: Yale University Press.

Thaler, R. 1991. *Quasi-Rational Economics.* New York: Russell Sage Foundation.

The Union of Concerned Scientists. 1977. "The Risks of Nuclear Power Reactors, Memorandum." Cambridge, Mass.

Thomas, K., D. Maurer, M. Fishbein, H.J. Otway, R. Hinkle, and D.A. Simpson. 1980. "Comparative Study of Public Beliefs About Five Energy Systems." : IIASA (International Institute for Applied Systems Analysis).

Thomas, L.M. 1987. "Why We Must Talk About Risk." Pp 19–25 in *Risk Communication*, edited by J.C. Davies, V.T. Covello, and F.W. Allen. Washington, D.C.: The Conservation Foundation.

Thompson, M. 1980. "An Outline of the Cultural Theory of Risk." Laxenburg, Austria: International Institute for Applied Systems Analysis (IIASA).

Thompson, M., and M. Warburton. 1985. "Decision Making Under Contradictory Certainties: How to Save the Himalayas When You Can't Find Out What's Wrong with Them." *Journal of Applied Systems Analysis* 12:3–34.

Tilly, C. 1978. *From Mobilization to Revolution.* Reading, Mass.: Addison Wesley.

Tilly, Ch (Ed.). 1992. "Foreword." In J.F. Short and L. Clarke (Eds) *Organizations, Uncertainties, and Risk*, Westview Press, Boulder, CO

Trabasso, T., and G. Bower. 1964. "Presolution Reversal and Dimensional Shifts in Concept Identification." *Journal of Experimental Psychology* 67:398–399.

Treml, A.K. 1990. "Über den Zufall." *Universitas* 45:826–837.

Trivers, R. 1985. *Social Evolution.* Menlo Park, Cal.: The Benjamin Cummings Publishing Company.

Turner, B.A. 1995. "The Role of Flexibility and Improvisation in Emergency Response." Pp 463–475 in *Natural Risk and Civil Protection*, edited by T. Horlick-Jones, A. Amendola, and R. Casale. London: E & FN Spon, Chapman and Hall.

Tversky, A. 1972. "Elimination by Aspects: A Theory of Choice." *Psychological Review* 79:281–299.

Tversky, A., and D. Kahneman. 1974. "Judgment Under Uncertainty: Heuristics and Biases." *Science* 185:1124–1131.

Tversky, A., and D. Kahneman. 1981. "The Framing of Decisions and the Psychology of Choice." *Science* 211:453–458.

U.S. Atomic Energy Commission. 1975. "Reactor Safety Study – An Assessment of Accident Risks in U.S. Commercial Nuclear Power Plants." Washington, D.C.: U.S. Nuclear Regulatory Commission (WASH-1400).

United States Presidential Commission on the Space Shuttle Challenger Accident. 1986. *Report to the President.* Washington, D.C.: The Commission.

United States President's Commission on the Accident at Three Mile Island, and J.G. Kemeny. 1979. *Report of the President's Commission on the Accident at Three Mile Island.* Washington, D.C.: The Commission.

Urmson, J.O., and J. Ree (Eds.). 1991. *The Concise Encyclopedia of Western Philosophy and Philosophers.* London: Routledge.

Useem, B. 1980. "Solidarity Model, Breakdown Model and the Boston Anti-Busing Movement." *American Sociological Review* 45:357–369.

Vaughan, D. 1996. *The Challenger Launch Decision: Risky Technology, Culture, and Deviance at NASA.* Chicago, IL: University of Chicago Press.

Vayda, A.P., and R. Rappaport. 1968. "Ecology, Cultural and Noncultural." Pp 477–497 in *Introduction to Cultural Anthropology*, edited by J. Clifton. Boston: Houghton Mifflin.

Vlek, C.A. 1996. "A Multi-Level, Multi-Stage and Multi-Attribute Perspective on Risk Assessment, Decision-Making, and Risk Control." *Risk Decision and Policy* 1: 9–31.

von Neumann, J. 1935–36/1945. "A model of general economic equilibrium." *Review of Economic Studies* 13:1–9.

von Neumann, J., and O. Morgenstern. 1944. *Theory of Games and Economic Behavior.* Princeton, NJ: Princeton University Press.

von Winterfeldt, D. 1987. "Value Tree Analysis: An Introduction and an Application to Offshore Oil Drilling." Pp 439–377 in *Insuring and Managing Hazardous risks: From Seveso to Bhopal and Beyond*, edited by P.R. Kleindorfer and H.C. Kunreuther. Berlin: Springer.

von Winterfeldt, D., and W. Edwards. 1986. *Decision Analysis and Behavioral Research.* Cambridge: Cambridge University Press.

Wald, A. 1936. "Über einige Gleichungssysteme der mathematischen Ökonomie." *Zeitschrift für Nationalökonomie* 7:637–670.

Walley, P. 1991. *Statistical Reasoning with Imprecise Probabilities.* London: Chapman and Hall.

Walras, L. 1874–1877. *Elements d'economie politique pure.* Lausanne: Corbaz.

Walsh, E., and R.H. Warland. 1983. "Social Movement Involvement in the Wake of a Nuclear Accident: Activists and Free Riders in the Three Mile Island Area." *American Sociological Review* 48:764–781.

Walsh, E.J. 1981. "Resource Mobilization and Citizen Protest in Communities around Three Mile Island." *Social Problems* 29:1–21.
Watson, S.R. 1982. "Multi-attribute Utility Theory for Measuring Safety." *European Journal of Operational Research* 10:77–81.
Watts, N. 1983. "On the Poverty of Theory." Pp 231–262 in *Interpretations of Calamity*, edited by K. Hewitt. Boston: Allen and Unwin.
Weber, M. 1904/05/1993. *Die protestantische Ethik und der "Geist" des Kapitalismus.* Bodenheim: Athenaeum Hain Hanstein.
Weber, M. 1920–1921/1988. *Gesammelte Aufsätze zur Religionssoziologie (Photomechanischer Nachdruck).* Tübingen: Mohr.
Weber, M. 1968. *Economy and Society, edited by Guenther Roth and Claus Wittich.* New York: Bedminster Press.
Weber, M. 1983. *Entscheidungen bei Mehrfachzielen.* Wiesbaden.
Weber, M., C.W. Mills, and H.H. Gerth. 1958. *From Max Weber: essays in sociology.* New York: Oxford University Press.
Webler, T. 1995. "'Right' Discourse in Citizen Participation. An Evaluative Yardstick." Pp 35–86 in *Fairness and Competence in Citizen Participation. Evaluating New Models for Environmental Discourse*, edited by O. Renn, T. Webler, and P. Wiedemann. Dordrecht and Boston: Kluwer.
Webler, T., D. Levine, H. Rakel, and O. Renn. 1991. "The Group Delphi: A Novel Attempt at Reducing Uncertainty." *Technological Forecasting and Social Change* 39:253–263.
Weimann, J. 1991. *Umweltökonomik: Eine theorieorientierte Einführung.* Berlin: Springer-Verlag.
Weinberg, A.M. 1972. "Science and Trans-Science." *Minerva* 10:209–222.
Weinberg, A.M. 1981. "Reflections on risk assessment." *Risk Analysis* 1:5–7.
Weingart, P. 1983. "Verwissenschaftlichung der Gesellschaft – Politisierung der Wissenschaft." *Zeitschrift für Soziologie* 12:225–241.
Weinstein, N.D. 1980. "Unrealistic Optimism about Future Life Events." *Journal of Personality and Social Psychology* 39:106–120.
Whipple, C. (Ed.). 1992. *Inconsistent Values in Risk Management.* Westport, CT: Praeger.
White, G.F. 1961. "The Choice of Use in Resource Management." *Natural Resources Journal* 1:23–40.
White, G.F. (Ed.). 1974. *Natural Hazards: Local, National, Global.* Oxford: Oxford University Press.
White, H.C. 1981. "Where do markets come from?" *American Journal of Sociology* 87: 517–547.
White, H.C. 1992. *Identity and control. A structural theory of social action.* Princeton, NJ: Princeton University Press.
Whitehead, A.N. 1929/1978. *Process and Reality: an Essay in Cosmology.* New York: The Free Press.
WHO. 1977. "Evaluation of Carcinogenic Risk of Chemicals to Man." Vol. 1. Lyon: International Agency for Research on Cancer.
Wicker, A.W. 1979. "Attitudes vs Actions: The Relationship of Verbal and Behavioral Responses to Attitude Objects." *Journal of Social Issues* 22:41–78.
Wiedemann, P.M. 1993. "Tabu, Sünde, Risiko: Veränderung der gesellschaftlichen Wahrnehmung von Gefährdungen." Pp 43–68 in *Risiko ist ein Konstrukt. Wahrnehmungen von Risikowahrnehmung*, edited by Bayerische Rückversicherung. München: Knesbeck.
Wigley, T.M.L., R. Richels, and J.A. Edmonds. 1996. "Economic and Environmental Choices in the Stabilization of Atmospheric CO_2 Concentrations." *Nature* 379:240–243.
Wildavsky, A. 1990. "No Risk is the Highest Risk of All." Pp 120–127 in *Readings in Risk*, edited by T.S. Glickman and M. Gough. Washington, D.C.: Resources for the Future.
Wilkins, L., and P. Patterson. 1987. "Risk Analysis and the Construction of News." *Communication* 37:80–92.
Williamson, O.E. 1979. "Transaction-Cost Economics: the Governance of Contractual Relations." *Journal of Law and Economics* 22:233–261.
Wilson, E.O. 1978. *On Human Nature.* Cambridge, MA: Harvard University Press.
Wilson, R. 1979. "Analyzing the Daily Risks of Life." *Technology Review* 81:40–46.
Wilson, R., and E.A.C. Crouch. 1987. "Risk Assessment and Comparisons: An Introduction." *Science* 236:267–270.
Winkler, R.L. 1968. "The Consensus of Subjective Probability Distributions." *Management Science* 15:B61–B75.
Wittgenstein, L. 1953. *Philosophical Investigations.* Oxford: Basil Blackwell.
Wittgenstein, L. 1979. *On Certainty.* Oxford: Basil Blackwell.
Wright, J.D., and P.H. Rossi (Eds.). 1981. *Social Science and Natural Hazards.* Cambridge, MA: ABT Associates.
Wynne, B. 1982a. "Institutional Mythodologies and Dual Societies in the Management of Risk." In *The Risk Analysis Controversy. An Institutional Perspective*, edited by H.C. Kunreuther and E.V. Levy. Berlin: Springer-Verlag.
Wynne, B. 1982b. *Rationality and Ritual: the Windscale Inquiry and Nuclear Decisions in Britain.* Chalfont St. Giles: British Society for the History of Science.
Wynne, B. 1984. "Public Perceptions of Risk." Pp 246–259 in *The Urban Transportation of Irradiated Fuel*, edited by J. Aurrey. London: Macmillan.

Wynne, B. 1991. "After Chernobyl: Science Made Too Simple?" *New Scientist* 26:44–46.

Wynne, B. 1992. "Risk and Social Learning: Reification to Engagement." Pp 275–297 in *Social Theories of Risk*, edited by S. Krimsky and D. Golding. Westport, CT: Praeger.

Zald, M.N., and J.D. McCarthy (Eds.). 1987. *Social Movements in an Organizational Society: Collected Essays.* New Brunswick, N.J.: Transaction Books.

Zimmermann, R. 1987. "A Process Framework for Risk Communication." *Science, Technology, and Human Values* 12:131–137.

INDEX

Printed and bound by CPI Group (UK) Ltd, Croydon, CR0 4YY

23/10/2024

01777668-0001